2

Microbiology Monographs

Series Editor: Alexander Steinbüchel

Complex Intracellular Structures in Prokaryotes

Volume Editor: Jessup M. Shively

With 85 Figures, 23 in color

Springer

Volume Editor:

Dr. Jessup M. Shively
Prof. Emeritus of Biochemistry
Department of Genetics and Biochemistry
Clemson University
Clemson SC 29634
USA
e-mail: shy53911@bellsouth.net

Series Editor:

Professor Dr. Alexander Steinbüchel
Institut für Molekulare Mikrobiologie und Biotechnologie
Westfälische Wilhelms-Universität
Corrensstraße 3
48149 Münster
Germany
e-mail: steinbu@uni-muenster.de

ISSN 1862-5576
ISBN 978-3-642-06893-5 e-ISBN 978-3-540-32526-0

DOI 10.1007/11497158

Springer is a part of Springer Science+Business Media

springer.com

Printed in Germany

Editor: Dr. Christina Eckey, Heidelberg
Desk Editor: Dr. Jutta Lindenborn, Heidelberg
Cover design: eStudio Calamar S.L., Spain

Printed on acid-free paper 149/3100 YL – 5 4 3 2 1 0

Preface

Categorizing organisms as either prokaryotes or eukaryotes was first introduced in 1937 and proposed again in 1957 (Beck 2000); however, it was some years later, prompted to a certain degree by a publication by Stanier and van Niel (1962), before the concept became more widely accepted. At this early stage in the late 1950s, our complete understanding of the ultrastructure of both cell types, but especially prokaryotes, was still in its infancy and awaited the more common and extensive use of electron microscopy as well as a multitude of other techniques and instrumentation. Prokaryotes were generally viewed as primitive simple cells with little intracellular structure. The existence of some storage bodies had been demonstrated (see Vol. 1 of this series), but little else was known about the internal organization (first chapter, this volume). The absence of mitochondria or chloroplasts appeared certain, but the state of the nucleic acid within the prokaryotic cell and the presence/absence of mitosis were still somewhat controversial subjects. That our concept of a prokaryote has undergone a remarkable transformation during the last 50 years should become readily apparent as one proceeds through the chapters in this volume. Furthermore, the reader should come to realize that the rigid classification of an organism as either a prokaryote or eukaryote is not always so clear-cut. We are quite sure that the next 50 years will provide a clearer understanding of the complex intracellular structures presently known as well as bring to light surprising new ones.

As we gathered information on the intracellular structures found in a seemingly endless variety of prokaryotes it became obvious to us that the components were so numerous that it would be impossible to include topics for all of them in one "reasonably sized" volume of *Microbiology Monographs.* After considering different alternatives, we decided to place those components with primarily a metabolic storage/reserve function in Vol. 1 and label them "inclusions" see *Microbiology Monographs* Vol. 1 "Inclusions in Prokaryotes." Those components with other functions were then considered in Vol. 2 "Complex Intracellular Structures of Prokaryotes." After eliminating those topics well covered in microbiology and/or biochemistry textbooks we selected 10 topics that we feel exemplify the spectrum of prokaryote intracellular structures: proteasomes, phycobilisomes, chlorosomes, gas vesicles, carboxysomes, magnetosomes, intracytoplasmic membranes, membrane bound nucleoids, anammoxosomes, and cytoarchitecture of *Epulopiscium* spp. As we secured authors for these topics, suggestions for other topics seemed to come from everywhere. In an effort to make the monograph somewhat more "complete," but still remain within the suggested book length, we added seven topics as mini (cameo) chapters. Even with the expanded coverage, worthy topics have likely been excluded. It is our hope that the omissions will not displease either the researchers investigating the excluded subjects or readers.

We would like to express our sincere thanks to all of the authors; without their contributions this book would not have been possible. We realize that the preparation of a lengthy manuscript is a difficult task in the face of the many other duties these authors are expected to perform. Their contributions are excellent and were submitted in a timely fashion. Our thanks also go to Springer for publishing this monograph. Special thanks are due to Christina Eckey and Jutta Lindenborn for their valuable suggestions and support.

Clemson and Münster, January, 2006

Jessup Shively
Alexander Steinbüchel

References

Beck RW (2000) A chronology of microbiology in historical context. ASM Press, Washington
Stanier RY, van Neil CB (1962) The concept of a bacterium. Archiv Microbiol 42:17–35

Contents

Part I
Complex Intracellular Structures in Prokaryotes

Microbiol Monogr (2)
J.M. Shively: Complex Intracellular Structures in Prokaryotes
DOI 10.1007/7171_018/Published online: 13 May 2006

Prokaryote Complex Intracellular Structures: Descriptions and Discoveries

Jessup M. Shively

Department of Genetics and Biochemistry, Clemson University, Clemson, SC 29634, USA
shy53911@bellsouth.net

Abstract The gas vacuole was first observed in 1895, but details of this structure (gas vesicles) as well as discovery of the other structures covered in this monograph (proteasomes, phycobilisomes, chlorosomes, carboxysomes and carboxysome-like inclusions, magnetosomes, intracytoplasmic membranes, membrane-bounded nucleoids, pirellulosomes, anammoxosomes and the cytoarchitecture of *Epulopiscium* spp.) awaited the availability of the transmission electron microscope and related technologies. Additional advancements in electron microscopy were required for the optimal visualization of some structures.

1 Introduction

It is quite interesting to peruse microbiology textbooks from the 1950s (for example Stanier et al. 1957). As a rule and although several different cell inclusions are briefly discussed, including glycogen, granulose, fat, volutin, and sulfur (see Vol. 1 of this series), little information is presented on any complex intracellular structures. DNA was known, but the presence/absence of a nucleus and mitotic figures in bacteria was still a controversial subject. The molecular mechanisms of protein synthesis are missing, including information on ribosomes. Chromatophores, recently isolated but not yet observed in the cell, containing the photosynthetic pigments are briefly described as "special structures" (see chapter by Niederman, this volume). Surprisingly, gas vacuoles discovered in cyanobacteria in 1895 and in bacteria in 1913 (Fig. 1) are not even mentioned (see chapter by Pfeifer, this volume). In defense of the omission, definitive work on the vacuoles did not occur until 1956 and later. Also, cyanobacteria, still considered to be

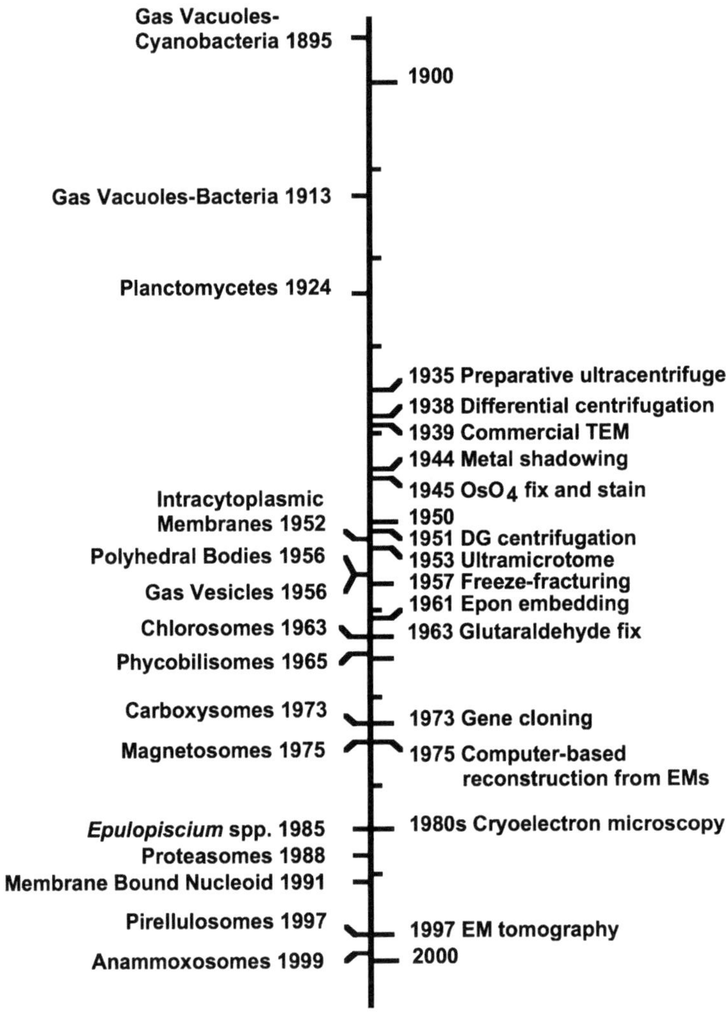

Fig. 1 Timeline of complex intracellular structure discovery (*left*) and selected technological advances (*right*). Abbreviations: *TEM* transmission electron microscope, *DG centrifugation* density gradient centrifugation, *SEM* scanning electron microscope, *EMs* electron micrographs, *EM tomography* electron microscope tomography. Technological advance dates taken from Alberts et al. (2002) and Beck (2002)

the "property" of plant scientists, were yet to become a part of the realm of microbiology. However, even botany textbooks provided poor coverage of the cyanobacteria; they were generally only well covered in separate monographs. Essentially, the only complex structure given much attention was the endospore found in certain bacteria. Thus, the "primary" complex intracellular structure topics (proteasomes, phycobilisomes, chlorosomes, gas vesicles, carboxysomes and carboxysome-like inclusions, magnetosomes, intracytoplasmic membranes, the membrane-bounded nucleoids and pirellulosomes, anammoxosomes, and cytoarchitecture of *Epulopiscium* spp.), presented as chapters in this volume, were unknown except for the preliminary information on gas vacuoles and chromatophores. This lack of information on the now-known complex intracellular structures was essentially due to "what you can't or don't see (resolve), you can't describe".

The gas vacuole is actually composed of gas vesicles and the chromatophore is a part of an extensive intracytoplasmic membrane system; neither of these structures can be resolved in the light microscope. The same is true for many of the other structures, and in some instances the structures are so uncommon that their serendipitous discovery was required even if they could have been resolved. Therefore, the discovery of complex intracellular structures necessitated the introduction of a new technology, namely the electron microscope and related technologies, as well as the careful ultrastructural characterization of known and newly isolated prokaryotes (Fig. 1). Admittedly, other technologies, encompassing many disciplines, are also of paramount importance and are required to fully understand a structure; this remains true today. The discovery and characterization of complex intracellular structures in prokaryotes continues into this century.

A brief description and the historical background of each "primary" intracytoplasmic structure mentioned above will be presented in this introductory chapter. Subsequent chapters (this volume) by Maupin-Furlow et al., Adir et al., Frigaard and Bryant, Pfeifer, Heinhorst et al., Scheffel and Schüler, Niederman, Fuerst, Fuerst et al., and Angert provide in-depth coverage of each structure.

After the "primary topics" for this volume were selected, colleagues and authors suggested several other interesting structures. In an effort to provide more complete coverage, but still remain within the established guidelines for the length of the monograph, seven structure topics were added as mini (cameo) chapters: see chapters authored by Vollmer, Kürner et al., Balish, Hoiczyk, Sergueev et al., van Keulen, and Corsaro and Venditti in this volume.

2
Complex Intracellular Structures: Descriptions and Discovery

The structures vary from those that exist as either naked protein or protein-lipid (membrane) complexes to those that are enclosed in either a monolayer protein coat or bilayer membrane(s), to those that represent intracellular "active" offspring.

2.1
Proteasomes

Proteasomes, intracellular nanocompartments, are energy-dependent proteases that degrade proteins into 3–30 amino acid oligopeptides (Baumeister et al. 1998; Bochtler et al. 1999; De Mot et al. 1999; Maupin-Furlow et al. 2000, 2001; Zwickl et al. 2000). The structures are ubiquitous in both eukaryotes and archaea but are rare in bacteria, having been identified only in actinomycetales. The 20S catalytic core, 11–12 nm in diameter by 15 nm in length, is a cylinder (four-stacked heptameric rings) with narrow openings at each end allowing limited substrate access to the central chamber where hydrolysis takes place (Zwickl et al. 2000; Maupin-Furlow et al. 2000, 2001; Kaczowka and Maupin-Furlow 2003). In both eukaryotes and archaea the 20S core associates with ATPase regulatory components, 19S cap and proteasome-activating nucleotidase (PAN), respectively, to form the energy-dependent complex (Zwickl et al. 1999; Maupin-Furlow et al. 2001; Kaczowka and Maupin-Furlow 2003). The bacterial 20S proteasome is thought to associate with ARC (ATPase forming ring-shaped complexes), but the ability of ARC to facilitate protein unfolding and/or the association with 20S proteasomes remains to be established (Wolf et al. 1998).

Although proteasome-like structures were first described in eukaryotes in the early 1960s (Baumeister et al. 1997, 1998), it was not until 1988 that enzymological studies revealed an array of proteolytic activities that lead to a consensus name "multicatalytic proteinase" (Dahlmann et al. 1988). However, that same year Arrigo et al. (1988), recognizing the structure as a functional compartment, suggested the name "proteasome". Dahlmann et al. (1989, 1992), noting that archaea possess some properties reminiscent of eukaryotes, investigated the archaebacterium, *Thermoplasma acidophilum*, and discovered the presence of proteolytically active particles (prosome/proteasome) similar in shape to those of eukaryotes. Although it was suggested that proteasomes were restricted in the Archaea to *Thermoplasma* (Pühler et al. 1994), Maupin-Furlow and Ferry (1995) demonstrated their presence in a methanogenic archaeon, *Methanosarcina thermophila*. Genome sequences now reveal that they are ubiquitous in the Archaea. The first bacterial 20S proteasome was found in the actinomycete, *Rhodococcus erythropolis*

(Tamura et al. 1995; Zuhl et al. 1997). A chapter by Maupin-Furlow et al., this volume, provides in-depth coverage of proteasomes.

2.2 Phycobilisomes

Phycobilisomes are supramolecular protein complexes present in both prokaryotic (cyanobacteria) and eukaryotic cells (cyanelles, red algae) (Shively 1974; Gantt 1980; Glazer 1985; Shively et al. 1988; Sidler 1994; Tandeau de Marsac 2003; Adir 2005). The phycobilisomes, attached to photosynthetic lamellae and functioning as the light-harvesting antennae for photosystem II, are primarily composed of intensely pigmented phycobiliproteins, and smaller amounts of non-chromophore bearing "linker peptides". There are three major phycobiliprotein families: phycocyanins, allophycocyanins, and phycoerythrins present in red algae and cyanobacteria (Sidler 1994; Tandeau de Marsac 2003). Electron microscopy reveals the morphology of the phycobilisome to be quite variable depending on the organism, with four different types having been described: hemi-ellipsoidal, hemi-discoidal, block shaped, and bundle shaped (Shively et al. 1988; Sidler 1994; Tandeau de Marsac 2003; Adir 2005).

Esenbeck (1836) reported on a water-soluble blue-colored pigment he labeled "saprocyanin" (later named phycocyanin) that was released from the cyanobacterium *Oscillatoria*. During the rest of the 19th century and well into the 20th century additional phycobiliproteins were discovered and their various properties described (Sidler 1994; Tandeau de Marsac 2003). Myers and Kratz (1955) and Myers et al. (1956) determined that phycobiliproteins were major constituents of cyanobacteria and red algae and theorized that the 22 nm granules observed between the thylakoids were actually phycobiliprotein aggregates. In an elegant series of studies on the ultrastructure of the red alga *Porphyridium cruentum*, Gantt and Conti (1965, 1966a,b) demonstrated that phycobiliproteins were present as 40 nm granules arranged in regular rows on the stromal surfaces of the thylakoids. They named the granules phycobilisomes. Lefort (1965) observed the structures in the cyanobacterial endosymbionts, but did not identify the structures as phycobilisomes until later (Bourdu and Lefort 1967). Gantt and collaborators as well as other research groups reported on the presence of phycobiliproteins in diverse red algae and cyanobacteria (Edwards et al. 1968; Gantt and Conti 1969; Edwards and Gantt 1971; Gantt 1980). Gantt and Lipschultz (1972) developed a method for the isolation of the structures from *P. cruentum* and later the method was modified for the isolation of phycobilisomes from cyanobacteria (Gray et al. 1973), thus paving the way for extensive studies of the phycobilisomes. The relative ease of isolation and purification of various phycobilisome components enabled researchers to perform very detailed spectroscopic and crystallographic studies, which afforded a precise molecular view of the structure and function of the complex (Huber 1989; Glazer 1989). A chapter by Adir et al., this volume, provides in-depth coverage of phycobilisomes.

2.3 Chlorosomes

Chlorosomes (originally chlorobium vesicles), the light-harvesting structures of the green photosynthetic bacteria, are elongated ellipsoid bodies 30–70 wide × 100–200 nm long, surrounded by a non-unit monolayer membrane ∼ 3 nm thick, and filled with bacteriochlorophyll *c*, *d*, or *e* (Shively 1974; Staehelin et al. 1980; Feick et al. 1982; Shively et al. 1988; Wahlund et al. 1991; Blankenship et al. 1995; Blankenship and Matsuura 2003; Frigaard and Bryant 2004). The structures, firmly attached to the inner surface of the cytoplasmic membrane, transfer absorbed light energy to the photosynthetic reaction centers in the cytoplasmic membrane (Blankenship and Matsuura 2003; Frigaard and Bryant 2004). Each cell may contain 200–250 chlorosomes (Blankenship and Matsuura 2003; Frigaard and Bryant 2004).

While studying the ultrastructure of several strains of "green bacteria" via electron microscopy Cohen-Bazire (1963) noted in every strain the presence of oblong structures, she called chlorobium vesicles, 100–150 nm long and 30–40 nm wide lining the "cortex of the cytoplasm between the surface membrane system and the ribosomal region". Fractionation of broken cells seemed to indicate that the bulk of the cell chlorophyll was present in vesicles of this size and shape. During the next several years the vesicles were reported to be present in still other green bacteria, and studies demonstrated the vesicles to be surrounded by a non-unit single layered membrane (Cohen-Bazire et al. 1964; Cohen-Bazire and Sistrom 1966; Holt et al. 1966; Pfennig and Cohen-Bazire 1967). The chlorobium vesicles were partially purified from *Chlorobium* strains and shown to be highly enriched in bacteriochlorophyll, to contain large amounts of monoglactosyl diglyceride and to be low in phospholipids (Cruden and Stainer 1970; Cruden et al. 1970). Some years later, Staehelin et al. (1978) studied the supramolecular architecture of chlorobium-type vesicles in freeze-fractured cells of *Chloroflexus aurantiacus* and proposed changing the name to chlorosomes. A chapter by Frigaard and Bryant, this volume, provides in-depth coverage of chlorosomes.

2.4 Gas Vesicles

Gas vacuoles, composed of a variable number of gas vesicles, are compound organelles found mainly in certain aquatic prokaryotes. The gas vesicles are gas-filled cylinders with conical ends that, depending on the organism, vary in width from 45 to greater than 100 nm, and in length from 100 nm to over 1.0 mm (Bowen and Jensen 1965a,b; Cohen-Bazire et al. 1969; Walsby 1972,

1994). The vesicle protein wall, approximately 2 nm thick, has a ribbed construction, the rib periodicity being between 4 and 5 nm.

Ahlborn (1895), Klebahn (1895), and Strodtman (1895) independently discovered the gas vacuoles in waterbloom-forming cyanobacteria. Klebahn (1895, 1922, 1925, 1929) showed that they were filled with gas, provided cells with buoyancy, and disappeared when cells were placed under pressure. Based on his observations, Klebahn proposed that the gas vacuoles were surrounded by a membrane. Numerous other early studies dealt with the gas vacuoles of planktonic cyanobacteria (Molisch 1903; Canabaeus 1929; Fogg 1941). Lauterborn (1913, 1915) expanded the list of gas vacuole producers to include aquatic heterotrophic bacteria. Many years later Bowen and Jensen (1965a,b), observing thin sections in the electron microscope, found that the gas vacuoles of five cyanobacteria were composed of stacks of cylindrical, hollow structures they called "gas vesicles". The cylinders had conical ends and were bounded by a membrane only 2 nm thick. Although the majority of the early research was accomplished on cyanobacterial gas vacuoles, they were first observed in the electron microscope by Houwink (1956) in *Halobacterium halobium*. He noted that the large gas vacuoles observed in the light microscope were composed of much smaller, hollow structures (vesicles). Without thin sections he was able to demonstrate the three dimensional structure of the vesicles based on stereo-pairs. He also noted that the vesicles collapsed due to the pressure created during centrifugation. His elegant work was confirmed and extended by Larsen et al. (1967) and Stoeckenius and Kunau (1968). Gas vesicles are commonly limited to planktonic bacteria and halophilic archaea, and have now been reported in well over 150 species (Shively et al. 1988; Walsby 1994). Recently, gas vesicle gene clusters have been determined to be present in *Bacillus megaterium* and some actinomycete genera (Li and Cannon 1998, van Keulen et al. 2005). A chapter by Pfeifer, this volume, provides in-depth coverage of gas vesicles. See also a cameo chapter by van Keulen, this volume.

2.5
Carboxysomes and Carboxysome-Like Inclusions

Carboxysomes, referred to as polyhedral bodies before being isolated and partially characterized, are simple organelles that carry out the fixation of carbon dioxide via the enzyme ribulose-1,5-bisphosphate carboxylase/oxygenase (RuBisCO) (Cannon et al. 2001).

Inclusions with polygonal profiles were first observed in the cyanobacteria (Drews and Niklowitz 1956; Lefort 1960a,b; Hagedorn 1961) and based on their characteristic shape, Jensen and Bowen (1961) labeled the inclusions of *Nostoc pruniforme* "polyhedral bodies." During the next 12 years additional cyanobacteria (Pankratz and Bowen 1963; Gantt and Conti 1969; Wolk 1973) as well as several autotrophic bacteria (Murray and Watson 1965; van Gool et al. 1969; Wang and Lundgren 1969; Wang et al. 1970; Shively et al. 1970; Shively 1974; Shively et al. 1988) were shown to possess polyhedral bodies. Shively et al. (1973a,b) isolated the bodies from *Halothiobacillus (Thiobacillus) neapolitanus*, demonstrated that they were surrounded by a monolayer shell, possessed the enzyme RuBisCO, were functional in carbon dioxide fixation, and proposed the name carboxysomes. Subsequently the polyhedral bodies of two cyanobacteria and *Nitrobacter agilis* were shown to be carboxysomes (Codd and Stewart 1976; Lanaras and Codd 1981; Shively et al. 1977). It is now widely accepted that polyhedral bodies, i.e., carboxysomes, surrounded by a monolayer shell and varying in size from approximately 100 to 700 nm in diameter, are ubiquitous in cyanobacteria, but otherwise are found only in a limited number of chemoauthotrophic bacteria (Jensen 1993; Cannon et al. 2001; Badger and Price 2003).

More recently it has been reported (Walter et al. 1997; Bobik et al. 1997) that the operons containing the genes encoding the proteins for propanediol and ethanolamine utilization by *Salmonella typhimurium* also contain genes that are homologous to certain of the carboxysome shell genes. Subsequently, Shively et al. (1998) confirmed the presence of the gene homologs in *Salmonella typhimurium* and showed that several enteric bacteria had polyhedral bodies when grown on propanediol/ethanolamine. A year later Bobik et al. (1999) demonstrated that the polyhedral bodies were directly involved in propanediol utilization. With the recent progress in genome sequencing, it has now been shown that a considerable number of prokaryotic organisms have the potential of producing polyhedral bodies that could be involved in other metabolic processes. Thus, the presence of polyhedral organelles appears to be more widespread than originally considered. A chapter by Heinhorst et al., this volume, provides in-depth coverage of carboxysome and carboxysome-like inclusions.

2.6 Magnetosomes

Many Gram-negative, motile, aquatic bacteria are capable of aligning themselves along geomagnetic field lines and thus are said to be magnetotactic (Blakemore 1982; Ofer et al. 1984; Frankel and Blakemore 1988, 1989; Bazylinski 1995; Bazylinski and Frankel 2004). Magnetotaxis, along with aerotaxis, guide bacteria to an environment of appropriate oxygen tension (Bazylinski and Frankel 2004). The bacteria accomplish the geomagnetic field alignment by creating intracellular membrane-bound organelles called magnetosomes that contain either magnetite or gerigite crystals. The magnetosome is morphologically diverse depending on the bacterium, but they are commonly polygonal shaped structures 30–200 nm in width (Balkwill et al. 1980; Schüler 2002, 2004; Keim et al. 2005).

Blakemore (1975) described motile bacteria in marine sediments that were directed by a local geomagnetic field as a magnetotactic response. He noted that the bacteria had flagella and possessed iron-rich particles contained within membrane vesicles, and postulated that the particles might be responsible for the magnetic cell movement. Four years later Frankel et al. (1979) established the presence of magnetite crystals in bacteria. Balkwill et al. (1980) published on the ultrastructure of a magnetotactic spirillum wherein they described in detail the electron-dense, enveloped, iron-rich particles. The term "magnetosome" was proposed for the particles found in this spirillum as well as those found in other magnetotactic bacteria. A chapter by Scheffel and Schüler, this volume, provides in-depth coverage of magnetosomes.

2.7 Intracytoplasmic Membranes

Many prokaryotic organisms, including photosynthetic bacteria, cyanobacteria, aerobic chemolithoautotrophic nitrifying bacteria, and methanotrophs, possess extensive intracytoplasmic bilayer membranes (Echlin and Morris 1965; Murray and Watson 1965; Watson and Mandel 1971; Wolk 1973; Pfennig 1977; Remsen 1978; Golecki and Drews 1982; Nierzwicki-Bauer et al. 1983; Hanson and Hanson 1996; Schmidt et al. 2001). These "extra" membranes appear as various morphological types, e.g., vesicles, tubules, and single, paired (thylakoids), or stacked lamellae, depending on the organism and the culture conditions. They can exist at the periphery or in the interior of the cytoplasmic compartment (Pankratz and Bowen 1963; Echlin and Morris 1965; Lang 1968: Remsen 1978; Golecki and Drews 1982; Jensen 1993). In most, if not all of these cases, the membranes originate from, and arise by invagination of, the cytoplasmic membrane and serve to increase the membrane surface area for the enzymes of metabolic processes (Lang 1968; Remsen 1978; Golecki and Drews 1982; Jensen 1993; Zak et al. 2001). Although the above represent the well known intracytoplasmic membranes, others with completely different functions do exist (see chapters by Fuerst, and Fuerst et al., this volume).

Schachman et al. (1952) and Pardee et al. (1952) isolated a particulate fraction from disrupted cells of the photosynthetic bacterium *Spirillum rubrum* (*Rhodospirillum rubrum*) that contained all of the pigments of the cell. The particles, with a sedimentation coefficient of 190 S, were either flattened disks (110 nm diameter) or vesicles (60 nm diameter). Schachman named the particles chromatophores. Two years later, Frenkel (1954) demonstrated that the chromatophores, in the presence of ADP, inorganic phosphate and light, could carry out photophosphorylation. Thus, the particles had all of the bacterial chlorophyll and carotenoids of the cell and were photochemically active. When the membrane-bound vesicles (chromatophores) were first observed inside cells, it was suggested that they were discrete cytoplasmic organelles (Vatter and Wolfe 1958). However, it was clearly shown by subsequent researchers that the vesicles were the result of tubular, subsequently sequentially constricted, invaginations of the cytoplasmic membrane (Cohen-Bazire and Kunisawa 1963; Drews and Giesbrecht 1963; Boatman 1964; Holt and Marr 1965; Remsen 1978). The photosynthetic lamellae of the cyanobacteria were seen first with the electron microscope by Calvin and Lynch (1952) and later observed and described in more detail by Niklowitz and Drews (1956, 1957). The lamellae appear as closed discs termed thylakoids (Wolk 1973). A chapter by Niederman, this volume, provides in-depth coverage of intracytoplasmic membranes.

2.8 Planctomycetes

Based on phylogenetic analyses, the planctomycetes, order Planctomycetales, comprise a separate kingdom-level division of the Domain Bacteria, at present classified as a separate phylum of this Domain (Jenkins et al. 2002; Fuerst 2005). Their distinctive phylogeny is mirrored in their phenotype; budding reproduction, proteinaceous cell wall without peptidoglycan, and compartmentation by intracytoplasmic membranes (Lindsay et al. 2001). Several species, reminiscent of eukaryotes, have nucleoids enveloped by either a single (pirellulosome) or double membrane (Fuerst and Webb 1991; Lindsay et al. 2001). In addition, at least three different genera possess a bilayer-bound, anaerobic ammonia-oxidizing compartment called the anammoxosome (Strous et al. 1999; Schmid et al. 2000; Lindsay et al. 2001; van Niftrik et al. 2004).

Historically, it was Gimesi (1924) who first described a budding organism, *Planctomyces bekefii*, but it was identified as an aquatic fungus because of its appearance, i.e., threadlike forms bearing spherical structures. Some years later Henrici and Johnson (1935) observed stalked and budding bacteria, morphologically similar to what Giemsi had reported; these bacteria were later given the name *Blastobacter henricii*. For many years a number of planctomycetes from diverse aquatic habitats were discovered, but it was not until 1973 that the first isolation and species description of a budding bacterium, *Pasteuria ramosa*, was reported (Staley 1973). The organism was later shown by molecular sequencing to be a member of the phylum *Planctomycetes* and renamed *Pirellula staleyi* (Schlesner and Hirsch 1984; Schlesner and Hirsch 1987). To date, through the efforts of various researchers many planctomycetes have been described from numerous environmental sources culminating in the establishment of nine genera, six of which have been cultured (Staley et al. 1976; Schlesner 1986, 1989, 1994; Ward et al. 1995; Jenkins et al. 2002; Wang et al. 2002; Chouari 2003, Fuerst 2005).

2.8.1 Membrane-Bounded Nucleoids and Pirellulosomes

Franzmann and Skerman (1984) published a detailed description of a new budding bacterium, *Gemmata obscuriglobus*. A micrograph in their publication showed, what appeared to be, "packaged" nuclear material. This micrograph prompted Fuerst and Webb (1991) to undertake a more detailed examination of the bacterium. Using several different electron microscopy techniques, including thin sectioning of chemically fixed cells, thin sectioning of freeze-substituted cells, freeze-fracture/freeze-etch, and immunogold labeling, the researchers were able to demonstrate double-stranded DNA associated with a fibrillar nucleoid that was surrounded by two nuclear membranes. Lindsay et al. (1997) used essentially the same techniques plus serial sectioning and computer simulations to examine two additional planctomycetes, *Pirellula marina* (now known as *Blastopirellula marina*) and *Pirellula staleyi*. In these bacteria the condensed fibrillar nucleoid was also enclosed in a membrane-enveloped compartment labeled the pirellulosome, however, in this instance there was only a single membrane. A chapter by Fuerst, this volume, provides in-depth coverage of membrane-bounded nucleoids and pirellulosomes.

2.8.2 Anammoxosomes

Broda (1977) theorized, based on thermodynamic calculations, the existence of an ammonia-oxidizing chemolithotrophic bacterium; the oxidant being nitrate, oxygen, or carbon dioxide with the end product being dinitrogen gas. Broda's conclusions were shown to be correct when the microbiological anaerobic ammonium oxidation (the anammox process) was demonstrated with nitrite serving as the electron acceptor, producing dinitrogen gas (Mulder et al. 1995; van de Graaf 1995). Strous et al. (1999) identified a new lithotrophic planctomycete as the organism responsible for the oxidation. The organism, tentatively named "*Candidatus* Brocadia anammoxidans" was demonstrated to have a ribosome-free cytoplasm compartment, termed the anammoxosome, bounded by a single membrane and possessing hydroxylamine oxidoreductase, an enzyme unique to anaerobic ammonium oxidation (Lindsay et al. 2001). A chapter by Fuerst et al., this volume, provides in-depth coverage of anammoxosomes.

2.9 Cytoarchitecture of *Epulopiscium* spp.

Epulopiscium spp. and related intestinal symbionts, belong to the bacterial Phylum Firmicutes, within the Clostridiales. The cells of the largest *Epulop-*

iscium spp. are cigar-shaped and reach lengths in excess of 600 μm. They exhibit an unusual mode of cellular reproduction where an individual may form multiple, intracellular "active" offspring (Angert 2005).

Researchers, studying the biology of reef fish in the Red Sea, discovered enormous intestinal microorganisms within the brown surgeonfish *Acanthurus nigrofuscus* (Fishelson et al. 1985). This publication documented several characteristics exhibited by these microbes: (1) an extensive internal tubular membrane system, (2) a lack of typical eukaryotic organelles, and (3) an extensive array of cilia-like filaments covering the cells, but lacking the structure characteristics of eukaryotic cilia. Based on their size and mode of reproduction, Montgomery and Pollack (1988) described the organisms as a new genus and species of protist, "*Epulopiscium fishelsoni*". A survey of surgeonfish in the Pacific revealed that similar intestinal symbionts were widespread among surgeonfish (Clements et al. 1989). Further ultrastructural studies provided additional evidence that all of these unusual organisms possessed features more typical of prokaryotes than eukaryotes (Clements and Bullivant 1991). The surface filaments appeared to be bacterial-type flagella, and freeze-fractures revealed that the internal membranes contained no nuclear-type pores or other prominent features. The DNA of the cells was described as "coagulated" in appearance, like that of bacterial DNA, when cells were fixed and prepared for electron microscopy by conventional methods (Clements and Bullivant 1991). Finally, based on their small subunit rRNA gene sequence, these enormous symbionts were placed within the Domain Bacteria (Angert et al. 1993). A meticulous study using light and electron microscopy described the location and appearance of cellular DNA of large *Epulopiscium* cells and suggested that the "islands" of DNA in a single cell may represent numerous, typical bacterial nucleoids (Robinow and Angert 1998).

Recent studies have focused on the process of internal offspring formation. The close evolutionary relationship between endospore-forming bacteria and the *Epulopiscium* group lead to the hypothesis that vivipary in *Epulopiscium* may have evolved from endosporulation (Angert et al. 1996). Numerous studies support this hypothesis (Angert and Losick 1998; Angert and Clements 2004; Flint et al. 2005). A chapter by Angert, this volume, provides in-depth coverage of the cytoarchitecture of *Epulopiscium* spp.

References

Adir N (2005) Elucidation of the molecular structures of components of the phycobilisome: reconstructing a giant. Photosynthesis Res 85:15–32

Ahlborn F (1895) Über die Wasserblüte *Byssus flosaquae* und ihr Verhalten gegen Druck. Verh Naturwiss Ver Hamburg III 2:25

Alberts B, Johnson A, Lewis J, Raff M, Roberts K, Walter P (2002) Molecular biology of the cell, 4th edn. Garland Science, New York

Angert ER (2005) Alternatives to binary fission in bacteria. Nat Rev Microbiol 3:214–224
Angert ER, Clements KD (2004) Initiation of intracellular offspring in *Epulopiscium*. Mol Microbiol 51:827–835
Angert ER, Losick RM (1998) Propagation by sporulation in the guinea pig symbiont *Metabacterium polyspora*. Proc Natl Acad Sci USA 95:10218–10223
Angert ER, Clements KD, Pace NR (1993) The largest bacterium. Nature 362:239–241
Angert ER, Brooks AE, Pace NR (1996) Phylogenetic analysis of *Metabacterium polyspora*: clues to the evolutionary origin of daughter cell production in *Epulopiscium* species, the largest bacteria. J Bacteriol 178:1451–1456
Arrigo AP, Tanaka K, Goldberg AL, Welch WJ (1988) Identity of the 19S 'prosome' particle with the large multifunctional protease complex of mammalian cells (the proteasome). Nature 331:192–194
Badger MR, Price GD (2003) CO_2 concentrating mechanisms in cyanobacteria: molecular components, their diversity and evolution. J Exptl Bot 54:609–622
Balkwill DL, Maratea D, Blakemore RP (1980) Ultrastructure of a magnetotactic spirillum. J Bacteriol 141:1399–1408
Baumeister W, Cejka Z, Kania M, Seemüller E (1997) The proteasome: a macromolecular assembly designed to confine proteolysis to a nanocompartment. Biol Chem 378:121–130
Baumeister W, Walz J, Zühl F, Seemüller E (1998) The proteasome: paradigm of a self-compartmentalizing protease. Cell 92:367–380
Bazylinski DA (1995) Structure and function of the bacterial magnetosome. ASM News 61:337–343
Bazylinski DA, Frankel RB (2004) Magnetosome formation in prokaryotes. Nat Rev Microbiol 2:217–230
Beck RW (2000) A chronology of microbiology in historical context. ASM, Washington
Blakemore R (1975) Magnetotactic bacteria. Science 190:377–379
Blakemore RP (1982) Magnetotactic bacteria. Annu Rev Microbiol 36:217–238
Blankenship RE, Matsuura K (2003) Antenna complexes from green photosynthetic bacteria. In: Green BR, Parson WW (eds) Light-harvesting antennas in photosynthesis. Kluwer, Dordrecht, The Netherlands, p 195–217
Blankenship RE, Olson JM, Miller M (1995) Antenna complexes from green photosynthetic bacteria. In: Blankenship RE, Madigan MT, Bauer CE (eds) Anoxygenic photosynthetic bacteria. Kluwer, Dordrecht, p 399–435
Boatman ES (1964) Observations of the fine structure of spheroplasts of *Rhodospirillum rubrum*. J Cell Biol 20:297–311
Bobik TA, Xu Y, Jeter RM, Otto KE, Roth JR (1997) Propanediol utilization genes (pdu) of *Salmonella typhimurium*: Three genes for the propanediol dehydratase. J Bacteriol 179:6633–6639
Bobik TA, Havemann GD, Busch RJ, Williams DS, Aldrich HC (1999) The propanediol utilization (*pdu*) operon of *Salmonella enterica* serovar typhimurium LT2 includes genes necessary for formation of polyhedral organelles involved in coenzyme B12-dependent 1,2-propanediol degradation. J Bacteriol 181:5967–5975
Bochtler M, Ditzel L, Groll M, Hartmann C, Huber R (1999) The proteasome. Annu Rev Biophys Biomol Struct 28:295–317
Bourdu R, Lefort M (1967) Structure fine, observee en cryode capage, des lamelles photosynthetiques des yanophycees endosymbiotiques: *Glaucocystis nostochinearum* Itzigs, et *Cyanophora paradoxa*. Compt Rend 265:37–40
Bowen CC, Jensen TE (1965a) Blue-green algae: fine structure of the gas vacuoles. Science 147:1460–1462

Bowen CC, Jensen TE (1965b) Fine structure of gas vacuoles in blue-green algae. Amer J Bot 52:641
Broda E (1977) Two kinds of lithotrophs missing in nature. Z Allg Mikrobiol 17:491–493
Calvin M, Lynch V (1952) Grana-like structures of *Synechococcus cedorum*. Nature 169:455
Canabaeus L (1929) Über die Heterocysten und Gasvakuolen der Blaualgen und ihre Beziehung zueinander. Pflanzenforschung, vol 13. Jena
Cannon GC, Bradburne CE, Aldrich HC, Baker SH, Heinhorst S, Shively JM (2001) Microcompartments in prokaryotes: carboxysomes and related polyhedra. Appl Environ Microbiol 67:5351–5361
Chouari R, Paslier D, Daegelen P, Ginestet P, Weissenbach J, Sghir A (2003) Molecular evidence for novel planctomycete diversity in a municipal wastewater treatment plant. Appl Environ Microbiol 69:7365–7363
Clements KD, Bullivant S (1991) An unusual symbiont from the gut of surgeonfishes may be the largest known prokaryote. J Bacteriol 173:5359–5362
Clements KD, Sutton DC, Choat JH (1989) Occurrence and characteristics of unusual protistan symbionts from surgeonfishes Acanthuridae of the Great Barrier Reef Australia. Marine Biol 102:403–412
Codd GA, Stewart WDP (1976) Polyhedral bodies and ribulose 1,5-diphosphate carboxylase of the blue-green alga *Anabaena cylindrica*. Planta 130:323–326
Cohen-Bazire G (1963) Some observations on the organization of the photosynthetic apparatus in purple and green bacteria. In: Gest H, San Pietro A, Vernon LP (eds) Bacterial photosynthesis. Antioch, Yellow Springs, Ohio, pp 89–119
Cohen-Bazire G, Kunisawa R (1963) The fine structure of *Rhodospirillum rubrum*. J Cell Biol 16:401–419
Cohen-Bazire G, Sistrom WR (1966) The procaryotic photosynthetic apparatus. In: Vernon LP, Seeley GR (eds) The chlorophylls. Academic, New York, p 313–341
Cohen-Bazire G, Pfennig N, Kunisawa R (1964) The fine structure of green bacteria. J Cell Biol 22:207–225
Cohen-Bazire G, Kunisawa R, Pfennig N (1969) Comparative study of the structure of gas vacuoles. J Bacteriol 100:1049–1061
Cruden DL, Stanier RY (1970) The characterization of chlorobium vesicles and membranes isolated from green bacteria. Arch Mikrobiol 72:115–134
Cruden Dl, Cohen-Bazire G, Stanier RY (1970) Chlorobium vesicles: The photosynthetic organelles of green bacteria. Nature 228:1345–1347
Dahlmann B, Kuehn L, Ishiura S, Tsukahara T, Sugita H, Tanaka K, Rivett AJ, Hough RF, Rechsteiner M, Mykles DI et al. (1988) The multicatalytic proteinase: a high-Mr endopeptidase. Biochem J 255:750–751
Dahlmann B, Kopp F, Kuehn L, Niedel B, Pfeifer G, Hegerl R, Baumeister W (1989) The multicatalytic proteinase (prosome) is ubiquitous from eukaryotes to archaebacteria. FEBS Lett 251:125–131
Dahlmann B, Kuehn L, Grziwa A, Zwickl P, Baumeister W (1992) Biochemical properties of the proteasome from *Thermoplasma acidophilum*. Eur J Biochem 208:789–797
De Mot R, Nagy I, Walz J, Baumeister W (1999) Proteasomes and other self-compartmentalizing proteases in prokaryotes. Trends Microbiol 7:88–92
Drews G, Giesbrecht P (1963) On the morphogenesis of bacterial chromatophores (thylakoids) and on the synthesis of bacteriochlorophyll in *Rhodopseudomonas spheroides* and *Rhodospirillum rubrum*. Zentralbl Bakteriol 190:508–535
Drews G, Niklowitz W (1956) Beiträge zur Cytologie der Blaualgen. II. Zentroplasma und granulare Einschlüsse von *Phormidium uncinatum*. Arch Mikrobiol 24:147–162

Echlin P, Morris I (1965) The relationship between blue-green algae and bacteria. Biol Rev 40:143–187

Edwards MR, Gantt E (1971) Phycobilisomes of the thermophilic blue-green alga *Synechococcus lividus*. J Cell Biol 50:896–900

Edwards MR, Berns DS, Holt SC, Ghiorse WC (1968) Ultrastructure of the thermophilic blue-green alga *Synechococcus lividus* Copeland. J Phycol 4:283–298

Esenbeck N (1836) Ueber einen blau-rothen Farbstoff, der sich bei der Zersetzung von *Oscillatorien* bildet. Liebigs Ann Chem XVII:75–82

Feick RG, Fitzpatrick M, Fuller RC (1982) Isolation and characterization of cytoplasmic membranes and chlorosomes from the green bacterium *Chloroflexus aurantiacus*. J Bacteriol 150:905–915

Fishelson L, Montgomery WL, Myrberg AA (1985) A unique symbiosis in the gut of a tropical herbivorous surgeonfish (Acanthuridae: Teleostei) from the Red Sea. Science 229:49–51

Flint JF, Drzymalski D, Montgomery WL, Southam G, Angert ER (2005) Nocturnal production of endospores in natural populations of *Epulopiscium*-like surgeonfish symbionts. J Bacteriol 187:7460–7470

Fogg GE (1941) The gas-vacuoles of the Myxophyceae (Cyanophyceae). Biol Rev Cambridge Phil Soc 16:205–217

Frankel RB, Blakemore RP (1988) Magnetite and magnetotaxis in microorganisms. Adv Exp Med Biol 238:321–330

Frankel RB, Blakemore RP (1989) Magnetite and magnetotaxis in microorganisms. Bioelectromagnetics 10:223–237

Frankel RB, Blakemore RP, Wolfe RS (1979) Magnetite in freshwater magnetotactic bacteria. Science 203:1355–1356

Franzmann PD, Skerman VB (1984) Gemmata obscuriglobus, a new genus and species of the budding bacteria. Antonie Van Leeuwenhoek 50:261–268

Frenkel AW (1954) Light-induced photophosphorylation by cell-free preparations of photosynthetic bacteria. J Am Chem Soc 76:5568–5569

Frigaard N-U, Bryant DA (2004) Seeing green bacteria in a new light: genomics-enabled studies of the photosynthetic apparatus in green sulfur bacteria and filamentous anoxygenic phototrophic bacteria. Arch Microbiol 182:265–276

Fuerst JA (2005) Intracellular compartmentation in planctomycetes. Annu Rev Microbiol 59:299–328

Fuerst JA, Webb RI (1991) Membrane-bounded nucleoid in the Eubacterium *Gemmata obscuriglobus*. Proc Natl Acad Sci USA 88:8184–8188

Gantt E (1980) Structure and function of phycobilisomes: light-harvesting pigment complexes in red and blue-green algae. Int Rev Cytol 66:45–80

Gantt E, Conti SF (1965) The ultrastructure of *Porphyridium cruentum*. J Cell Biol 26:365–375

Gantt E, Conti SF (1966a) Granules associated with the chloroplast lamellae of *Porphyridium cruentum*. J Cell Biol 9:423–434

Gantt E, Conti SF (1966b) Phycobiliprotein localization in algae. Brookhaven Symp Biol 19:393–405

Gantt E, Conti SF (1969) Ultrastructure of blue-green algae. J Bacteriol 97:1486–1493

Gantt E, Lipschultz CA (1972) Phycobilisomes of *Porphyridium cruentum*. I. Isolation. J Cell Biol 54:313–324

Gimesi N (1924) Hydrobiologiai tanulma ìnyok (Hydrobiologische Studien). I. Planctomyces bekefii Gim nov gen et sp., p 1–8. Kiadja a Magyar Ciszterci Rend, Budapest. (In Hungarian with German translation.)

Glazer AN (1985) Light harvesting by phycobilisomes. Annu Rev Biophys Biophys Chem 14:47–77
Glazer AN (1989) Light guides. Directional energy transfer in a photosynthetic antenna. J Biol Chem 264:1–4
Golecki JR, Drews G (1982) Supramolecular organization and composition of membranes. In: Carr NG, Whitton BA (eds) The biology of cyanobacteria. University of California, Berkeley, p 125–141
Gray BH, Lipschultz CA, Gantt E (1973) Phycobilisomes from a blue-green alga *Nostoc* species. J Bacteriol 116:471–478
Hagedorn H (1961) Untersuchungen über die Feinstruktur der Blaualgen. Z Naturforsch 16b:825–829
Hanson RS, Hanson TE (1996) Methanotrophic bacteria. Microbiol Rev 60:439–471
Henrici AT, Johnson DE (1935) Studies of freshwater bacteria. II. Stalked bacteria, a new order of schizomycetes. J Bacteriol 30:61–93
Holt SC, Marr AG (1965) Location of chlorophyll in *Rhodospirillum rubrum*. J Bacteriol 89:1402–1412
Holt SC, Conti SF, Fuller RC (1966) Effect of light intensity on the formation of the photochemical apparatus in the green bacterium *Chloropseudomonas ethylicum*. J Bacteriol 91:349–355
Houwink AL (1956) Flagella, gas vacuoles and cell-wall structure in *Halobacterium halobium*; an electron microscope study. J Gen Microbiol 15:146–150
Huber R (1989) Nobel lecture. A structural basis of light energy and electron transfer in biology. EMBO J 8:2125–2147
Jenkins C, Kedar V, Fuerst JA (2002) Gene discovery within the planctomycete division of the domain Bacteria using sequence tags from genomic DNA libraries. Genome Biol 3:1–11
Jensen TE (1993) Cyanobacterial ultrastructure. In: Berner T (ed) Ultrastructure of microalgae. CRC, Boca Raton FL, p 7–51
Jensen TE, Bowen CC (1961) Organization of the centroplasm in *Nostoc pruniforme*. Iowa Acad Sci Proc 68:86–89
Kaczowka SJ, Maupin-Furlow JA (2003) Subunit topology of two 20S proteasomes from *Haloferax volcanii*. J Bacteriol 185:165–174
Keim CN, Solórzano G, Farina M, Lins U (2005) Intracellular inclusions of uncultured magnetotactic bacteria. Int Microbiol 8:111–117
Klebahn H (1895) Gasvacuolen, ein Bestandteil der Zellen der Wasserblüte bildenden Phycochromaceen. Flora 80:241–282
Klebahn H (1922) Neue Untersuchungen über die Gasvakuolen. Jahrb Wiss Bot 61:535–589
Klebahn H (1925) Weitere Untersuchungen über die Gasvakuolen. Ber Deut Bot Ges 43:143–159
Klebahn H (1929) Über die Gasvakuolen der Cyanophyceen. Verh Int Ver Theoret Angew Limnologie 4:408–414
Lanaras T, Codd GA (1981) Ribulose 1,5-bisphosphate carboxylase and polyhedral bodies of *Chlorogloeopsis fritschii*. Planta 153:279–285
Lang NJ (1968) The fine structure of the blue-green algae. Ann Rev Microbiol 22:15–46
Larsen H, Omang S, Steensland H (1967) On the gas vacuoles of the halobacteria. Arch Mikrobiol 59:197–203
Lauterborn R (1913) Zur Kenntnis einiger sapropelischer Schizomyceten. Allg Bot Z 19:97–100

Lauterborn R (1915) Die sapropelische Lebewelt. Naturhist-med. Vereins Z Heidelberg, NF 13:396–480

Lefort M (1960a) Structure inframicroscopique du chromatoplasma de quelques cyanophycees. C R Acad Sci 250:1525–1527

Lefort M (1960b) Nouvelles recherches sur l'infrastructure du chromatoplasma des cyanophycees. C R Acad Sci 251:3046–3048

Lefort M (1965) Sur le chromatoplasma d'une cyanophycde endosymbiotique: *Glaucocystis nostochinearum* Itzigs. Compt Rend 261:233–236

Li N, Cannon M (1998) Gas vesicle genes identified in *Bacillus megaterium* and functional expression in *Escherichia coli*. J Bacteriol 180:2450–2458

Lindsay MR, Webb RI, Fuerst JA (1997) Pirellulosomes: a new type of membrane-bounded cell compartment in planctomycete bacteria of the genus *Pirellula*. Microbiol 143:739–748

Lindsay MR, Webb RI, Strous M, Jetten MS, Butler MK, Forde RJ, Fuerst JA (2001) Cell compartmentalisation in planctomycetes: novel types of structural organisation for the bacterial cell. Arch Microbiol 175:413–429

Maupin-Furlow JA, Ferry JG (1995) A proteasome from the methanogenic archaeon *Methanosarcina thermophila*. J Biol Chem 270:28617–28622

Maupin-Furlow JA, Wilson HL, Kaczowka SJ, Ou MS (2000) Proteasomes in the archaea: From structure to function. Front Biosci 5:837–865

Maupin-Furlow JA, Kaczowka SJ, Ou MS, Wilson HL (2001) Archaeal proteasomes: proteolytic nanocompartments of the cell. Adv Appl Microbiol 50:279–338

Myers A, Preston RD, Ripley GW (1956) Fine structure in the red algae – I. X-ray and electronmicroscope investigation of *Griffithsia flosculosa*. Proc R Soc London Ser B 144:450–459

Myers J, Kratz WA (1955) Relation between pigment content and photosynthetic characteristics in a blue-green alga. J Gen Physiol 39:11–22

Molisch H (1903) Die sogennanten Gasvakuolen und das Schweben gewisser Phycochromaceen. Bot Z 61:47

Montgomery WL, Pollak PE (1988) *Epulopiscium fishelsoni* n.g., n.s., a protist of uncertain taxonomic affinities from the gut of an herbivorous reef fish. J Protozool 35:565–569

Mulder A, van de Graaf AA, Robertson LA, Kuenen JG (1995) Anaerobic ammonium oxidation discovered in a denitrifying fluidized bed reactor. FEMS Microbiol Ecol 16:177–184

Murray RGE, Watson SW (1965) Structure of *Nitrosocystis oceanus* and comparison with *Nitrosomonas* and *Nitrobacter*. J Bacteriol 89:1594–1609

Nierzwicki-Bauer SA, Balkwill DL, Stevens SE Jr (1983) Three-dimensional ultrastructure of a unicellular cyanobacterium. J Cell Biol 97:713–722

Niklowitz W, Drews G (1956) Beiträge zur Cytologie der Blaualgen. I. Untersuchungen zur Substruktur von *Phormidium uncinatum* Gom. Arch Mikrobiol 24:134–146

Niklowitz W, Drews G (1957) Beiträge zur Cytologie der Blaualgen. IV. Vergleichende elektronenmikroskopische Untersuchungen zur Substruktur einiger Hormogonales. Arch Mikrobiol 27:150–165

Ofer S, Nowik I, Bauminger ER, Papaefthymiou GC, Frankel RB, Blakemore RP (1984) Magnetosome dynamics in magnetotactic bacteria. Biophys J 46:57–64

Pankratz HS, Bowen CC (1963) Cytology of blue-green algae. I. The cells of *Symploca muscorum*. Amer J Bot 50:387–399

Pardee AB, Schachman HK, Stanier RY (1952) Chromatophores of *Rhodospirillum rubrum*. Nature 169:282–283

Pfennig N (1977) Phototrophic green and purple bacteria: a comparative systematic survey. Annu Rev Microbiol 31:275–290
Pfennig N, Cohen-Bazire G (1967) Some properties of the green bacterium *Pelodictyon clathratiforme*. Arch Mikrobiol 59:226–236
Pühler G, Pitzer F, Zwickl P, Baumeister W (1994) Proteasomes: multisubunit proteinases common to *Thermoplasma* and eukaryotes. System Appl Microbiol 16:734–741
Remsen CC (1978) Comparative subcellular architecture of photosynthetic bacteria. In: Clayton RK, Sistrom WR (eds) The photosynthetic bacteria. Plenum, New York, p 31–60
Robinow C, Angert ER (1998) Nucleoids and coated vesicles of *Epulopiscium* spp. Arch Microbiol 170:227–235
Schachman HK, Pardee AB, Stanier RY (1952) Studies on the macromolecular organization of microbial cells. Arch Biochem Biophys 38:245–260
Schlesner H (1986) *Pirella marina* sp. nov, a budding, peptidoglycanless bacterium from brackish water. Syst Appl Microbiol 8:177–180
Schlesner H (1989) *Planctomyces brasiliensis* sp. nov, a halotolerant bacterium from a salt pit. Syst Appl Microbiol 12:159–161
Schlesner H (1994) The development of media suitable for microorganisms morphologically resembling *Planctomyces* spp., *Pirellula* spp., and other Planctomycetales from various aquatic habitats using dilute media. Syst Appl Microbiol 17:135–145
Schlesner H, Hirsch P (1984) Assignment of ATCC 27377 to *Pirella* gen nov comb nov. Int J Syst Bacteriol 34:492–495
Schlesner H, Hirsch P (1987) Rejection of the genus name *Pirella* for pear-shaped budding bacteria and proposal to create the genus *Pirellula* gen nov. Int J Syst Bacteriol 37:441
Schmid M, Twachtmann U, Klein M, Strous M, Juretschko S, Jetten M, Metzger JW, Schleifer K-H, Wagner M (2000) Molecular evidence for genus level diversity of bacteria capable of catalyzing anaerobic ammonium oxidation. Syst Appl Microbiol 23:93–106
Schmidt I, Zart D, Bock E (2001) Effects of gaseous NO_2 on cells of *Nitrosomonas eutropha* previously incapable of using ammonia as an energy source. Antonie van Leeuwenhoek 79:39–47
Schüler D (2002) The biomineralization of magnetosomes in *Magnetospirillum gryphiswaldense*. Int Microbiol 5:209–214
Schüler D (2004) Molecular analysis of a subcellular compartment: the magnetosome membrane in *Magnetospirillum gryphiswaldense*. Arch Microbiol 181:1–7
Shively JM (1974) Inclusion bodies of prokaryotes. Ann Rev Microbiol 28:167–187
Shively JM, Decker GL, Greenawalt JW (1970) Comparative ultrastructure of the thiobacilli. J Bacteriol 101:618–627
Shively JM, Ball F, Brown DH, Saunders RE (1973) Functional organelles in prokaryotes: polyhedral inclusions (carboxysomes) of *Thiobacillus neapolitanus*. Science 182:584–586
Shively JM, Ball FL, Kline BW (1973) Electron microscopy of the carboxysomes (polyhedral bodies) of *Thiobacillus neapolitanus*. J Bacteriol 116:1405–1411
Shively JM, Bock E, Westphal K, Cannon GC (1977) Icosahedral inclusions (carboxysomes) of *Nitrobacter agilis*. J Bacteriol 132:673–675
Shively JM, Bryant DA, Fuller RC, Konopka AE, Stevens SE Jr, Strohl WR (1988) Functional inclusions in prokaryotic cells. Int Rev Cytol 113:35–100

Shively JM, Bradburne CE, Aldrich HC, Bobik TA, Mehlman JA, Jin S, Baker SH (1998) Sequence homologs of the carboxysomal polypeptide CsoS1 of the thiobacilli are present in cyanobacteria and enteric bacteria that form carboxysomes-polyhedral bodies. Can J Bot 76:906–916

Sidler WA (1994) Phycobilisome and phycobiliprotein structures. In: Bryant DA (ed) The molecular biology of cyanobacteria. Kluwer Academic, Dordrecht, The Netherlands, p 139–216

Staehelin LA, Golecki JR, Fuller RC, Drews G (1978) Visualization of the supramolecular architecture of chlorosomes (chlorobium type vesicles) in freeze-fractured cells of *Chloroflexus aurantiacus.* Arch Microbiol 119:269–277

Staehelin LA, Golecki JR, Drews G (1980) Supramolecular organization of chlorosomes (chlorobium vesicles) and of their membrane attachment sites in *Chlorobium limicola.* Bioch Biophys Acta 589:30–45

Stanier RY, Doudoroff M, Adelberg EA (1957) The microbial world. Prentice-Hall, Englewood Cliffs NJ

Staley JT (1973) Budding bacteria of the Pasteuria-Blastobacter group. Can J Microbiol 19:609–614

Staley JT, Bont JA, Jonge K (1976) *Prosthecobacter fusiformis* nov gen et sp., the fusiform caulobacter. Antonie Van Leeuwenhoek 42:333–342

Stoeckenius W, Kunau WH (1968) Further characterization of particulate fractions from lysed cell envelopes of *Halobacterium halobium* and isolation of gas vacuole membranes. J Cell Biol 38:336–357

Strodtman S (1895) Die Ursache des Schwebevermögens bei den Cyanophyceen. Biol Zentralbl 15:113–115

Strous M, Fuerst JA, Kramer EHM, Logemann S, Muyzer G, van de pas-Schoonen KT, Webb R, Kuenen JG, Jetten MSM (1999) Missing lithotroph identified as new planctomycete. Nature 400:446–449

Tamura T, Nagy I, Lupas A, Lottspeich F, Cejka Z, Schoofs G, Tanaka K, De Mot R, Baumeister W (1995) The first characterization of a eubacterial proteasome: the 20S complex of *Rhodococcus.* Curr Biol 5:766–774

Tandeau de Marsac N (2003) Phycobiliproteins and phycobilisomes: the early observations. Photosynthesis Res 76:197–205

van de Graaf AA, Mulder A, de Bruijn P, Jetten MS, Robertson LA, Kuenen JG (1995) Anaerobic oxidation of ammonium is a biologically mediated process. Appl Environ Microbiol 61:1246–1251

van Gool AP, Lambert R, Laudelout H (1969) The fine structure of frozen etched *Nitrobacter* cells. Arch Mikrobiol 69:281–293

van Keulen G, Hopwood DA, Dijkhuizen L, Sawers RG (2005) Gas vesicles in actinomycetes: Old buoys in novel habitats? Trends Microbiol 13:350–354

van Niftrik LA, Fuerst JA, Sinninghe Damste JS, Kuenen JG, Jetten MS, Strous M (2004) The anammoxosome: an intracytoplasmic compartment in anammox bacteria. FEMS Microbiol Lett 233:7–13

Vatter AE, Wolf RS (1958) The structure of photosynthetic bacteria. J Bacteriol 75:480–488

Wahlund TM, Woese CR, Castenholz RW, Madigan MT (1991) A thermophilic green sulfur bacterium from New Zealand hot springs, *Chlorobium tepidum* sp nov. Arch Microbiol 156:81–90

Walter D, Ailion M, Roth J (1997) Genetic characterization of the pdu operon: Use of 1,2-propanediol in *Salmonella typhimurium.* J Bacteriol 179:1013–1022

Walsby AE (1994) Gas vesicles. Microbiol Rev 58:94–144

Walsby AE (1972) Structure and function of gas vacuoles. Bacteriol Rev 36:1–32

Wang J, Jenkins C, Webb RI, Fuerst JA (2002) Isolation of Gemmata-like and Isosphaera-like planctomycete bacteria from soil and freshwater. Appl Environ Microbiol 68:417–422

Wang WS, Lundgren DG (1969) Poly-3-hydroxybutyrate in the chemolithotrophic bacterium *Ferrobacillus ferrooxidans*. J Bacteriol 97:947–950

Wang WS, Korczynski MS, Lundgren DG (1970) Cell envelope of an iron-oxidizing bacterium: studies of lipopolysaccharide and peptidoglycan. J Bacteriol 104:556–565

Ward N, Rainey FA, Stackebrandt E, Schlesner H (1995) Unraveling the extent of diversity within the order Planctomycetales. Appl Environ Microbiol 61:2270–2275

Watson SW, Mandel M (1971) Comparison of the morphology and deoxyribonucleic acid composition of 27 strains of nitrifying bacteria. J Bacteriol 107:563–569

Wolf S, Nagy I, Lupas A, Pfeifer G, Cejka Z, Muller SA, Engel A, De Mot R, Baumeister W (1998) Characterization of ARC, a divergent member of the AAA ATPase family from *Rhodococcus erythropolis*. J Mol Biol 277:13–25

Wolk CP (1973) Physiology and cytological chemistry of blue-green algae. Bacteriol Rev 37:32–101

Zak E, Norling B, Maitra R, Huang F, Andersson B, Pakrasi HB (2001) The initial steps of biogenesis of cyanobacterial photosystems occur in plasma membranes. Proc Natl Acad Sci USA 98:13443–13448

Zuhl F, Tamura T, Dolenc I, Cejka Z, Nagy I, De Mot R, Baumeister W (1997) Subunit topology of the *Rhodococcus* proteasome. FEBS Lett 400:83–90

Zwickl P, Ng D, Woo KM, Klenk HP, Goldberg AL (1999) An archaebacterial ATPase, homologous to ATPases in the eukaryotic 26S proteasome, activates protein breakdown by 20S proteasomes. J Biol Chem 274:26008–26014

Zwickl P, Baumeister W, Steven A (2000) Dis-assembly lines: The proteasome and related ATPase-assisted proteases. Curr Op Struct Biol 10:242–250

Microbiol Monogr (2)
J.M. Shively: Complex Intracellular Structures in Prokaryotes
DOI 10.1007/7171_019/Published online: 25 May 2006

Proteasomes and Other Nanocompartmentalized Proteases of Archaea

Julie A. Maupin-Furlow (✉) · Malgorzata A. Gil · Matthew A. Humbard · P. Aaron Kirkland · Wei Li · Christopher J. Reuter · Amy J. Wright

Department of Microbiology and Cell Science, University of Florida, Gainesville, Florida 32611-0700, USA
jmaupin@ufl.edu

Abstract A growing number of proteases and peptidases have been identified that form large nanocompartmentalized structures in the cytosol, membrane, and extramembrane of cells. In archaea, these include the intracellular energy-dependent proteasomes and the membrane-associated Lon protease as well as the intracellular energy-independent tetrahedral aminopeptidase (TET), tricorn peptidase (TRI), and PfpI-like proteases. Homologs of HtrA proteins are also distributed in some archaea and may form nanocompartments that switch function from chaperone to protease with increasing temperature. The location of these latter homologs remains to be determined.

1 Introduction

Nano- or self-compartmentalized proteases and peptidases are widely distributed in all three domains of life. These enzymes include the energy-dependent regulatory proteases of the proteasome (HslV), Lon, ClpP and FtsH families (Lupas et al. 1997) as well as the energy-independent tricorn peptidase (TRI) (Walz et al. 1999), tetrahedral aminopeptidase (TET) (Franzetti et al. 2002), PfpI protease/peptidase (Du et al. 2000), HtrA chaperone/protease (Kim and Kim 2005), bleomycin hydrolase (Gal6) (Joshua-Tor et al. 1995), leucine aminopeptidase (LAP) (Burley et al. 1990), and diaminopeptidase (DppA) (Remaut et al. 2001). Although members of this large group share a compartmentalized architecture and are able to hydrolyze peptide bonds, these enzymes are otherwise unrelated and, thus, appear to have arisen by convergent evolution.

The self-compartmentalized structure of proteases and peptidases is likely to serve a variety of roles. For energy-dependent proteases, this architecture minimizes the uncontrolled degradation of proteins in the cytosol and membrane by confining the proteolytic active sites to only those protein substrates that are unfolded or that display degradation signals. Likewise, the narrow channels of some peptidases appear to restrict access of proteins and large peptides. A compartmentalized structure also allows additional layers of control to be added to the proteolytic system by coupling the proteases to various regulatory enzymes including those that modulate covalent modification (e.g. ubiquitin, phosphorylation) and unfolding of substrate proteins. Other advantages to compartmentalizing the active sites of peptide hydrolysis include the proper orientation of substrate as well as the enhanced catalytic efficiency and processivity of the enzyme. The shared architecture of the large peptidase complexes with energy-dependent proteases may also enhance their physical and functional interaction and, thus, allow channeling of protein substrate to ultimately generate the free amino acids needed for de novo protein synthesis. Self-compartmentalization also offers the flexibility to deploy proteases and peptidases to different locations of the cell.

Several self-compartmentalized proteases and peptidases have been studied in archaea. These include the energy-dependent proteasomes and Lon protease as well as the intracellular energy-independent tetrahedral aminopeptidase (TET), tricorn peptidase (TRI), and PfpI-like proteases. Some archaea also encode homologs of the HtrA family, members of which in eukaryotes and bacteria form protein compartments outside the cell membrane that switch with increasing temperature from chaperone to protease. The ClpP and FtsH proteases, which are widespread in most bacteria and eukaryotic organelles, have not been identified in archaea.

2 Proteasomes

Proteasomes are proteolytic nanocompartments or nanomachines distributed in all three domains of life: *Bacteria*, *Archaea*, and *Eucarya* (Volker and Lupas 2002). These enzymes maintain protein quality control by degrading misfolded and denatured proteins in response to cell stress (Esser et al. 2004). They also mediate general protein turnover and play central roles in regulating cellular activities including cell division, metabolism, DNA repair, translation, and transcription (Wolf and Hilt 2004).

2.1 Catalytic 20S Core Proteasome

Proteasomes consist of a 20S proteolytic core particle (CP) which is a large, cylindrical complex of four stacked heptameric rings in an $\alpha_7\beta_7\beta_7\alpha_7$ stoichiometry (Fig. 1). The outermost α-rings form a gate at each end of the cylinder which restricts folded proteins from accessing the central channel connecting two antechambers to a central chamber formed by the two β-rings. Proteolysis occurs within the central chamber at active centers formed by autocatalytic processing of β-preproteins to expose the *N*-terminal nucleophile (Ntn) threonine (Seemüller et al. 1996; Maupin-Furlow et al. 1998). The high concentration of active sites (hundreds of mM) within the CP ensures multicleavage of most polypeptides into fragments of 3 to 24 amino acids in length (Kisselev et al. 1998). The small cleavage products apparently diffuse out of

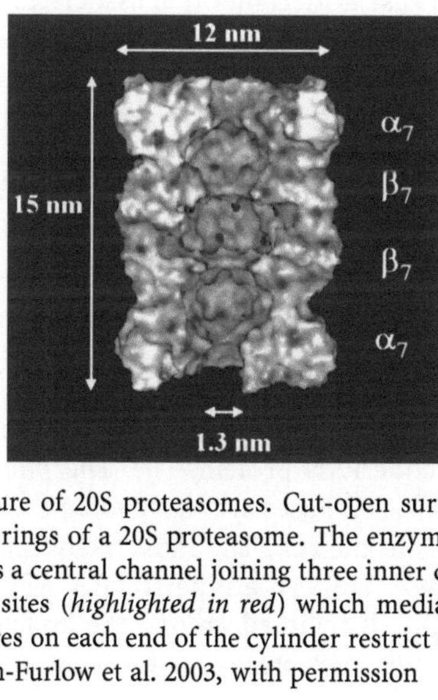

Fig. 1 Cylindrical structure of 20S proteasomes. Cut-open surface representation of the four stacked heptameric rings of a 20S proteasome. The enzyme is in an $\alpha7\beta7\beta7\alpha7$ configuration which includes a central channel joining three inner cavities. The central cavity is *lined* by the catalytic sites (*highlighted in red*) which mediate the hydrolysis of peptide bonds. The axial pores on each end of the cylinder restrict access of substrate. Figure reproduced from Maupin-Furlow et al. 2003, with permission

the proteolytic chamber while the export of longer peptides may require an ATPase partner (Kim et al. 2000).

2.2
Proteasome-associated Regulators

Proteins of the AAA family associate with 20S proteasomes and serve multiple roles in regulating the degradation of proteins in the cell (Wolf and Hilt 2004; Römisch 2005). In archaea, these AAA proteins include the proteasome-activating nucleotidases (PANs) which associate with 20S proteasomes to catalyze the energy-dependent and processive degradation of proteins (Maupin-Furlow et al. 2004). In addition, archaea encode Cdc48 homologs of which the analogous protein in eukaryotes has been shown to facilitate the ATP-dependent movement of substrates to proteasomes for degradation (Wang et al. 2004; Richly et al. 2005; Elsasser and Finley 2005).

2.2.1
Proteasome-activating Nucleotidase (PAN) Proteins

Most archaea encode homologs of the eukaryal Rpt (regulatory particle triple-A) subunits of 26S proteasomes. These eukaryal Rpt proteins form the base component of the 19S cap required for energy-dependent protein degradation by 26S proteasomes. In archaea, these Rpt-like proteins have been denoted as PAN for proteasome-activating nucleotidase based on their function. The PAN protein of *Methanocaldococcus jannaschii* (*Mj*PAN, MJ1176) has been characterized from recombinant *E. coli* and shown to form an irregular ring-shaped ATPase dodecamer that associates with CPs (Zwickl et al. 1999; Wilson et al. 2000). The purified *Mj*PAN stimulates CP-mediated degradation of proteins (e.g. casein, GFP-SsrA) in the presence of ATP or CTP. Substrate binding to *Mj*PAN activates nucleotide hydrolysis which successively promotes substrate unfolding and opening of the axial gate (Benaroudj and Goldberg 2000; Benaroudj et al. 2003). PAN proteins may also facilitate substrate translocation into the CP.

2.2.2
CDC48 Homologs

Although proteasomal CPs are universally distributed among archaea, not all of these organisms encode PAN proteins (*i.e. Thermoplasma* and *Pyrobaculum* sp.). Thus, other AAA ATPases are likely to be required for CP-mediated degradation of proteins in the cell. The most likely candidates are homologs of Cdc48 (VCP, VAT, p97, TER94), a type II AAA ATPase found in all three domains and universally distributed among the archaea. Single particles of the *T. acidophilum* VCP/Cdc48 protein have been analyzed by electron tomography and cryo-electron microscopy and reveal a hexameric barrel-like structure with a central pore (Rockel et al. 1999, 2002). This VCP/Cdc48 protein binds denatured protein substrates and mediates the energy-dependent refolding and unfolding of proteins at low and high concentrations of Mg^{2+}, respectively (Golbik et al. 1999; Gerega et al. 2005). Similar to other AAA+ proteins, the catalytic mechanism of Cdc48 is thought to be mediated by threading substrate proteins through its central pore which is lined with hydrophobic residues (Wang et al. 2001b).

In eukaryotes, Cdc48/p97 acts as a molecular chaperone in a wide variety of cellular activities, many of which are directly or indirectly regulated by the ubiquitin-proteasome system (Wang et al. 2004). In particular, Cdc48 is closely linked to the endoplasmic reticulum-associated degradation (ERAD) pathway of the Ub-proteasome (Neuber et al. 2005; Schuberth and Buchberger 2005; Elsasser and Finley 2005).

2.3 Proteasomal Subtypes

Complete genome sequences reveal a wide variety of organisms, including archaea, carry duplicated genes predicted to encode proteasomal CP and regulatory particle AAA subtypes (Maupin-Furlow et al. 2004; Smalle and Vierstra 2004). Differential expression of these genes is likely to diversify the functional capacity of proteasomes.

Biochemical evidence supporting the synthesis of proteasomal subtypes ranges from vertebrates to archaea. This includes the IFN-γ-inducible immunoproteasome that facilitates immune responses in vertebrates (Van den Eynde and Morel 2001), the spermatogenesis-specific proteasome synthesized in insects (Ma et al. 2001), and the COP9 signalsome (CSN), a distant relative of the 19S cap which regulates proteasome-mediated protein turnover in higher eukaryotes (Harari-Steinberg and Chamovitz 2004). Similarly, the halophilic archaeon *Haloferax volcanii* synthesizes CP and PAN subtypes of different subunit composition (Wilson et al. 1999; Kaczowka and Maupin-Furlow 2003; Reuter et al. 2004). The subunit topology of the CP subtypes has been determined and includes two complexes, one of $\alpha 1$ and β composition and the other of $\alpha 1$, $\alpha 2$ and β (Kaczowka and Maupin-Furlow 2003). This latter complex is asymmetric with rings of $\alpha 1$ forming one end of the cylinder and $\alpha 2$ forming the other. In addition, PAN complexes of PanA, PanB, and PanAB composition have been purified from *H. volcanii* (Reuter et al., personal communication, 2006).

The role of proteasomal subtypes in archaeal physiology has not yet been elucidated. It is known that in *H. volcanii* the levels of $\alpha 1$, β and PanA are relatively constant and abundant during "normal" growth suggesting these proteasomal proteins may play a role in housekeeping functions of the cell (Reuter et al. 2004). In contrast, the levels of $\alpha 2$ and PanB are relatively low in log phase but increase several-fold as cells enter stationary phase suggesting these paralogs provide an ancillary role that may optimize the proteasome system for starvation and/or other stress. Although the two α-type proteins are closely related, there are structural differences that include residues predicted to precede the *N*-terminal α-helix at each end of the cylinder and the loop that restricts the channel opening. Thus, modulating the ratio of these α subunits may control distinct structural domains at the ends of the CP. This in turn would be expected to influence CP gating as well as the type of AAA regulatory protein (i.e. PanA, PanB or PanAB complexes) which recognizes substrate and associates with the CP.

2.4 Proteasomal Substrate Recognition

How proteins are targeted for proteasome-mediated destruction in archaea remains to be elucidated. In eukaryotes, many proteasomal substrates are covalently linked to chains of polyubiquitin (poly-Ub) (Ciechanover 1994). The formation of these chains is often modulated by other post-translational modifications of protein substrate including protein phosphorylation (Harari-Steinberg and Chamovitz 2004), glycosylation (Yoshida et al. 2002) and acetylation (Li et al. 2002a; Giandomenico et al. 2003).

A combination of structural and sequence similarity approaches predicts several proteins with Ub-like folds and Ub-association domains are encoded by prokaryotic genomes (Rudolph et al. 2001; Wang et al. 2001a; Bienkowska et al. 2003; Spreter et al. 2005). The homologs with Ub-like folds classify to two superfamilies: 2Fe-2S ferredoxins and MoaD/ThiS proteins of the molybdenum and thiamin cofactor biosynthetic pathway (Bienkowska et al. 2003). The structural homology of these proteins to Ub combined with similar sulfur chemistry suggests that they share a common ancestor with eukaryotic Ub. Other evidence for prokaryotic Ub chemistry is based on the *C*-terminal domain of an archaeal nascent polypeptide associated complex (i.e. MTH177), which is structurally related to eukaryal Ub-associated proteins (Spreter et al. 2005). Whether any of these proteins participate in proteasome-mediated proteolysis in archaea remains to be determined.

To understand the role of archaeal proteasomes within the context of whole cells, irreversible inhibitors of the CP proteolytic active site have been used. The tri-peptide carboxybenzyl-leucyl-leucyl-leucine vinyl sulfone (Z-L_3VS) was used in an early study to inhibit proteasomes in *Thermoplasma acidophilum* (Ruepp et al. 1998). The Z-L_3VS tri-peptide is a potent inhibitor of CPs and was shown to modify up to 80% of the proteasomal β-subunits in cell culture. Although inhibition of the CPs arrested growth under heat shock conditions, it had only a marginal effect on growth under normal conditions. A more recent study used *clasto*-lactacystin β-lactone (CLβL) to inhibit the CPs of *H. volcanii* in cell culture (Reuter and Maupin-Furlow 2004). In contrast to the earlier study, addition of CLβL significantly reduced the "normal" growth rate of these cells. These results suggest that the physiological requirement for CP activity may vary among the archaea.

In addition to proteasome inhibitors, green fluorescent protein (GFP) reporter proteins with degrons (proteolytic tags) have been used in cell culture to examine the physiological role of proteasomes. This was made possible by the finding that soluble, modified, red-shifted derivatives of GFP could be synthesized and readily detected in recombinant *H. volcanii* cells (Reuter and Maupin-Furlow 2004). Amino acid residues of various sequence and length from 1 to 11 residues, which were not expected to influence overall protein structure, were added to the *C*-terminus of this GFP reporter. The inclusion of hydrophobic residues, not length, reduced the levels of this reporter protein in the cell. Proteasomes were found to be responsible, at least in part, for modulating these reporter protein levels.

3
Lon Protease

The proteolytic domains of Lon are highly conserved and widely distributed among archaea, bacteria, and eukaryotic organelles (Iyer et al. 2004). Of these, the LonA-type proteases of *Escherichia coli* and yeast mitochondria have been most thoroughly studied. These enzymes are formed by a polypeptide chain that contains an *N*-terminal LAN domain, central AAA+ domain and *C*-terminal proteolytic (P-) domain with a Ser – Lys catalytic dyad responsible for the hydrolysis of peptide bonds (Rotanova et al. 2004). The bacterial LonA polypeptide assembles into hexameric ring-like particles in the cytosol (Botos et al. 2004), and the yeast orthologue forms ring-like heptamers in the mitochondrial matrix (Stahlberg et al. 1999). Both function as energy-dependent proteases in the processive degradation of proteins.

The archaeal Lon homologs are highly diverse in domain organization. Most lack the *N*-terminal LAN domain, with the rare exception of the LonA homologs of methanosarcina (e.g. MA1862, MM3118). One common theme among archaeal Lon homologs, however, is the presence of an *N*-terminal signal sequence and/or transmembrane spanning motifs suggesting most are membrane-associated. Many homologs contain both AAA+ and P-domains, but some have only a single ATPase or protease domain. These latter homologs are not predicted to independently hydrolyze folded proteins and may instead associate with partners to facilitate energy-dependent proteolysis.

The biological role of the archaeal Lon homologs has not been established. Some (e.g. PAB1313, PH0452, PF0467) contain inteins and thus are likely to be vital for host survival (Pietrokovski 2001). Recent global analysis of *Halobacterium* NRC-1 reveals the level of *lon*-specific transcript increases after exposure to UV irradiation coupled with shifts in temperature and nutrient conditions, thus, suggesting the protease may function in stress response (Baliga et al. 2004). On the basis of analogy to other organisms, it is possible that the archaeal Lon protease functions not only in proteolysis but also in the catalysis of protein folding (Rep et al. 1996) and in the binding of inorganic polyphosphate, DNA, and RNA (Nomura et al. 2004; Liu et al. 2004).

Of the archaeal Lon homologs that have been studied at the biochemical level, all have similar domain organization including an *N*-terminal AAA+ domain with two transmembrane spanning helices and a *C*-terminal P-domain. These proteins have all been purified from recombinant *E. coli* and include the full-length *Thermococcus kodakaraensis* TK1264 (*Tk*LonB) (Fukui et al. 2002), *Thermoplasma acidophilum* Ta1081 (*Ta*LonB) (Besche and Zwickl 2004; Besche et al. 2004), and *Archaeoglobus fulgidus* AF0364 (*Af*LonB) (Botos et al. 2005) as well as the P-domains of *Af*LonB (Botos et al. 2005) and *Methanocaldococcus jannaschii* MJ1417 (*Mj*LonB) (Im et al. 2004). Of these, the *Ta*LonB has been shown to form membrane-associated, hexameric ring-shaped particles (Fig. 2) (Besche et al. 2004). Likewise, *Tk*LonB and *Ta*LonB have been shown to associate with the membrane fractions of their native host based on Western blot (Fukui et al. 2002; Besche et al. 2004). All three of the purified, full-length LonB proteins catalyze the energy-dependent degradation of proteins (Fukui et al. 2002; Besche et al. 2004; Botos et al. 2005). However, the isolated proteolytic (P-) domains of the Lon proteases do not exhibit detectable enzyme activity.

To further understand the catalytic mechanism of LonB, site-directed mutagenesis of *Ta*LonB (Besche and Zwickl 2004; Besche et al. 2004) and *Af*LonB (Botos et al. 2005) has been performed. In addition, atomic resolution crystal structures of a dimeric form of the *Mj*LonB P-domain (Im et al. 2004) and

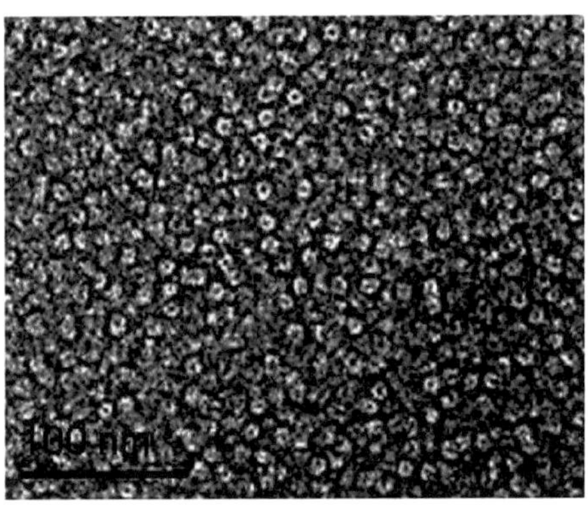

Fig. 2 LonB-type protease of archaea forms a ring-shaped complex. The 510-kDa *T. acidophilum* LonB was purified from detergent solubilized membranes of a recombinant *E. coli* strain expressing Ta1081. Electron micrograph of the negatively stained protein reveals ring-shaped particles likely to be of hexameric configuration. Figure reproduced from Besche and Zwickl 2004, with permission

hexameric *Af*LonB P-domain (Botos et al. 2005) have been determined and compared to *Ec*LonA (Botos et al. 2004). From these results, a Ser – Lys – Asp catalytic triad distinct from the canonical catalytic dyad of LonA proteases was proposed (Im et al. 2004). However, future studies which examine the structure of proteolytically active derivatives of LonB are needed to fully confirm and understand this catalytic mechanism.

4
Peptidases Functioning Downstream of Energy-dependent Proteases

Energy-dependent proteases, such as the proteasome, release oligopeptide products which vary from 3 to 24 amino acids with an average length distribution of 7 to 8 amino acids (Niedermann et al. 1996; Kisselev et al. 1998, 1999; Emmerich et al. 2000). These oligopeptides are rapidly degraded in vivo (Reits et al. 2003, 2004) by mechanisms that are not fully characterized. Oligopeptide hydrolysis is needed to provide amino acid precursors for the biosynthesis of nascent polypeptides and to avoid accumulation of peptide fragment intermediates that may otherwise aggregate and/or interfere with proper protein–protein interaction (Saric et al. 2004).

In eukaryotes, it is believed that the bulk of proteasome products (especially those of 6 to 17 residues) are hydrolyzed to single amino acids by the combined activities of thimet oligopeptidase (TOP) and aminopeptidases (Saric et al. 2004). In addition, the giant tripeptidyl peptidase II (TPP-II) formed by a stack of eight ring-shaped segments (Geier et al. 1999) appears to be the primary peptidase responsible for degrading products greater than 15 residues (Tomkinson 1999; Wang et al. 2000; Reits et al. 2004).

Recently, it was proposed that tricorn peptidase (TRI) and tetrahedral aminopeptidase (TET) act as functional analogues in the degradation of oligopeptides in prokaryotes (Borissenko and Groll 2005). The TRI has a narrow distribution, being encoded only by the genomes of a select number of prokaryotes; whereas, TET is more widespread. Among archaeal species, homologs of the tricorn core (i.e. Ta1490) are found in *Thermoplasma*, *Ferroplasma*, *Picrophilus*, *Pyrobaculum* and *Sulfolobus*. In contrast, most other archaea encode homologs of TET with the exception of *Pyrobaculum aerophilum*, which encodes both TRI and TET homologs, and *Nanoarchaeum equitans*, which does not encode either homolog.

4.1
Tricorn Peptidase (TRI)

Tricorn (TRI) is a giant tricorn-shaped peptidase that exceeds the size of 20S proteasomes (Fig. 3). The enzyme complex has been extensively characterized from *T. acidophilum* and is composed of a 121-kDa subunit (i.e. Ta1490) that forms a 720-kDa homohexamer. This further assembles to form a 14.6-MDa icosahedral capsid composed of 20 hexamers (Brandstetter et al. 2001).

Tricorn displays trypsin and chymotrypsin specificities with preferential di- and tricarboxypeptidase activities (Tamura et al. 1996, 1998). The crystal structure of the tricorn hexamer has been determined in the presence and absence of chloromethyl ketone-based inhibitors (Brandstetter et al. 2001). This, in combination with site-directed mutagenesis, reveals peptide bond hydrolysis proceeds via a classical typsin-like serine protease mechanism. A catalytic Ser – His – Glu – Ser tetrad with an oxyanion hole formed by the amide nitrogens of Asp966 and Gly918 is responsible for peptide bond hydrolysis (Brandstetter et al. 2001). The tetrad includes a Ser965 nucleophile which is activated by a His746 residue that is oriented by Ser745 that in turn is polarized by Glu1023. The enzyme is proposed to cleave the peptide substrates by an electrostatically driven processive mechanism (Kim et al. 2002). Substrate peptides are thought to enter the catalytic chamber through a β-propeller tunnel (β7) that widens from the capsid surface toward the active site (Brandstetter et al. 2001) similar to dipeptidyl peptidase (DP) IV and prolyl oligopeptidase (POP) (Fülöp et al. 1998; Engel et al. 2003). Unlike the other proteases, tricorn has an additional conical β-propeller tunnel (β6) that widens toward the solvent (Brandstetter et al. 2001). This latter tunnel appears to serve as an "exhaust system" to channel products to downstream peptidases (Kim et al. 2002; Brandstetter et al. 2002).

Tricorn associates with three additional peptidases denoted as tricorn interacting factors F1 (Ta0830), F2 (Ta0301) and F3 (Ta0815) (Tamura et al. 1996, 1998). These accessory proteins are aminopeptidases that cooperate with tricorn to degrade oligopeptides to free amino acids. Indeed, reconsti-

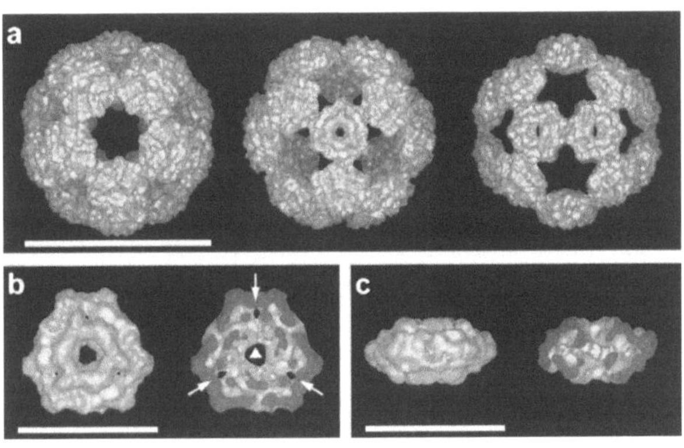

Fig. 3 Icosohedral tricorn capsid of *T. acidophilum*. Views of **a** the caspid down five-, three- and two-fold symmetry axes (bar = 50 nm), **b** a single tricorn hexamer with the main opening on the threefold axis and side windows indicated by *arrows* (bar = 20 nm), **c** a hexamer side-on (*left*) and cut-open (*right*) (bar = 20 nm). Figure reproduced from Walz et al. 1997, with permission

tution of tricorn with these interacting factors completes the proteasomal degradation pathway by generating free amino acids from protein substrates (Tamura et al. 1998). The 14.6-MDa capsid apparently acts as an organizing center or scaffold for the F1, F2 and F3 aminopeptidases. Interacting with the tricorn core is believed to enhance the intrinsic activities of the aminopeptidases and elicit new activities not catalyzed by the individual components. Thus, the compartmentalized capsid arrangement is thought to mediate allosteric control of peptidase activity.

Homologs of the F1 proline aminopeptidase are widespread among archaea. In contrast, archaea encoding F1 in combination with F2, F3 and the tricorn core are limited to *Thermoplasma*, *Sulfolobus*, and *Pyrobaculum* sp. Thus, most archaea do not encode a complete tricorn complex and are likely to use other peptidases to process the oligopeptides generated by energy-dependent proteases to free amino acids. Among these, the widespread tetrahedral aminopeptidase (TET) is one likely candidate for recycling amino acids (see section below).

4.2 TET Peptidase

TET peptidase was named based on its large tetrahedral-shape. It is a dodecameric, energy-independent peptidase of the M42 family in the MH clan of metallopeptidases (Fig. 4). The enzyme was isolated from the haloarchaeon *Haloarcula marismortui* (HmTET) where it was found to have a broad aminopeptidase activity and process peptides up to 30 to 35 amino acids in length to free amino acids (Franzetti et al. 2002). The general architecture of this complex is different from ring- or barrel-shaped proteases, such as the proteasome, in that it is tetrahedral in shape and has a central cavity accessible through four narrow and four wider channels (i.e. less than 17 Å and 21 Å in diameter, respectively) (Franzetti et al. 2002). Protein sequence, as determined by nanoelectrospray tandem mass spectrometry of tryptic peptide fragments, revealed that the HmTET is a member of the M42 aminopeptidase family. Members of this family are found in all archaea that do not encode the TRI core. Exceptions include *Nanoarchaeum equitans*, which does not encode either TRI or TET homologs, and *P. aerophilum*, which encodes both.

Many archaeal members of the M42 aminopeptidase family have been characterized including the commercially available deblocking aminopeptidases (DAP or TET1) of *Pyrococcus furiosus* (Tsunasawa et al. 1998) and *P. horikoshii* (Ando et al. 1999; Onoe et al. 2002) as well as the *P. horikoshii* PH1527 paralog denoted as PhTET2 or FrvX (Russo and Baumann 2004; Durá et al. 2005; Borissenko and Groll 2005). All form dodecameric complexes, which have a relatively broad aminopeptidase activity, but differ in their preferences for amino acids in the S1 position. Unlike the DAPs, neither HmTET nor PhTET2 catalyze deblocking and instead require oligopeptides

(a)

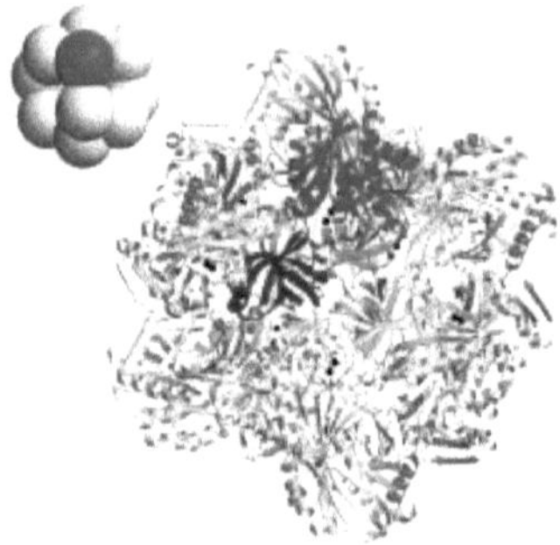

(b)

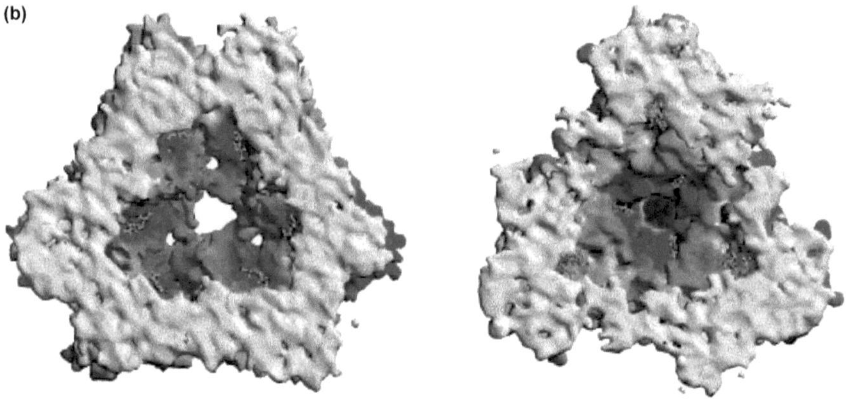

Fig. 4 Intracellular tetrahedral aminopeptidase (TET). **a** Ribbon representation of PhTET2 along the two-fold symmetry axis with a ball model illustrating the orientation of the particle. **b** Surface representation of two halves of PhTET2 cut-open and viewed along the three-fold axis. *Color coded* by electrostatic potential from − 15 to 15 kT/e (*red* to *blue*). Figure reproduced from Borissenko and Groll 2005, with permission

with unmodified *N*-termini for processing (Franzetti et al. 2002; Durá et al. 2005). Although these enzymes sequentially degrade peptides from the *N*-termini, they are not processive with release of peptide intermediates. Short peptides of 9 to 12 amino acids are more readily cleaved than longer peptides (greater than 19 amino acids), and cleavage is not detected for those peptides of 40 amino acids or longer.

Steric factors are believed to disfavor access of long peptides to the active sites of TET. The compartmentalized structure of TET is in support of this, as determined by single particle analysis of negatively stained electron micrographs of HmTET (Franzetti et al. 2002) and X-ray diffraction analysis of PhTET2 (Russo and Baumann 2004; Borissenko and Groll 2005). On the basis of a combination of these studies, TET was found to be a compartmentalized tetrahedron built from six dimers with each of the 12 active sites located on the interior side of the complex. The two catalytic divalent metal ions of each active site are coordinated by the conserved residues Asp235,

His68, Asp182, Glu213, and His323 (based on PhTET2 residues). Substrate access to these sites is highly restricted by four narrow channels (10 to 18 Å diameter) located at the center of each triangular face. These openings should allow passage of α-helical or double β-stranded peptides but restrict the access of large peptides and proteins which have additional structure. In one of the PhTET2 structures (Borissenko and Groll 2005), three smaller openings (9 Å diameter) are positioned in close proximity to the central pore on each face and are proposed to facilitate product exit. However, these smaller channels do not appear in the earlier structure of PhTET2 and are instead blocked by a Phe residue (Phe224) (Russo and Baumann 2004).

In the cell, TET is believed to act downstream of proteasomes to generate free amino acids. Consistent with this, a minimal protein degradation system was reconstituted in vitro from the initial unfolding of protein substrates to the release of free amino acids (Borissenko and Groll 2005). The assay included the protein substrate GFP-SsrA and the recombinant enzymes *M. jannaschii* PAN, *A. fulgidus* 20S proteasome, and *P. horikoshii* TET2. The proteolytic system was found to generate free amino acids and require the presence of 20S proteasomes and PhTET2. Release of free amino acids was also observed when PhTET2 was replaced by the TRI core and interacting factor F1 of *T. acidophilum*. Although this system is reconstituted from a wide variety of archaea and a tmRNA SsrA-tagging system has not been identified in this domain, the study does demonstrate the ability of these enzymes to coordinate the generation of free amino acids from a folded protein substrate.

5 Proteases of the DJ-1/ThiJ/PfpI Superfamily

The PfpI protease was first discovered in the hyperthermophilic archaeon *Pyrococcus furiosus* based on its gelatin hydrolyzing activity which is stable even after incubating the enzyme for 24 h at nearly boiling temperatures in relatively high concentrations of detergent (i.e. 1% SDS) (Blumentals et al. 1990; Halio et al. 1996). Early evidence suggested a compartmentalized structure for this protease based on the tendency of the inactive 18.8-kDa monomers to self-associate into higher-order oligomers (some greater than 200 kDa) that were proteolytically active (Halio et al. 1996). Although the polypeptide was immunologically related to 20S proteasomes, reverse genetics revealed the encoding *pfpI* gene defined a new protease family (Halio et al. 1996).

Homologs of PfpI are widely distributed in archaeal genomes suggesting this protease serves an important role in cell physiology. Although this biological role is still unclear, Kelly and colleagues (Blumentals et al. 1990; Snowden et al. 1992) provided insight when they demonstrated that PfpI protein levels are regulated in response to growth conditions. PfpI was one of the most predominant proteins in cells grown under peptide-limiting conditions.

The PfpI levels diminished when cultures were supplemented with maltose or high levels of peptides. This pattern of PfpI regulation combined with the universal distribution of PfpI homologs among archaea suggest that the enzyme is important during starvation and/or other stressful conditions. PfpI may be needed to mediate a high turnover of oligopeptides and/or proteins to replenish the free amino pool for de novo protein biosynthesis. In addition, PfpI may be used to meet the metabolic requirements of those archaea that use peptides as growth substrates.

The isolation and characterization of the *pfpI* gene facilitated the structural characterization of this new protease family. On the basis of homology to PfpI, the *P. horikoshii* PH1704 was targeted for purification from recombinant *E. coli* (Du et al. 2000). Similar to PfpI, a variety of SDS-resistant oligomers of the PH1704 protein were functional in the ATP-independent hydrolysis of gelatin. The structure of the hexameric form of the recombinant PH1704 enzyme was solved by X-ray diffraction analysis and found to be a ring-shaped trimer of dimers with a central channel of 24 Å diameter (Fig. 5) (Du et al. 2000; Halio et al. 1996). A conserved cysteine residue (Cys100) was located in a sharp turn of a "nucleophile elbow", a motif common to the active sites of α/β hydrolases. Thus, a Cys – His – Glu catalytic triad is proposed to form at each dimer interface of the PfpI-like proteases and would be consistent with the dependence of proteolytic activity on oligomerization. Access of even the smallest globular protein substrate to the three active sites of this hexameric complex appears to be hindered by the compartmentalized structure.

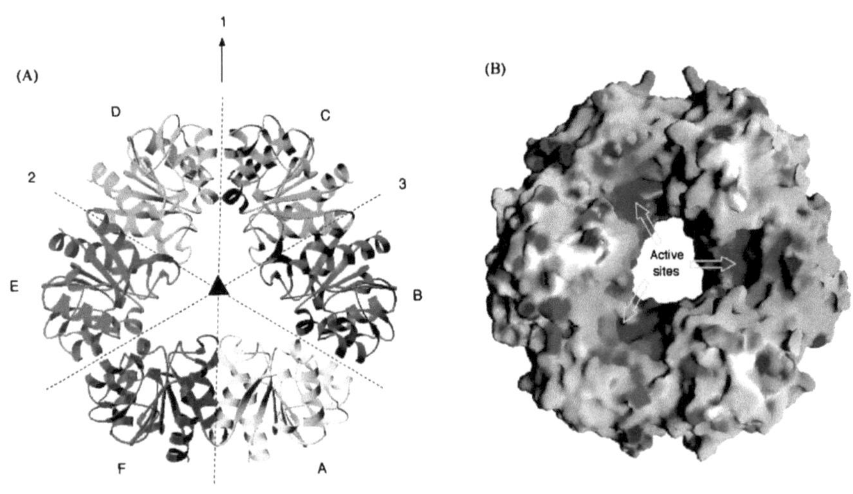

Fig. 5 Intracellular PfpI-related protease PH1704. **a** Ribbon diagram and **b** surface representation of the PfpI-like hexamer. Putative active sites indicated. *Color-coded* by electrostatic potential. Figure reproduced from Du et al. 2000, with permission

PfpI is now known to be a member of a large superfamily (DJ-1/ThiH/PfpI) of proteins with diverse function. Members include the PfpI-like proteases (Du et al. 2000; Halio et al. 1996) as well as protein chaperones (Quigley et al. 2003), catalases (Horvath and Grishin 2001), ThiJ kinases of thiamine biosynthesis (Mizote et al. 1996, 1999), and AraC-type transcriptional regulators such as AdpA (Ohnishi et al. 2005). DJ-1 has received particular interest based on the association of loss of function mutations of this protein with recessively inherited Parkinson's disease (Bonifati et al. 2003; Miller et al. 2003). Consensus maximum likelihood tree analysis of the DJ-1/ThiH/PfpI superfamily reveals a number of distinct clads (Bandyopadhyay and Cookson 2004). It also suggests that the PfpI-like intracellular proteases of archaea are close relatives of the bacterial Hsp31 chaperones. Both the proteases and chaperone have retained the consensus sequence [aliphatic]-[aliphatic]-Cys-His-[Ser, Ala, or Gly]. In the PH1704 structure, the Cys and His residues of this consensus form the tentative catalytic triad with a conserved Asp or Glu from an adjacent monomer (Du et al. 2000). The Hsp31 chaperone, which harbors these conserved residues, also catalyzes the hydrolysis of peptide bonds (Lee et al. 2003).

6 HtrA Protease/Chaperone Family

Homologs of the high temperature requirement A protein (HtrA) family are distributed in all domains of life. Members form cage-like proteins (Fig. 6) that differ from other compartmentalized proteases and peptidases in their extracytoplasmic location and extreme flexibility with the potential to change overall shape and internal structure from a chaperone to a serine protease (Jeffery 2004; Schlieker et al. 2004; Groll et al. 2005). The direction of this switch is reversible and depends on temperature and the folded state of substrate protein, with an increase in temperature and unfolded protein substrate enhancing protease activity.

HtrA proteins perform many cellular roles, most notable of which is the quality control of proteins in the extracytoplasmic space where these enzymes function as both molecular chaperones and proteases in the absence of ATP (Kim and Kim 2005). HtrA proteins have also been implicated in virulence (Bäumler et al. 1994; Elzer et al. 1996; Yorgey et al. 2001; Jones et al. 2001; Stack et al. 2005), biogenesis of secreted proteins (Poquet et al. 2000; Lyon and Caparon 2004), and cell signaling via the destruction of key regulatory proteins (Alba and Gross 2004). Of the three HtrA proteins that are synthesized in *E. coli* (DegP, DegQ, DegS), DegP is the most thoroughly studied. It is heat shock inducible, essential at high temperatures, and proposed to be key in the digestion of abnormal periplasmic proteins (Strauch and Beckwith 1988; Lipinska et al. 1988, 1990; Strauch et al. 1989). DegP is

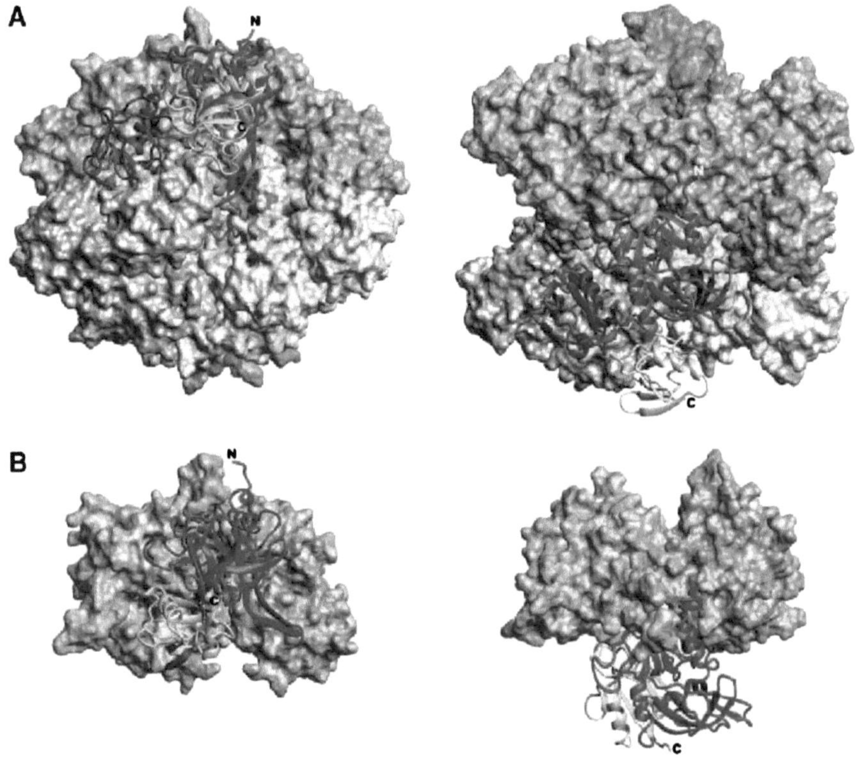

Fig. 6 Proteases of the HtrA family. The right images are rotated 90° about the horizontal axis of the left images. The monomers are individually colored with one monomer of each complex represented as a ribbon diagram (*N*- and *C*-termini indicated by N and C, respectively) and the remaining as surface representations. **A** *E. coli* DegP hexamer of two PDZ domains (*purple and light pink*) and one protease domain (*red*) with the catalytic serine obscured in these views. **B** Human mitochondrial HtrA2/Omi trimer of one PDZ domain (*light pink*) and one protease domain (*red*). The catalytic serine is indicated (S). Figure reproduced from Day and Hinds 2002, with permission

also implicated in the specific targeting of proteins, such as colicin A lysis protein (Cavard et al. 1989), MalS α-amylase (Spiess et al. 1999), and the PapA subunit of the P pilus (Jones et al. 2002). In contrast to the soluble DegP, DegS is tethered to the cytoplasmic membrane in the periplasm and, although not essential, is required for robust growth at all temperatures (Waller and Sauer 1996; Bass et al. 1996). Interaction with outer membrane protein (OMP) *C*-termini which are exposed after stress, triggers the activation of DegS as a protease in the σ-E stress pathway (Walsh et al. 2003). Activated DegS cleaves the periplasmic domain of the anti-σ factor RseA which triggers YaeL (a site-2 type intramembrane cleaving protease) to successively degrade the cytoplasmic domain of RseA (Alba and Gross 2004). This action results in the release of active σ-E, which activates transcription of stress-response genes.

HtrA proteins contain a trypsin-like protease domain and one to two *C*-terminal PDZ domains. Similar PDZ domains are found in other proteases, including tail-specific proteases and site-2 type intramembrane cleaving proteases (e.g. YaeL), and have been shown to mediate protein–protein interactions such as the preferential binding of PDZ to the *C*-termini (3 to 4 residues) of target proteins (Saras and Heldin 1996; Schlieker et al. 2004). To further understand how the trypsin-like and PDZ domains function in HtrA catalysis, a variety of proteins from this family have been analyzed by X-ray diffraction including an *E. coli* DegP hexamer (Krojer et al. 2002), human HtrA2 trimer (Li et al. 2002b), *Thermotoga maritima* DegQ protease domain trimer (Kim et al. 2003), and *E. coli* DegS trimer in the presence and absence of an OMP signaling peptide (Wilken et al. 2004). These crystal structures provide insight into the structure and function of HtrA proteins. On the basis of the *E. coli* DegS structures, the binding of signaling peptides to the PDZ domains induces a series of conformational changes that activates protease function. Upon peptide binding, the Ser – His – Asp catalytic triad of the trypsin-like domain (e.g. Ser201, His96, and Asp126 of *E. coli* DegS), which is formed by two β-barrel lobes with a *C*-terminal helix, undergoes reorientation to an active state. Although the mechanism of active site rearrangement differs among HtrA family members, all of the enzymes undergo a reversible switch from an inactive to active protease.

HtrA homologs with conserved residues of the His – Asp – Ser catalytic triad are found in archaea spanning from the *Euryarchaeota* (e.g. haloarchaea and *Methanothermobacter* sp.) to *Crenarchaeota* (e.g. *Sulfolobus*, *Ferroplasma*, and *Pyrobaculum* sp.). The trypsin-like protease domain common to all of these homologs is followed by a single PDZ domain similar to *E. coli* DegS. Only the haloarchaeal homologs are likely to be extracytoplasmic based on the conservation of *N*-terminal signal sequences common to proteins translocated by the Sec system. The remaining archaeal HtrA homologs do not contain predicted signal sequences of the Sec or TAT pathways nor do they contain predicted transmembrane spanning domains. Thus, these latter homologs may be cytosolic. Although archaeal HtrA proteins have not been characterized to date, *Halobacterium* NRC-1 *htrA*-specific transcripts have been observed to increase after exposure to stress, similar to the increases observed for *lon*-specific transcripts (Baliga et al. 2004).

7
Concluding Remarks

Compartmentalized proteases and peptidases now appear quite common among organisms including the archaea and may have arisen by convergent evolution. This shared architecture is likely to serve many fundamental roles in facilitating protein quality control and regulating proteolysis in the cell. It will be interesting to see what proteins and pathways are regulated by proteases, how these substrate proteins are recognized for destruction, and whether proteases and peptidases physically associate in facilitating protein degradation to free amino acids in archaeal cells.

Acknowledgements This research was funded in part by a grant from the National Institutes of Health (R01 GM057498) and Department of Energy (DE-FG02-05ER15650).

References

Alba BM, Gross CA (2004) Regulation of the *Escherichia coli* σ^E-dependent envelope stress response. Mol Microbiol 52:613–619

Ando S, Ishikawa K, Ishida H, Kawarabayasi Y, Kikuchi H, Kosugi Y (1999) Thermostable aminopeptidase from *Pyrococcus horikoshii.* FEBS Lett 447:25–28

Baliga NS, Bjork SJ, Bonneau R, Pan M, Iloanusi C, Kottemann MC, Hood L, DiRuggiero J (2004) Systems level insights into the stress response to UV radiation in the halophilic archaeon *Halobacterium* NRC-1. Genome Res 14:1025–1035

Bandyopadhyay S, Cookson MR (2004) Evolutionary and functional relationships within the DJ1 superfamily. BMC Evol Biol 4:6

Bass S, Gu Q, Christen A (1996) Multicopy suppressors of Prc mutant *Escherichia coli* include two HtrA (DegP) protease homologs (HhoAB), DksA, and a truncated R1pA. J Bacteriol 178:1154–1161

Bäumler AJ, Kusters JG, Stojiljkovic I, Heffron F (1994) *Salmonella typhimurium* loci involved in survival within macrophages. Infect Immun 62:1623–1630

Benaroudj N, Goldberg AL (2000) PAN, the proteasome-activating nucleotidase from archaebacteria, is a protein-unfolding molecular chaperone. Nat Cell Biol 2:833–839

Benaroudj N, Zwickl P, Seemuller E, Baumeister W, Goldberg AL (2003) ATP hydrolysis by the proteasome regulatory complex PAN serves multiple functions in protein degradation. Mol Cell 11:69–78

Besche H, Tamura N, Tamura T, Zwickl P (2004) Mutational analysis of conserved AAA$^+$ residues in the archaeal Lon protease from *Thermoplasma acidophilum.* FEBS Lett 574:161–166

Besche H, Zwickl P (2004) The *Thermoplasma acidophilum* Lon protease has a Ser – Lys dyad active site. Eur J Biochem 271:4361–4365

Bienkowska JR, Hartman H, Smith TF (2003) A search method for homologs of small proteins. Ubiquitin-like proteins in prokaryotic cells? Protein Eng 16:897–904

Blumentals II, Robinson AS, Kelly RM (1990) Characterization of sodium dodecyl sulfate-resistant proteolytic activity in the hyperthermophilic archaebacterium *Pyrococcus furiosus.* Appl Environ Microbiol 56:1992–1998

Bonifati V, Rizzu P, van Baren MJ, Schaap O, Breedveld GJ, Krieger E, Dekker MC, Squitieri F, Ibanez P, Joosse M, van Dongen JW, Vanacore N, van Swieten JC, Brice A, Meco G, van Duijn CM, Oostra BA, Heutink P (2003) Mutations in the DJ-1 gene associated with autosomal recessive early-onset parkinsonism. Science 299:256–259

Borissenko L, Groll M (2005) Crystal structure of TET protease reveals complementary protein degradation pathways in prokaryotes. J Mol Biol 346:1207–1219

Botos I, Melnikov EE, Cherry S, Kozlov S, Makhovskaya OV, Tropea JE, Gustchina A, Rotanova TV, Wlodawer A (2005) Atomic-resolution crystal structure of the proteolytic domain of *Archaeoglobus fulgidus* Lon reveals the conformational variability in the active sites of Lon proteases. J Mol Biol 351:144–157

Botos I, Melnikov EE, Cherry S, Tropea JE, Khalatova AG, Rasulova F, Dauter Z, Maurizi MR, Rotanova TV, Wlodawer A, Gustchina A (2004) The catalytic domain of *E. coli* Lon protease has a unique fold and a Ser – Lys dyad in the active site. J Biol Chem 279:8140–8148

Brandstetter H, Kim JS, Groll M, Gottig P, Huber R (2002) Structural basis for the processive protein degradation by tricorn protease. Biol Chem 383:1157–1165

Brandstetter H, Kim JS, Groll M, Huber R (2001) Crystal structure of the tricorn protease reveals a protein disassembly line. Nature 414:466–470

Burley SK, David PR, Taylor A, Lipscomb WN (1990) Molecular structure of leucine aminopeptidase at 2.7-Å resolution. Proc Natl Acad Sci USA 87:6878–6882

Cavard D, Lazdunski C, Howard SP (1989) The acylated precursor form of the colicin A lysis protein is a natural substrate of the DegP protease. J Bacteriol 171:6316–6322

Ciechanover A (1994) The ubiquitin-proteasome proteolytic pathway. Cell 79:13–21

Day CL, Hinds MG (2002) HtrA—a renaissance protein. Structure (Camb) 10:737–739

Du X, Choi IG, Kim R, Wang W, Jancarik J, Yokota H, Kim SH (2000) Crystal structure of an intracellular protease from *Pyrococcus horikoshii* at 2-Å resolution. Proc Natl Acad Sci USA 97:14079–14084

Durá MA, Receveur-Brechot V, Andrieu JP, Ebel C, Schoehn G, Roussel A, Franzetti B (2005) Characterization of a TET-like aminopeptidase complex from the hyperthermophilic archaeon *Pyrococcus horikoshii*. Biochemistry 44:3477–3486

Elsasser S, Finley D (2005) Delivery of ubiquitinated substrates to protein-unfolding machines. Nat Cell Biol 7:742–749

Elzer PH, Phillips RW, Robertson GT, Roop RM 2nd (1996) The HtrA stress response protease contributes to resistance of *Brucella abortus* to killing by murine phagocytes. Infect Immun 64:4838–4841

Emmerich NP, Nussbaum AK, Stevanovic S, Priemer M, Toes RE, Rammensee HG, Schild H (2000) The human 26S and 20S proteasomes generate overlapping but different sets of peptide fragments from a model protein substrate. J Biol Chem 275:21140–21148

Engel M, Hoffmann T, Wagner L, Wermann M, Heiser U, Kiefersauer R, Huber R, Bode W, Demuth HU, Brandstetter H (2003) The crystal structure of dipeptidyl peptidase IV (CD26) reveals its functional regulation and enzymatic mechanism. Proc Natl Acad Sci USA 100:5063–5068

Esser C, Alberti S, Höhfeld J (2004) Cooperation of molecular chaperones with the ubiquitin/proteasome system. Biochim Biophys Acta 1695:171–188

Franzetti B, Schoehn G, Hernandez JF, Jaquinod M, Ruigrok RW, Zaccai G (2002) Tetrahedral aminopeptidase: a novel large protease complex from archaea. EMBO J 21:2132–2138

Fukui T, Eguchi T, Atomi H, Imanaka T (2002) A membrane-bound archaeal Lon protease displays ATP-independent proteolytic activity towards unfolded proteins and ATP-dependent activity for folded proteins. J Bacteriol 184:3689–3698

Fülöp V, Böcskei Z, Polgár L (1998) Prolyl oligopeptidase: an unusual b-propeller domain regulates proteolysis. Cell 94:161–170

Geier E, Pfeifer G, Wilm M, Lucchiari-Hartz M, Baumeister W, Eichmann K, Niedermann G (1999) A giant protease with potential to substitute for some functions of the proteasome. Science 283:978–981

Gerega A, Rockel B, Peters J, Tamura T, Baumeister W, Zwickl P (2005) VAT, the *Thermoplasma* homolog of mammalian p97/VCP, is an N domain regulated protein unfoldase. J Biol Chem 280:42856–42862

Giandomenico V, Simonsson M, Gronroos E, Ericsson J (2003) Coactivator-dependent acetylation stabilizes members of the SREBP family of transcription factors. Mol Cell Biol 23:2587–2599

Golbik R, Lupas AN, Koretke KK, Baumeister W, Peters J (1999) The Janus face of the archaeal Cdc48/p97 homologue VAT: protein folding versus unfolding. Biol Chem 380:1049–1062

Groll M, Bochtler M, Brandstetter H, Clausen T, Huber R (2005) Molecular machines for protein degradation. Chembiochem 6:222–256

Halio SB, Blumentals II, Short SA, Merrill BM, Kelly RM (1996) Sequence, expression in *Escherichia coli*, and analysis of the gene encoding a novel intracellular protease (PfpI) from the hyperthermophilic archaeon *Pyrococcus furiosus*. J Bacteriol 178:2605–2612

Harari-Steinberg O, Chamovitz DA (2004) The COP9 signalosome: mediating between kinase signaling and protein degradation. Curr Protein Pept Sci 5:185–189

Horvath MM, Grishin NV (2001) The *C*-terminal domain of HPII catalase is a member of the type I glutamine amidotransferase superfamily. Proteins 42:230–236

Im YJ, Na Y, Kang GB, Rho SH, Kim MK, Lee JH, Chung CH, Eom SH (2004) The active site of a Lon protease from *Methanococcus jannaschii* distinctly differs from the canonical catalytic dyad of Lon proteases. J Biol Chem 279:53451–53457

Iyer LM, Leipe DD, Koonin EV, Aravind L (2004) Evolutionary history and higher order classification of AAA+ ATPases. J Struct Biol 146:11–31

Jeffery CJ (2004) Molecular mechanisms for multitasking: recent crystal structures of moonlighting proteins. Curr Opin Struct Biol 14:663–668

Jones CH, Bolken TC, Jones KF, Zeller GO, Hruby DE (2001) Conserved DegP protease in gram-positive bacteria is essential for thermal and oxidative tolerance and full virulence in *Streptococcus pyogenes*. Infect Immun 69:5538–5545

Jones CH, Dexter P, Evans AK, Liu C, Hultgren SJ, Hruby DE (2002) *Escherichia coli* DegP protease cleaves between paired hydrophobic residues in a natural substrate: the PapA pilin. J Bacteriol 184:5762–5771

Joshua-Tor L, Xu HE, Johnston SA, Rees DC (1995) Crystal structure of a conserved protease that binds DNA: the bleomycin hydrolase, Gal6. Science 269:945–950

Kaczowka SJ, Maupin-Furlow JA (2003) Subunit topology of two 20S proteasomes from *Haloferax volcanii*. J Bacteriol 185:165–174

Kim DY, Kim DR, Ha SC, Lokanath NK, Lee CJ, Hwang H-Y, Kim KK (2003) Crystal structure of the protease domain of a heat-shock protein HtrA from *Thermotoga maritima*. J Biol Chem 278:6543–6551

Kim DY, Kim KK (2005) Structure and function of HtrA family proteins, the key players in protein quality control. J Biochem Mol Biol 38:266–274

Kim JS, Groll M, Musiol HJ, Behrendt R, Kaiser M, Moroder L, Huber R, Brandstetter H (2002) Navigation inside a protease: substrate selection and product exit in the tricorn protease from *Thermoplasma acidophilum*. J Mol Biol 324:1041–1050

Kim YI, Burton RE, Burton BM, Sauer RT, Baker TA (2000) Dynamics of substrate denaturation and translocation by the ClpXP degradation machine. Mol Cell 5:639–648

Kisselev AF, Akopian TN, Goldberg AL (1998) Range of sizes of peptide products generated during degradation of different proteins by archaeal proteasomes. J Biol Chem 273:1982–1989

Kisselev AF, Akopian TN, Woo KM, Goldberg AL (1999) The sizes of peptides generated from protein by mammalian 26 and 20S proteasomes. Implications for understanding the degradative mechanism and antigen presentation. J Biol Chem 274: 3363–3371

Krojer T, Garrido-Franco M, Huber R, Ehrmann M, Clausen T (2002) Crystal structure of DegP (HtrA) reveals a new protease-chaperone machine. Nature 416:455–459

Lee SJ, Kim SJ, Kim IK, Ko J, Jeong CS, Kim GH, Park C, Kang SO, Suh PG, Lee HS, Cha SS (2003) Crystal structures of human DJ-1 and *Escherichia coli* Hsp31, which share an evolutionarily conserved domain. J Biol Chem 278:44552–44559

Li M, Luo J, Brooks CL, Gu W (2002a) Acetylation of p53 inhibits its ubiquitination by Mdm2. J Biol Chem 277:50607–50611

Li W, Srinivasula SM, Chai J, Li P, Wu JW, Zhang Z, Alnemri ES, Shi Y (2002b) Structural insights into the pro-apoptotic function of mitochondrial serine protease HtrA2/Omi. Nat Struct Biol 9:436–441

Lipinska B, Sharma S, Georgopoulos C (1988) Sequence analysis and regulation of the *htrA* gene of *Escherichia coli*: a σ^{32}-independent mechanism of heat-inducible transcription. Nucleic Acids Res 16:10053–10067

Lipinska B, Zylicz M, Georgopoulos C (1990) The HtrA (DegP) protein, essential for *Escherichia coli* survival at high temperatures, is an endopeptidase. J Bacteriol 172:1791–1797

Liu T, Lu B, Lee I, Ondrovicová G, Kutejová E, Suzuki CK (2004) DNA and RNA binding by the mitochondrial Lon protease is regulated by nucleotide and protein substrate. J Biol Chem 279:13902–13910

Lupas A, Flanagan JM, Tamura T, Baumeister W (1997) Self-compartmentalizing proteases. Trends Biochem Sci 22:399–404

Lyon WR, Caparon MG (2004) Role for serine protease HtrA (DegP) of *Streptococcus pyogenes* in the biogenesis of virulence factors SpeB and the hemolysin streptolysin S. Infect Immun 72:1618–1625

Ma J, Katz E, Belote JM (2001) Expression of proteasome subunit isoforms during spermatogenesis in *Drosophila melanogaster*. Insect Mol Biol 11:627–639

Maupin-Furlow JA, Aldrich HC, Ferry JG (1998) Biochemical characterization of the 20S proteasome from the methanoarchaeon *Methanosarcina thermophila*. J Bacteriol 180:1480–1487

Maupin-Furlow JA, Gil MA, Karadzic IM, Kirkland PA, Reuter CJ (2004) Proteasomes: perspectives from the archaea [update 2004]. Front Biosci 9:1743–1758

Maupin-Furlow JA, Kaczowka SJ, Reuter CJ, Zuobi-Hasona K, Gil MA (2003) Archaeal proteasomes: potential in metabolic engineering. Metabol Eng 5:151–163

Miller DW, Ahmad R, Hague S, Baptista MJ, Canet-Aviles R, McLendon C, Carter DM, Zhu PP, Stadler J, Chandran J, Klinefelter GR, Blackstone C, Cookson MR (2003) L166P mutant DJ-1, causative for recessive Parkinson's disease, is degraded through the ubiquitin-proteasome system. J Biol Chem 278:36588–36595

Mizote T, Tsuda M, Nakazawa T, Nakayama H (1996) The *thiJ* locus and its relation to phosphorylation of hydroxymethylpyrimidine in *Escherichia coli*. Microbiology 142:2969–2974

Mizote T, Tsuda M, Smith DD, Nakayama H, Nakazawa T (1999) Cloning and characterization of the *thiD/J* gene of *Escherichia coli* encoding a thiamin-synthesizing bifunctional enzyme, hydroxymethylpyrimidine kinase/phosphomethylpyrimidine kinase. Microbiology 145:495–501

Neuber O, Jarosch E, Volkwein C, Walter J, Sommer T (2005) Ubx2 links the Cdc48 complex to ER-associated protein degradation. Nat Cell Biol 7:993–998

Niedermann G, King G, Butz S, Birsner U, Grimm R, Shabanowitz J, Hunt DF, Eichmann K (1996) The proteolytic fragments generated by vertebrate proteasomes: structural rela-

tionships to major histocompatibility complex class I binding peptides. Proc Natl Acad Sci USA 93:8572–8577
Nomura K, Kato J, Takiguchi N, Ohtake H, Kuroda A (2004) Effects of inorganic polyphosphate on the proteolytic and DNA-binding activities of Lon in *Escherichia coli*. J Biol Chem 279:34406–34410
Ohnishi Y, Yamazaki H, Kato JY, Tomono A, Horinouchi S (2005) AdpA, a central transcriptional regulator in the A-factor regulatory cascade that leads to morphological development and secondary metabolism in *Streptomyces griseus*. Biosci Biotechnol Biochem 69:431–439
Onoe S, Ando S, Ataka M, Ishikawa K (2002) Active site of deblocking aminopeptidase from *Pyrococcus horikoshii*. Biochem Biophys Res Commun 290:994–997
Pietrokovski S (2001) Intein spread and extinction in evolution. Trends Genet 17:465–472
Poquet I, Saint V, Seznec E, Simoes N, Bolotin A, Gruss A (2000) HtrA is the unique surface housekeeping protease in *Lactococcus lactis* and is required for natural protein processing. Mol Microbiol 35:1042–1051
Quigley PM, Korotkov K, Baneyx F, Hol WGJ (2003) The 1.6-Å crystal structure of the class of chaperones represented by *Escherichia coli* Hsp31 reveals a putative catalytic triad. Proc Natl Acad Sci USA 100:3137–3142
Reits E, Griekspoor A, Neijssen J, Groothuis T, Jalink K, van Veelen P, Janssen H, Calafat J, Drijfhout JW, Neefjes J (2003) Peptide diffusion, protection, and degradation in nuclear and cytoplasmic compartments before antigen presentation by MHC class I. Immunity 18:97–108
Reits E, Neijssen J, Herberts C, Benckhuijsen W, Janssen L, Drijfhout JW, Neefjes J (2004) A major role for TPPII in trimming proteasomal degradation products for MHC class I antigen presentation. Immunity 20:495–506
Remaut H, Bompard-Gilles C, Goffin C, Frère JM, Van Beeumen J (2001) Structure of the *Bacillus subtilis* D-aminopeptidase DppA reveals a novel self-compartmentalizing protease. Nat Struct Biol 8:674–678
Rep M, van Dijl JM, Suda K, Schatz G, Grivell LA, Suzuki CK (1996) Promotion of mitochondrial membrane complex assembly by a proteolytically inactive yeast Lon. Science 274:103–106
Reuter CJ, Kaczowka SJ, Maupin-Furlow JA (2004) Differential regulation of the PanA and PanB proteasome-activating nucleotidase and 20S proteasomal proteins of the haloarchaeon *Haloferax volcanii*. J Bacteriol 186:7763–7772
Reuter CJ, Maupin-Furlow JA (2004) Analysis of proteasome-dependent proteolysis in *Haloferax volcanii* cells using short-lived green fluorescent proteins. Appl Envir Microbiol 70:7530–7538
Richly H, Rape M, Braun S, Rumpf S, Hoege C, Jentsch S (2005) A series of ubiquitin binding factors connects CDC48/p97 to substrate multiubiquitylation and proteasomal targeting. Cell 120:73–84
Rockel B, Jakana J, Chiu W, Baumeister W (2002) Electron cryo-microscopy of VAT, the archaeal p97/CDC48 homologue from *Thermoplasma acidophilum*. J Mol Biol 317:673–681
Rockel B, Walz J, Hegerl R, Peters J, Typke D, Baumeister W (1999) Structure of VAT, a CDC48/p97 ATPase homologue from the archaeon *Thermoplasma acidophilum* as studied by electron tomography. FEBS Lett 451:27–32
Römisch K (2005) Endoplasmic reticulum-associated degradation. Annu Rev Cell Dev Biol 21:435–456
Rotanova TV, Melnikov EE, Khalatova AG, Makhovskaya OV, Botos I, Wlodawer A, Gustchina A (2004) Classification of ATP-dependent proteases Lon and compar-

ison of the active sites of their proteolytic domains. Eur J Biochem 271:4865–4871

Rudolph MJ, Wuebbens MM, Rajagopalan KV, Schindelin H (2001) Crystal structure of molybdopterin synthase and its evolutionary relationship to ubiquitin activation. Nat Struct Biol 8:42–46

Ruepp A, Eckerskorn C, Bogyo M, Baumeiste5r W (1998) Proteasome function is dispensable under normal but not under heat shock conditions in *Thermoplasma acidophilum*. FEBS Lett 425:87–90

Russo S, Baumann U (2004) Crystal structure of a dodecameric tetrahedral-shaped aminopeptidase. J Biol Chem 279:51275–51281

Saras J, Heldin CH (1996) PDZ domains bind carboxy-terminal sequences of target proteins. Trends Biochem Sci 21:455–458

Saric T, Graef CI, Goldberg AL (2004) Pathway for degradation of peptides generated by proteasomes: a key role for thimet oligopeptidase and other metallopeptidases. J Biol Chem 279:46723–46732

Schlieker C, Mogk A, Bukau B (2004) A PDZ switch for a cellular stress response. Cell 117:417–419

Schuberth C, Buchberger A (2005) Membrane-bound Ubx2 recruits Cdc48 to ubiquitin ligases and their substrates to ensure efficient ER-associated protein degradation. Nat Cell Biol 7:999–1006

Seemüller E, Lupas A, Baumeister W (1996) Autocatalytic processing of the 20S proteasome. Nature 382:468–470

Smalle J, Vierstra RD (2004) The ubiquitin 26S proteasome proteolytic pathway. Annu Rev Plant Biol 55:555–590

Snowden LJ, Blumentals II, Kelly RM (1992) Regulation of proteolytic activity in the hyperthermophile *Pyrococcus furiosus*. Appl Environ Microbiol 58:1134–1141

Spiess C, Beil A, Ehrmann M (1999) A temperature-dependent switch from chaperone to protease in a widely conserved heat shock protein. Cell 97:339–347

Spreter T, Pech M, Beatrix B (2005) The crystal structure of archaeal nascent polypeptide-associated complex (NAC) reveals a unique fold and the presence of a ubiquitin-associated domain. J Biol Chem 280:15849–15854

Stack HM, Sleator RD, Bowers M, Hill C, Gahan CG (2005) Role for HtrA in stress induction and virulence potential in *Listeria monocytogenes*. Appl Environ Microbiol 71:4241–4247

Stahlberg H, Kutejová E, Suda K, Wolpensinger B, Lustig A, Schatz G, Engel A, Suzuki CK (1999) Mitochondrial Lon of *Saccharomyces cerevisiae* is a ring-shaped protease with seven flexible subunits. Proc Natl Acad Sci USA 96:6787–6790

Strauch KL, Beckwith J (1988) An *Escherichia coli* mutation preventing degradation of abnormal periplasmic proteins. Proc Natl Acad Sci USA 85:1576–1580

Strauch KL, Johnson K, Beckwith J (1989) Characterization of *degP*, a gene required for proteolysis in the cell envelope and essential for growth of *Escherichia coli* at high temperature. J Bacteriol 171:2689–2696

Tamura N, Lottspeich F, Baumeister W, Tamura T (1998) The role of tricorn protease and its amino peptidase-interacting factors in cellular protein degradation. Cell 95:637–648

Tamura T, Tamura N, Cejka Z, Hegerl R, Lottspeich F, Baumeister W (1996) Tricorn protease —the core of a modular proteolytic system. Science 274:1385–1389

Tomkinson B (1999) Tripeptidyl peptidases: enzymes that count. Trends Biochem Sci 24:355–359

Tsunasawa S, Nakura S, Tanigawa T, Kato I (1998) Pyrrolidone carboxyl peptidase from the hyperthermophilic archaeon *Pyrococcus furiosus*: cloning and overexpression in

Escherichia coli of the gene, and its application to protein sequence analysis. J Biochem (Tokyo) 124:778–783

Van den Eynde BJ, Morel S (2001) Differential processing of class-I-restricted epitopes by the standard proteasome and the immunoproteasome. Curr Opin Immunol 13:147–153

Volker C, Lupas AN (2002) Molecular evolution of proteasomes. Curr Top Microbiol Immunol 268:1–22

Waller PR, Sauer RT (1996) Characterization of *degQ* and *degS*, *Escherichia coli* genes encoding homologs of the DegP protease. J Bacteriol 178:1146–1153

Walsh NP, Alba BM, Bose B, Gross CA, Sauer RT (2003) OMP peptide signals initiate the envelope-stress response by activating DegS protease via relief of inhibition mediated by its PDZ domain. Cell 113:61–71

Walz J, Koster AJ, Tamura T, Baumeister W (1999) Capsids of tricorn protease studied by electron cryomicroscopy. J Struct Biol 128:65–68

Walz J, Tamura T, Tamura N, Grimm R, Baumeister W, Koster AJ (1997) Tricorn protease exists as an icosahedral supermolecule in vivo. Mol Cell 1:59–65

Wang C, Xi J, Begley TP, Nicholson LK (2001a) Solution structure of ThiS and implications for the evolutionary roots of ubiquitin. Nat Struct Biol 8:47–51

Wang EW, Kessler BM, Borodovsky A, Cravatt BF, Bogyo M, Ploegh HL, Glas R (2000) Integration of the ubiquitin-proteasome pathway with a cytosolic oligopeptidase activity. Proc Natl Acad Sci USA 97:9990–9995

Wang J, Song JJ, Franklin MC, Kamtekar S, Im YJ, Rho SH, Seong IS, Lee CS, Chung CH, Eom SH (2001b) Crystal structures of the HslVU peptidase-ATPase complex reveal an ATP-dependent proteolysis mechanism. Structure (Camb) 9:177–184

Wang Q, Song C, Li CC (2004) Molecular perspectives on p97-VCP: progress in understanding its structure and diverse biological functions. J Struct Biol 146:44–57

Wilken C, Kitzing K, Kurzbauer R, Ehrmann M, Clausen T (2004) Crystal structure of the DegS stress sensor: How a PDZ domain recognizes misfolded protein and activates a protease. Cell 117:483–494

Wilson HL, Aldrich HC, Maupin-Furlow JA (1999) Halophilic 20S proteasomes of the archaeon *Haloferax volcanii*: purification, characterization, and gene sequence analysis. J Bacteriol 181:5814–5824

Wilson HL, Ou MS, Aldrich HC, Maupin-Furlow JA (2000) Biochemical and physical properties of the *Methanococcus jannaschii* 20S proteasome and PAN, a homolog of the ATPase (Rpt) subunits of the eucaryal 26S proteasome. J Bacteriol 182:1680–1692

Wolf DH, Hilt W (2004) The proteasome: a proteolytic nanomachine of cell regulation and waste disposal. Biochim Biophys Acta 1695:19–31

Yorgey P, Rahme LG, Tan MW, Ausubel FM (2001) The roles of *mucD* and alginate in the virulence of *Pseudomonas aeruginosa* in plants, nematodes and mice. Mol Microbiol 41:1063–1076

Yoshida Y, Chiba T, Tokunaga F, Kawasaki H, Iwai K, Suzuki T, Ito Y, Matsuoka K, Yoshida M, Tanaka K, Tai T (2002) E3 ubiquitin ligase that recognizes sugar chains. Nature 418:438–442

Zwickl P, Ng D, Woo KM, Klenk H-P, Goldberg AL (1999) An archaebacterial ATPase, homologous to ATPases in the eukaryotic 26S proteasome, activates protein breakdown by 20S proteasomes. J Biol Chem 274:26008–26014

Microbiol Monogr (2)
J.M. Shively: Complex Intracellular Structures in Prokaryotes
DOI 10.1007/7171_020/Published online: 20 May 2006

Assembly and Disassembly of Phycobilisomes

Noam Adir (✉) · Monica Dines · Merav Klartag · Ailie McGregor · Meira Melamed-Frank

Department of Chemistry, Technion, Israel Institute of Technology, Technion City, 32000 Haifa, Israel
nadir@tx.technion.ac.il

Abstract The process of photosynthesis is initiated by the absorption of light energy by large arrays of pigments bound in an ordered fashion within protein complexes called antennas. These antennas transfer the absorbed energy at almost 100% efficiency to the reaction centers that perform the photochemical electron transfer reactions required for the conversion of the light energy into useful and storable chemical energy. In prokaryotic cyanobacteria, eukaryotic red algae and cyanelles, the major antenna complex is called the phycobilisome, an extremely large (3–7 MDa) multi subunit complex found on the stromal side of the thylakoid membrane. Phycobilisomes are assembled in an ordered sequence from similarly structured units that covalently bind a variety of linear tetrapyrolle pigments called bilins. Phycobilisomes have a broad cross-section of absorption (500–680 nm) and mainly transfer the absorbed energy to photosystem II. They can, however, function as an antenna of photosystem I, and their composition can be altered as a result of changes in the environmental light quality. The phycobilisome is structurally and functionally different from other classes of photosynthetic antenna complexes. In this review, we will describe the important structural and functional characteristics of the phycobilisome complex and its components, especially with respect to its assembly and disassembly.

Abbreviations

APC	allophycocyanin
CCA	complementary chromatic adaptation
LHC	light harvesting complex
PB(s)	phycobilin(s)
PBS(s)	phycobilisome(s)
PBP(s)	phycobiliprotein(s)
PC	phycocyanin
PCB	phycocyanobilin
PE	phycoerythrin
PEB	phycoerythrobilin
PUB	phycourobilin
PXB	phycoviobilin
PSI	photosystem I
PSII	photosystem II
RC(s)	reaction center(s)
TEM	transmission electron microscopy

1 Introduction

A major priority of any photosynthetic organism is the efficient absorption of the light energy available in the specific environmental niche that it occupies. In search of efficient light absorption, different forms of light harvesting complex (LHC) antennas have evolved, with structural characteristics that lead to specific functional attributes (Adir 2005; Blankenship et al. 1995; Cogdell et al. 2004; Dekker and Boekema 2005; Frigaard et al. 2001, 2003; Glazer 1985; Huber 1989; Melkozernov and Blankenship 2005; Ting et al. 2002; Xiong and Bauer 2002). In all cases these antennas are composed of proteins that bind various pigment molecules in a fashion that affords both extremely efficient energy absorption and energy transfer to the photochemical reaction centers (RCs). The sizes of these antennas vary from organism to organism, and the ratio of antenna pigments to RCs can be in the hundreds. The synthesis of a large excess of LHC pigments could be considered an energetic burden on the photosynthetic organism and is thus under tight regulation. The large sizes of antenna complexes indicate that it is imperative that the LHCs provide enough excitation energy to the RCs in order to perform photosynthesis at the maximum efficiency and to avoid possibly deleterious back reactions (Adir et al. 2003). Many LHCs are made up of transmembrane proteins located within the thylakoid membranes adjacent to the RC (Ben-Shem et al. 2003; Cogdell et al. 2004; Dekker and Boekema 2005). In these cases, the two-dimensionality of the membrane limits the size of each LHC, and one finds multiple oligomeric forms in either condensed complexes (such as in LHCII and LHCI of plants), or in ring form (as found in purple non-sulfur bacteria).

Two LHC types are located outside the membrane plane and are thus essentially less limited in their size. The largest of all LHCs is the chlorosome found in green sulfur bacteria (Blankenship et al. 1995; Frigaard et al. 2003). This structure is described in detail (Frigaard and Bryant 2006, in this volume) of this monograph. The second large non-membranous LHC is called the phycobilisome (PBS) (Fig. 1), found in cyanobacteria, red-algae and cyanelles (Adir 2005; Glazer 1989). In the following review we will describe the present state of knowledge on the structure and function of different PBS forms, with special emphasis on the mechanisms leading to the formation and disassembly of the PBS.

The PBS successfully covers a wide range of excitation wavelengths using variations on a single structural principle. All PBS co-factor containing components (Table 1) are initially formed by a basic building block recognized in the literature as a monomer (Fig. 2). These monomers are in fact heterodimers of two subunits (α and β) that are significantly homologous on both sequence and structural levels. Their structures and mode of association lead to a very even and highly symmetric structure; however, we will describe here how the differences in local environment, coupled with the manner of complex assembly, allows the PBS to efficiently perform energy absorption and transfer. Each ($\alpha\beta$) monomer unit covalently binds up to three linear *tetrapyrrole* chromophore molecules called phycobilins (PBs) (Brown et al. 1990; MacColl 1998). The ($\alpha\beta$) monomers then self-assemble in a number of apparently sequential steps (Fig. 3) into (i) trimeric $(\alpha\beta)_3$ disks with a diameter of ~ 110 Å, a width of 30 Å and an interior uneven aperture with a diameter of 15–50 Å; (ii) $(\alpha\beta)_6$ hexameric double-disks; and finally into (iii) one of two major substructures, known as cores and rods. While both of these larger assemblies appear as stacks of disks aligned in different directions, it is recognized today that slight variations in the packing of these disks lead to markedly different properties. The number of core components

Fig. 1 Electron micrograph of a *Synechococcus lividus* cell showing PBS particles attached to the thylakoid membranes (*arrows*). Magnification, ×60 000. The figure was kindly provided by Prof. E. Gantt and first appeared in Edwards, Gantt, J Cell Biol (1971), 50:896–900

Table 1 Phycobilisome protein subunits and pigments

Protein component	Function	Bilin type[3]	Number of bilins (per monomer)	Absorption maximum (nm)[5]	Gene(s)
Allophycocyanin	Major core antenna protein	PCB	2	652	*apcA, apcB*
Allophycocyanin α^B	Minor core antenna protein	PCB	1	671–679	*apcD*
Phycocyanin	Rod antenna protein	PCB	3	620	*cpcA, cpcB*
$Phycocyanin_{612}$[1]	Minor rod antenna protein	PCB	3	612	*cpcA, cpcB*
Phycoerythro-cyanin	Rod antenna protein	PCB PVB	2 1	580–600	*pccA, pccB*
Phycoerythrin	Rod antenna protein	PEB PUB	4–5[4]	495–565	*cpeA, cpeB, mpeA, mpeB*
L_R[2]	Internal rod linker	–	–	–	*cpcC cpcH, cpcI, cpeC, cpeD, cpeE*
L_C	Internal core linker	–	–	–	*apcC*
L_{RC}[2]	Rod-core linker	–	–	–	*cpcD*
L_{CM}	Core-membrane linker	PCB	1	671–679	*apcE*
Phycoerythrin γ subunit	PE linker and antenna	PEB PUB	1 1–3		*mpeC*

[1] PC_{612} has been identified only in PBS from *T. vulcanus*
[2] A number of L_R, L_{RC} proteins can exist within each type of PBS, depending on species, and environmental conditions
[3] PCB, phycocyanoblin; PVB, phycoviolobilin; PEB, phycoerythrobilin; PUB, phycourobilin
[4] Different types of phycoerythrin contain 4–5 PEB, PUB bilins at different ratios
[5] Absorption spectra of $(\alpha\beta)_3$ trimeric forms

(typically called cylinders) and rods is species dependent. One of the most common PBS forms consists of a tricylindrical core surrounded by six rods that form a hemidiscoidal semi-circle (Adir 2005; Glauser et al. 1992; Glazer 1989; MacColl 1998). Other PBS forms include two or five cylinders and different numbers of rods (Ducret et al. 1998; Glauser et al. 1992). It is important to remember that while all of the different $(\alpha\beta)_3$ disks appear to be similar, the association of disks to form either core cylinders or rods is architecturally different than the forces that provide for rod-core assembly. The original discovery and description of the PBS emanated from transmission electron microscopy (TEM) studies performed in the 1960s and 1970s (Fig. 1) (Bryant

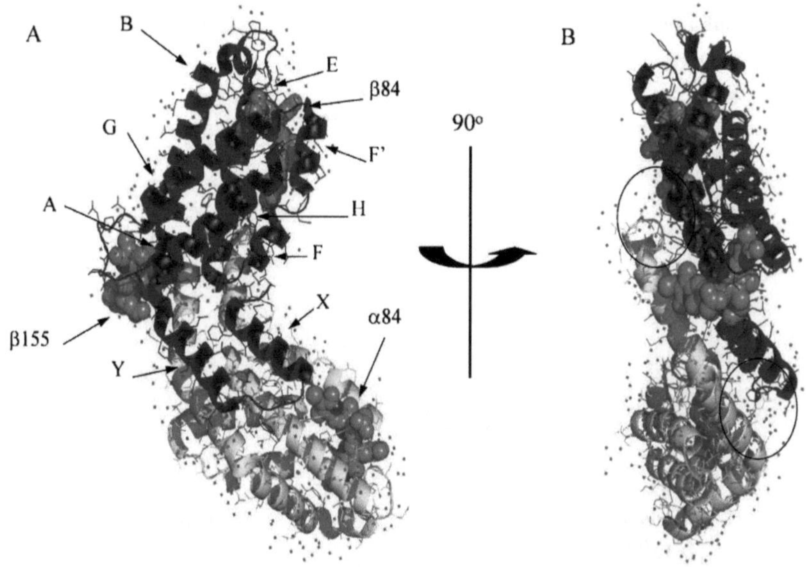

Fig. 2 Phycocyanin monomer. The phycocyanin monomer exemplifies all of the basic structural characteristics of PBPs. *Panel A*, the α and β subunits (*yellow* and *blue* respectively) are depicted in cartoons, the three PCB cofactors in CPK colored spheres, solvent molecules are in small *red balls*. The positions of helices A–Y of the β subunit are designated. *Panel B*, the structure has been rotated by 90°, revealing the monomer interface between the two subunits. Two *black rings* identify the position of two zones that contain strong polar interactions in all PBPs and stabilize the monomer. In some thermophilic cyanobacteria an additional zone exists in the vicinity of the β155 PCBs (for more details see [45])

et al. 1979; Gantt and Conti 1966a,b; Gantt and Lipschultz 1972; Glazer and Bryant 1975; Sidler 1994; Yamanaka et al. 1978). To date, much of what we know about the structure of entire PBS complexes is still based on these negatively stained and fixed TEM preparations.

The range of absorbed wavelengths covered by the PBS is species dependent due to the presence of different monomer constituents. While all PBSs contain both allophycocyanin (APC, $\lambda_{max} = 652$ nm) and phycocyanin (PC, $\lambda_{max} = 620$ nm), some species contain additional components: phycoerythrocyanin (PEC, $\lambda_{max} = 567$ nm) or phycoerythrin (PE, $\lambda_{max} = 560$ nm). Along with these PB-containing subunits are a number of unpigmented proteins of various lengths that are located within the central openings of the trimeric/hexameric disks. As a consequence of their position, these proteins are collectively called linker proteins (Liu et al. 2005). The actual absorption spectra are a result of the covalently bound bilin type, the protein identity, the associated linker protein and the level of organization. A list of the components of the PBS (biliproteins, bilins and linker proteins) is compiled in Table 1.

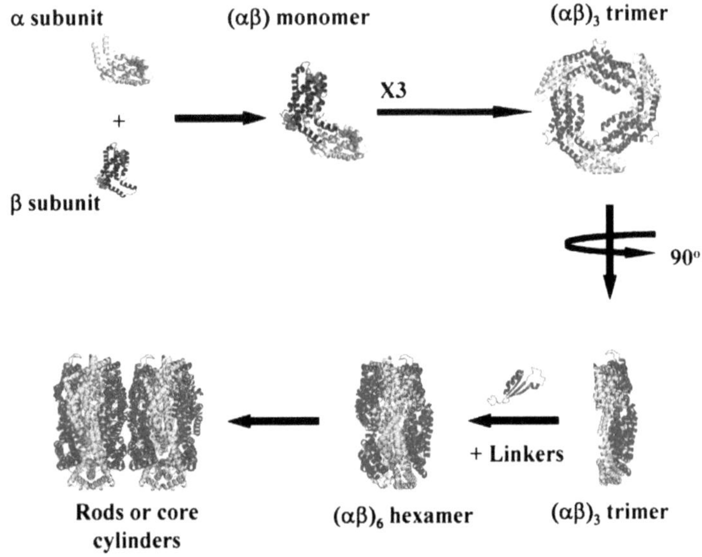

Fig. 3 Schematic model of the PBS assembly process. The $(\alpha\beta)$ monomer is formed from individual subunits already containing bilins. The linker shown is of the L_C type from the 1B33 APC structure, while in reality linkers associated with the process will be of different sizes and orientations. It is not clear whether APC forms tight double hexamers, or four trimers in looser association

2 Evolution and the Genetics of the Phycobilisome Antenna System

It is quite clear today that the evolution of life on earth quickly led to the development of species able to utilize the energy of the sun for driving metabolism (Blankenship and Hartman 1998). Fossil records appear to suggest that within the first billion years, photosynthetic bacteria had evolved the mechanisms required to abstract electrons from water and evolve oxygen (Awramik 1992; Buick 1992; De Marais 2000; Xiong and Bauer 2002). These primitive species appear to be quite similar to modern-day cyanobacteria, and it is quite likely that these organisms possessed antennas similar to the PBS. On the basis of amino acid homology and a similarity in protein folds, it was suggested that the phycobiliprotein (PBP) scaffold evolved from an ancient globin-type protein, which may already have embodied some of the structural characteristics needed for multi-subunit assembly (Holm and Sander 1993). All PBPs can be traced to a most recent common ancestor protein, which diverged, duplicated, fused and mutated in the development of the PBS. A detailed sequence comparison-based study and phylogenetic analysis of the evolutionary events leading to the PBS have been described by Grossman and co-workers (Apt et al. 1995).

Does the formation of the PBS require the concerted transcription of the various genes encoding for the different subunits? To date, the genomes of eight photosynthetic species/organelles containing PBSs have been fully determined, allowing us a sufficiently broad view of the positions of the PBS encoding genes in cyanobacteria and red algae. To the complete genomes we can add a number of other species whose PBS genes have been identified and characterized (Grossman et al. 1993). The primary PBS genes encode for the two subunits of each monomeric component, the various linker proteins, and accessory proteins involved in cofactor synthesis, cofactor attachment, and disassembly (Fig. 4 and Table 1). It is clear that in all of the genomes sequenced so far, a certain degree of clustering of the PBS genes appears to be advantageous to optimal control over PBS gene expression. This characteristic has been carried over from the cyanobacteria to the red algae whose PBS genes are almost all found within the plastid genome and remain clustered. Within the different clusters, the order of the genes is mostly, but not entirely, conserved apparently affording somewhat different control mechanisms over protein expression. For instance the *cpcEF* genes encoding for the PC lyase

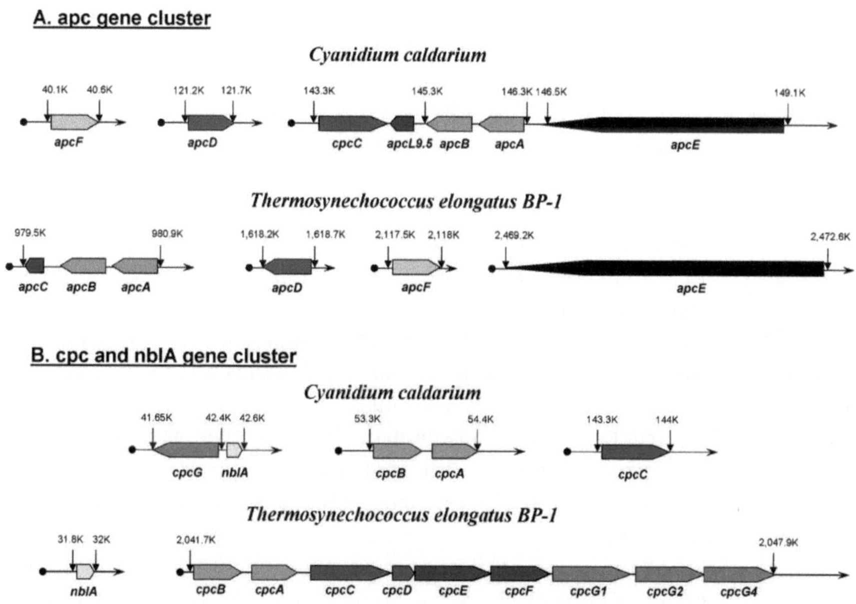

Fig. 4 Physical maps of the *apc* (**A**), and *cpc* and *nbl*A (**B**) genes and gene clusters from the red algae *Cyanidium caldarium* and the cyanobacterium *Thermosynechococcus elongatus.* The specific protein encoded by each of the genes is noted. The different sizes are relatively represented by the lengths of the *arrow-bars*; the *arrow* designates coding direction. The *numbers* denote the actual location of the genes within each genome as its annotated in the Entrez Genome subsection of the National Center for Biological Information (http://www.ncbi.nlm.nih.gov/Genomes/)

α and β subunits are contiguous within the *cpc* operon in *T. elongatus* but have been translocated to separate positions within the *Synechocystis* sp. PCC 6803 genome. Clustering most likely indicates the importance of polycistronic transcription, and indeed long transcripts encoding for the entire operon have been identified (Grossman et al. 1993). However, smaller transcripts have also been identified, indicating that gene expression can be modified under different environmental conditions. The use of either RT-PCR or microarray forms of analysis allow the quantification of different transcripts that appear under different conditions in vivo. Both methods were recently used to show changes in the expression pattern of a variety of cyanobacterial genes, due to changes in either light intensity or light quality (Huang et al. 2002; MacDonald et al. 2003). It is quite clear that PBS expression is tightly controlled at the transcriptional level, with all genes probed showing similar changes in transcript level. In the native environments of PBS-containing species, changes in the light regime may be less extreme, leading to gene specific rather than operon specific transcript changes. One of the most notable changes in gene expression occurs in those species (such as *Fremyella diplosiphon*) that can undergo complimentary chromatic adaptation (CCA) in which the expression of entire protein classes change (Grossman et al. 1993, 2001; MacColl 1998). Exposure of the cells to either red or green light induces changes in the expression of the operons encoding for both the PE and PC subunits. These operons can be switched on or off, thereby changing the action spectra for photosynthesis. Thus, we can conclude that optimal PBS assembly requires transcriptional regulation, coupled to species-specific modifications needed to obtain a great degree of fitness to its environment.

There is still little evidence that PBS assembly is controlled on the level of mRNA translation. Post-translational events are extremely important for the maturation of the subunits, and include attachment of the bilin cofactors and the site-specific methylation of βAsn72 of almost all subunits (see below; Duerring et al. 1988; Klotz et al. 1986; Swanson and Glazer 1990). The level of conservation in the sequences of the various proteins has been described in great detail (Apt et al. 1995) and strongly indicates that optimization of the overall PBS structure has been obtained. However, even small changes in the amino acid sequence allows for fine-tuning of absorption energies and/or for the addition of stabilization points, according to the requirements of the specie's environmental niche.

Comparison of the many PBP sequences available shows a remarkable level of conservation (Apt et al. 1995). Many of the conserved residues are functionally required to preserve the α-helical-based structure and obtain the optimal surroundings for bilin function. In addition, a large number of residues required for correct assembly are also highly conserved; emphasizing that PBS assembly is a primary prerequisite for its function. Conservation of these residues also indicates that the method of assembly may be quite similar in the evolutionary dissimilar PBS-containing species.

3 Characteristics of the Phycobilisome Components

The highest level of structural and functional analysis on the PBS has been performed on isolated components in a variety of oligomeric states. The following sections will describe the characteristics of the isolated components and how each component contributes to the overall PBS function.

3.1 The Bilin Cofactors

The wealth of structural data available today enables us to describe the PB cofactors with a high degree of precision. All PB types are linear tetrapyrolles, covalently bonded through a thioether linkage to free cysteines (Glazer 1989). The linked ring is typically called ring A, with only a subset of phycoerythrobilin (PEB) doubly linked through both rings A and D (MacColl 1998). The initial steps in the synthetic pathway for production of the different PB species are identical to those steps that produce heme (and chlorophyll), which are followed by degradation of heme by heme oxygenase (Migita et al. 2003; Sugishima et al. 2005), and reduction and isomerization steps (Frankenberg et al. 2001). The major difference between the various PBs is in the arrangement and number of conjugated double bonds, resulting in a change in the absorption maxima.

Intermediate resolution crystal structures have provided us with a means to visualize the minor twists and changes in ring planarity that affect PB absorption (Adir 2005). This is coupled with the identification of polar or charged group interactions, putative hydrogen bonds and solvent interactions that together complete the PB environment (Adir et al. 2002). A recent report of the high-resolution (0.97 Å) structure of PE545 (a non-PBS forming protein, see below), shows the five PBs in exquisite detail (Doust et al. 2004). Although the positions of hydrogen atoms were not included in the PDB depositions (PDB codes 1XF6 and 1XG0), the electron density maps allow the identification of protonation states of the different chromophores. Our group has recently collected high resolution data (1.35 Å) on PC from *T. vulcanus* (unpublished data), in which the visualization of the cofactor configuration and environment is greatly improved over the previous 1.6 Å data (PDB code 1KTP) (Adir et al. 2002). These details will provide the basis for a precise quantum mechanical description of the ground states of the PB and will help to predict the conformation and configuration of the excited states. The PB cofactors can also play structural roles in their association with the surrounding proteins. We identified in the PC structure of the thermophilic species *T. vulcanus* that a 60° shift in the position of the β155 phycocyanobilin (PCB) into the protein interior induced stabilization of the interface of the monomer and hexamer associations (see Sects. 4.1 and 4.3; Adir et al. 2001, 2002).

3.2 Allophycocyanin

Allophycocyanin (APC) is a general name for a number of biliproteins located in the core of the PBS and which are in contact with the photosynthetic membrane and in proximity to PSII (for a recent review see MacColl 2004). APC has been typically isolated in a trimeric $(\alpha\beta)_3$ form, and its α and β subunits are both highly conserved amongst different species (> 85%) and to a lesser extent between themselves (~ 35%) (Apt et al. 1995). Each α and β subunit covalently binds one PCB chromophore through conserved cysteine residues α84 and β84. The three-dimensional structures of APC have been obtained from two species of cyanobacteria and a red alga (Brejc et al. 1995; Liu et al. 1999; Reuter et al. 1999). X-ray diffraction studies on crystals of APC show the trimers as flattened rings or disks, having C_3 symmetry. The approximate diameter of a disk is 110 Å, and there is a central channel of uneven circumference of between 15–50 Å in diameter. Both subunits of APC monomers show almost identical structures, even though the subunit homology is only about 38%, and they are also highly homologous to PC and PE structures (MacColl et al. 2003; Sun and Wang 2003). In one case, a crystal structure of the APC–linker complex, $APL_C^{7.8}$ was obtained by Reuter et al. from the cyanobacterium *Mastigocladus luminosus* (Reuter et al. 1999). This structure revealed that the core linker is in contact with two of the three-APC β subunits, bringing the subunits closer together and directly interacting with the chromophores on these subunits. The presence of linkers in APC (and other PBPs) has indeed been shown to affect the absorption spectra of these chromophores. The distance between the two closest PCBs in the APC trimeric disk is 21 Å. The APC trimer has an unusual absorption maximum at 652 nm with a shoulder at 620 nm, while APC monomers have an absorption maximum at 615 nm (MacColl 2004). The fluorescence emission maxima of the trimer is 655 nm.

The most common form of the PBS consists of a central core formed by two, three or five cylinders of APC from which peripheral rods radiate in some form of hemidiscoidal arrangement. Despite extensive studies, it remains unclear how the APC core interacts with the thylakoid membrane and the photosystems within it. In a tricylindrical core, two of the APC cylinders lie in an anti-parallel orientation with respect to one another and appear to be in close contact with the thylakoid membrane (base cylinders). The third cylinder is located in the saddle formed by the base cylinders (Fig. 5), and is composed of a row of four APC $(\alpha\beta)_3$ disks associated with two L_C subunits. On the basis of low resolution TEM studies, models for the organization of the APC core envision that each of the membrane-associated base cylinders is composed of four trimeric disks with different biochemical compositions in the following order: (i) an APC $(\alpha\beta)_3$ disk associated with an L_C linker; (ii) an APC $(\alpha\beta)_3$ disk; (iii) $\alpha_2\beta_2\beta^{16}L_{CM}$—an APC $(\alpha\beta)_3$ disk with one α subunit re-

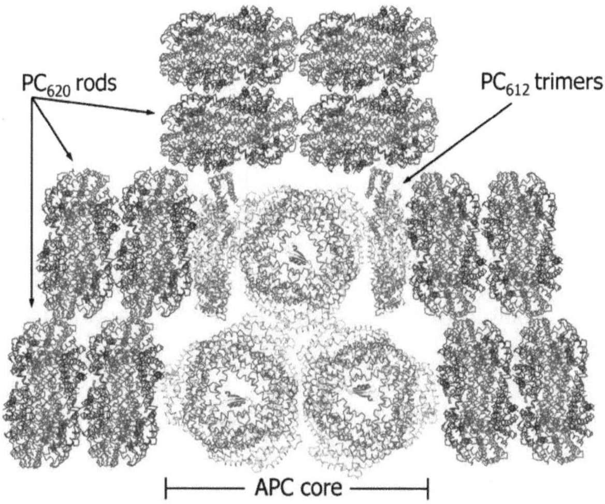

Fig. 5 Model of a tricylindrical PBS complex based on individual crystal structures. Symmetry related monomeric units of PC (PDB code 1KTP) assemble into trimers and hexamers in the crystals lattice. These hexamers were used to build the short rods. The three core cylinders are represented by three trimers of APC and the L_C linker protein from the 1B33 structure. The L_C is in the *magenta* cartoon. The two lower cylinders are rotated by 180° to approximate their anti-parallel orientation. The top cylinder was rotated to obtain optimal packing. The PC_{612} subunit (identified to date only in *T. vulcanus*, PDB code 1ON7), is proposed to fill in the gap between two rods and the upper APC cylinder. Cα backbone representations of the α and β subunits are colored: PC – *red* and *blue*, APC – *yellow* and *purple*, PC_{612} – *brown* and *cyan* respectively. The L_C subunits are represented in the *magenta* cartoon

placed by the PB domain of the L_{CM} and one β subunit replaced by the *apc*F protein (also known as APC A); and (iv) $\alpha^B\alpha_2\beta_3$—an APC $(\alpha\beta)_3$ disk with one α subunit replaced by the gene product of *apcD* which is also called APC α^B (Table 1).

All three APC crystal structures determined to date contain only $(\alpha\beta)_3$-type APC trimers, and are thus from upper level cylinders. Single particle analysis of the core complex revealed a mirror axis, indicative of two-fold symmetry (Barber et al. 2003; Brejc et al. 1995; Liu et al. 1999). It is suggested that the two cylinder nature of the APC core bottom layer matches the dimeric nature of the PSII complex so that each of the APC base cylinders interacts with the stromal surface of one of the monomers of the dimeric PSII (Barber et al. 2003; Ferreira et al. 2004; Loll et al. 2005). From these structures, possible interaction points can now be estimated between the APC core and the PSII core dimer. When viewed normal to the plane of the thylakoid membrane the densities corresponding to the interaction points of the two APC core base cylinders are located toward the periphery of the PSII dimer.

3.3 Phycocyanin

Phycocyanin (PC) is the major rod pigment-protein, found in all PBS forms. Unlike APC, most species possess only one set of PC genes, encoding for a single form of ($\alpha\beta$) monomer. This PBP is typically isolated in trimeric or hexameric form demonstrating the stability of these aggregation states. The molecular dimensions of the trimeric rings are similar to that of APC, but the disk circumference is uneven as a result of an added loop at the C-terminus of the β-subunit, to which a third PCB cofactor binds. As a result of the presence of this additional cofactor (which is attached to the βCys155 residue) the outer edges of the trimeric/hexameric rings contain a relatively solvent-exposed PCB. At this position, inter-rod energy transfer becomes likely, and new energy transfer pathways down and across the PBS are opened up (Padyana et al. 2001; Stec et al. 1999).

To date, PC crystal structures have been determined from nine different species, including both cyanobacteria (mesophiles and thermophiles) and red-algae (Adir 2005). The structures obtained are of relatively high resolution (Adir et al. 2002; Nield et al. 2003), and provide a well-defined molecular picture of the evolutionarily stable structural characteristics required for efficient energy transfer. These structures also afford insight into the fine-tuning required by species that have evolved in different surroundings, and afford clues on the process of PBS assembly.

Both α and β subunits in the monomer have thiol-linked PCB chromophores at symmetry-related positions (Cys84) located between helices E, F′ and G. The chemical surroundings of these two PCBs is, however, very different upon formation of the higher-order $(\alpha\beta)_3$ trimer. The α84 PCB is almost totally buried and hidden from the surrounding solvent unlike the β84 PCB which juts out into the $(\alpha\beta)_6$ hexameric ring interior. As noted above, the β-subunit has an additional PCB which is on the outside of the trimeric ring and may be important for intra and inter-rod energy transfer. The crystal structures of two thermophilic cyanobacterial species showed a significant change in the relative position of one of the pyrolle rings of the β155 PCB. The altered conformation of ring D was first identified in the *Thermosynecococcus vulcanus* structure determined at room temperature (Adir et al. 2001), and then further confirmed at cryogenic temperature in both *T. vulcanus* (Adir et al. 2002) and *T. elongatus* (Nield et al. 2003). This alteration in ring position was shown to be in a critical position that has the ability to stabilize both the ($\alpha\beta$) monomer interaction interface as well as the $(\alpha\beta)_6$ hexamer formation interface. Solvent molecules are located in almost identical positions in both structures, about equidistant between Aspα28, Asnβ35, and Argα'33 from the lower $(\alpha\beta)_3$ trimer. The presence of bound water in these positions at the typical growth temperatures of 55–60 °C is probably transient, but could add a significant amount of stabilization energy to both the ($\alpha\beta$) monomer and the $(\alpha\beta)_6$ hexamer. Additional solvent molecules were located in close contact (2.6–3.4 Å) with each of the three co-factors. Those solvent molecules associated with β84 PCB and β155 PCB interact with the molecular edge adjacent to the bulk solvent, while those associated with α84 PCB are positioned in the trimer association interface. These waters may have a role similar to that of the surrounding protein residues in stabilizing the PCB configuration to obtain optimal functionality (Adir et al. 2002).

The two internal PCB pyrolle rings contain propionic acids whose function is not clear, but that may be required to hold the PCB in an unfolded configuration that modifies their absorption spectra. Interactions between the propionic acids and polar residues have been previously identified for the PCBs in a number of crystal structures and these residues are highly conserved. In the *T. vulcanus* structure all three cofactors bind solvent molecules in the gap about equidistant between the two propionic carboxyl groups (Adir et al. 2002). When a superposition of the three co-factors is performed, these solvent molecules spatially coincide. It can be thus suggested that these solvent molecules are important for positioning of the propionic side chains in specific conformations. Indeed, molecular dynamics simulations of the propionic acid positions in the absence or presence of the intervening solvent molecules show that in their absence, the propionic acids repel each other and are attracted to adjacent positively charged residues. The coupling of the propionic acid self repulsion and their attraction to the positive residues would induce a greater degree of separation between the propionic acids, putting strain on the PCB molecule. Examination of other PC structures (*F. diplosiphon, T. elongatus*, etc.), show solvent molecules in similar positions, indicating that this requirement is not a characteristic trait limited to PC from thermophilic species. By obtaining their proper position, the propionic acid groups may maintain the two central pyrolle rings on almost the same plane and thus may be important for efficient energy transfer. The propionic acids may also interact with linker proteins, as was visualized in the *M. laminosus* APC structure (Reuter et al. 1999).

As indicated above, isolated PC maximal absorption is at 620 nm in its $(\alpha\beta)_3$ trimeric form (PC_{620}). PC monomer absorption is blue-shifted to 614 nm while the absorption of trimers isolated in the presence of linkers is red-shifted (629–638 nm) (MacColl 1998). These observations reaffirm the complexity of elucidating the mechanism of energy transfer in the PBS using isolated components. However, spectroscopic measurements made on isolated components, in conjunction with high resolution studies, continue to serve as the major source of information on PBS function. In the course of isolation of APC from *T. vulcanus*, our lab recently identified a minor fraction of PC that absorbed maximally at 612 nm, while in $(\alpha\beta)_3$ trimeric form (Adir and Lerner 2003). This form of PC (PC_{612}) was characterized crystallographically and appeared to have a number of unique structural traits. The difference in absorption was identified as caused by the lack of methylation of βAsn72 (Klotz et al. 1986). This residue has been found to be methylated in all PBP types, and this post-translational modification affects the absorption characteristics of the ubiquitous β84 PCB. Although crystallized under conditions similar to that of *T. vulcanus* PC_{620}, the PC_{612} did not form $(\alpha\beta)_6$ hexamers *in crystal*, as a result of changes in crystal packing. All previously determined PC structures have formed hexamers, and in most cases rod-like structures due to hexamer–hexamer interactions. The lack of assembly could be due to a small change in the nucleation process; however it revealed a number of characteristics that could have a bearing on PBS assembly. The lack of hexamer formation allowed a significantly higher degree of flexibility in the α subunit of each $(\alpha\beta)$ monomer, within the PC_{612} trimer. This flexibility could be advantageous in the formation of the interaction surface between the end of the rod and the circumference of the APC core cylinders. The presence of a trimer at the end of two of the rods could serve as a "filler" ring to make up for the differential in position between the lower base core cylinders and the upper cylinder (Fig. 5). This hypothesis is strengthened by the fact that the PC_{612} was only isolated in conjunction with APC cores, and thus it is probably not a small population of unmethylated PC. The lack of methylation could allow the rod linker (L_{RC}) present at the end of the rod to interact strongly with the β84 cofactor, affording a red-shift towards 630–640 nm. This in turn would thereby serve as a functional bridge between the bulk PC_{620} in the rods, and the APC in the cores.

3.4 Phycoerythrin

The phycoerythrin (PE) component of the PBS is found in only a subset of cyanobacterial species, and in red algae (Grossman et al. 1993; MacColl 1998). There exist quite a few variations of PE, depending on the types of bilin cofactors attached to the apo-proteins. These include not only PCB but also phycoerythrobilin (PEB), phycoviobilin (PXB) and phycourobilin (PUB), which differ in the number of conjugated double bonds (Glazer 1989). A reduction in the number of double bonds leads to a higher energy excited state, and a blue-shifted absorption spectra. PE typically contains five bilins, with the two extra bilins linked to αCys143 and a PEB doubly linked to βCys50 and βCys61. In addition, PEs assembled into trimers bind one of a number of special linker proteins that contain a single covalently bound bilin pigment, and are called γ subunits (Liu et al. 2005). Thus, the typical PE assembly is an $(\alpha\beta)_6\gamma$ hexameric disk. To date crystal structures of PE have not been able to visualize the γ subunit, although analysis of solubilized crystals has shown their presence (Chang et al. 1996).

PE is present in those species that have the ability to change their PBS composition as a result of the light quality, a process known as complimentary chromatic adaptation (CCA) (Grossman et al. 1993; MacColl 1998). In these species the PBS composition changes from rods containing only PC (in red-light grown cells) to rods containing both PE and PC (in green-light grown cells). This change in composition requires changes in gene expression of both the PBPs and linker proteins. It also requires in some cases a limited disassembly of already existing PBS complexes (see below).

3.5
Unusual Phycobiliprotein

A unique use of PBPs as antenna proteins is found in cryptophyte algae. These algae do not contain APC or an assembled PBS, rather they have either PC or PE in the lumen of the thylakoid membranes (Doust et al. 2004; Wilk et al. 1999). The PBPs are completely soluble, form tetrameric complexes with a distinctive α subunit which is unrelated to the PBP family of proteins. A very high resolution crystal structure of PE545 from *Rhodomonas CS24* has recently been determined (Doust et al. 2004; Montgomery et al. 2004), providing an outstanding view of the cofactor conformation and environment.

Another unusual use of the capabilities of PBPs are the "allophycocyanin-like" proteins (Apl) recently discovered in the laboratory of D. Kehoe (Montgomery et al. 2004). These members of the APC family contain bound cofactors, but are not found within assembled PBS, and contain sequence changes that indicate that they are indeed unable to assemble. On the basis of the positions of the genes that encode for the Apl proteins, it has been suggested that they may serve in a photoresponsive regulatory role, perhaps in the expression of the *cpe* and *cpc* operons. A similar form of regulatory role may be played by a form of PE in *Prochlorococcus*, a photosynthetic prokaryote that completely lacks PBS antennas (Steglich et al. 2005).

3.6
Linker Proteins

Whilst the brilliant colors of the PBPs aided their early identification, the colorless polypeptides that make up approximately 15% of the PBS mass were not characterized until much later. Tandeau de Marsac and Cohen-Bazir first recognized the added intricacy of the PBS structure, hypothesizing the involvement of these colorless polypeptides, now known as linker proteins, in the attachment of the PBS complex to the thylakoid membrane and positioning of the light-harvesting pigments within the PBS (Tandeau de Marsac and Cohen-Bazire 1977). As with the PBPs, the number and type of linker protein varies between species and can be altered as a consequence of changes in the environment. The linkers are classified into four sub categories according to their position in the PBS; rod linkers (L_{R}), rod-core linkers (L_{RC}), core linkers (L_{C}) and the core membrane linker (L_{CM}). There are different numbers of genes encoding for some of the linkers, depending on the species, and interestingly the level of sequence homology is low when compared to that of the pigmented PBS constituents. Linker proteins are known to be involved in assembly and stabilization of the PBS, as well as facilitating the efficiency and unidirectionality of energy flow down the complex. Trimeric or hexameric PBPs associated with a linker show spectroscopic properties distinct from those of isolated proteins (MacColl 1998; Pizarro and Sauer 2001; Yu and Glazer 1982). Zhang and co-workers have recently comprehensively reviewed the function of the linker proteins, and we will focus here on characteristics pertaining to the structure and assembly of the PBS (Liu et al. 2005).

The early low-resolution TEM studies that identified the PBS, could not visualize the linker proteins, indicating that these proteins are sequestered within the PE/PC/APC rings. This indicates that as in the case of the L_R or L_C proteins that are found within the trimeric or hexameric structures, the L_{RC} proteins are also mostly hidden from the outside environment. Looking at the various PBS models, this would imply that most of the L_{RC} is within the PC hexamer closest to the core and that the interaction with the core APC rings is relatively small. The channels of $\sim$ 30 Å diameter formed in PBS components that assemble into trimers, and by further stacking into hexamers, are quite sufficient for the placement of the linker protein in an extended conformation. Mimuro and co-workers have suggested that a three-fold redundancy in the primary structure of linkers is suggestive of equivalent binding or interaction with three monomers (Fuglistaller et al. 1987; Mimuro et al. 1999). It has further been suggested that the greater conservation of sequence around the N terminal may be needed for packing into a channel whereas the less conserved C terminal may provide an assembly interface between rod disks (Anderson and Toole 1998). The presence of linker proteins is apparently necessary for higher states of aggregation of PBPs in vivo (Kondo et al. 2005; Yu et al. 1981). The stoichiometry in the assembled PBS is usually one linker per hexamer (Liu et al. 2005; Lundell and Glazer 1983). By use of EM studies of the assembly of wild-type PBS vs. mutants, combined with reconstitution experiments of selected components, it has been possible to predict specific positioning of each linker within the complex (Ducret et al. 1998; Lundell and Glazer 1983). Rod extension is terminated by the presence of a small L_R linker that "caps" the rod. In it's absence in vitro, rods can form extended bundles of very long rods (Yu et al. 1981).

In contrast to PBPs, unassociated linker proteins have a basic pI, and have relatively low solubility in aqueous media (Fairchild et al. 1991; Mimuro et al. 1999). This would suggest that linker association with the other PBS components could be mediated by chaperones, or that association with the PBS components occurs concomitantly with the assembly of the PBS. The mode of linker association is unclear. For instance, both L_R and L_{RC} linkers associate with PC. There is no known difference between a PC hexamer adjacent to the core and PC hexamers in a more distal position in the rod. How then does the L_{RC} associate in the correct position? As mentioned in Sect. 3.3, the nonmethylated trimeric PC_{612} (Adir and Lerner 2003) could serve as the terminal unit of the rod. By virtue of its trimeric nature PC_{612} could present a significantly different environment than the PC hexamers, enabling the specific association of the L_{RC} linkers. Its additional flexibility could also assist in the matching between rod ends and the circumference of the core disks. However, this form of PC has not been seen in EM studies and further experimental studies are needed to address this question.

The possibility of post-translational modifications of the linkers has also been explored. Glazer and co-workers have provided evidence against the glycosylation of linkers (Fairchild et al. 1991). Recent research has indicated that linker proteins are phosphorylated in the assembled PSB and has suggested phosphorylation/dephosphorylation as a possible factor in controlling PBS degradation (Piven et al. 2005).

To date only a single linker protein has been characterized crystallographically—the L_C linker associated to an APC trimer from *M. laminosis* (Reuter et al. 1999). The linker structure is composed of a three-strand β-sheet and two short α-helices. The linker is positioned in the trimeric ring associating primarily at the β chromophores of two out of the three monomers. This association has little effect on the relative orientation of the α and β chromophores in relation to each other, however, an overall flattening of the APC trimer was seen. Whilst lack of further crystallographic data limits structural understanding of the linker proteins, spectroscopic measurements and functionality studies provide insights into their role within the PBS.

Perhaps the most intricate of all linkers is the L_{CM} (the product of the *apc*E gene) which anchors the entire PBS onto the membrane (Liu et al. 2005). This linker is 2–4 times larger than the L_R linkers depending on the number of APC disks in the core, and contains a number of identifiable domains. The *N*-terminal domain is homologous to APC α subunit (Apt et al. 1995), with a single covalently bound PCB cofactor. It was recently shown that PCBs can become covalently attached autocatalytically (Zhao et al. 2005). This domain pairs with a specific β subunit within the core, and directs the *C*-terminal out towards the membrane. The remaining part of the L_{CM} contains fused L_R-like domains, with spacing units. It has been proposed that the number of such domains should correlate to the number of APC disks in the core, however, there are a few discrepancies, and thus the overall mode of binding is unclear.

A second case of the use of a linker-like domain in a fusion protein is the case of ferredoxin-NADP+ oxidoreductase (FNR). This enzyme catalyzes the transfer of electrons from ferredoxin (the terminal electron acceptor of PSI), to $NADP^+$. Cyanobacterial FNR has an *N*-terminal extension that is homologous with the 9 kDa L_R (the *cpcD* gene product) linker, and has been found to associate with the PBS (Gomez-Lojero et al. 2003). The position of the bound FNR is at the distal end of the rods, and it has been suggested that the PBS may have the ability to associate with PSI through the rod ends, and not through the cores. Indeed, in a mutant lacking APC, and thus not containing assembled PBS, PC rods were found to be present, and energy transfer from these rods was primarily performed to PSI.

3.7
Phycobilisome Function

Spectroscopic, biochemical, biophysical, immunological, and X-ray crystallographic techniques have been used to establish the pathway for energy flow and the arrangement of the different PBPs within a PBS (Debreczeny et al. 1993; Glazer 1989; MacColl 1998, 2004; Pizarro and Sauer 2001; Sauer and Scheer 1988). Energy is transferred between pigments in a non-radiative fashion, from the highest energy absorbing pigment down the PBS funnel to the lowest energy absorbing pigment. The distances between pigments within $(\alpha\beta)_3$ trimers and $(\alpha\beta)_6$ hexamers are known to high precision as a result of the numerous crystal structures of the different PBPs (Duerring et al. 1991; Nield et al. 2003; Schirmer et al. 1985, 1986). Putative distances along the rods, within the cores, between adjacent rods or between rods and cores have been modeled using the various crystal packing modes associated with these same crystallographic studies (MacColl 1998, 2004). The distance between cofactors, coupled with experimental assays, support Förster resonance energy transfer with weak coupling (Forster 1948, 1965). In this regime, each chromophore maintains its individual absorption characteristics. These characteristics are fine-tuned by its three-dimensional conformation which is a result of its specific protein environment. Upon assembly, pigments come into closer contact with other pigments on adjacent monomers (typically no closer than about 20 Å). At these distances, spectral overlap occurs, resulting in energy transfer by exciton-coupling. This mechanism may provide most of the directionality of energy transfer in the PBS, as well as the large absorption shifts in the different components. For instance, monomeric APC absorbs maximally at 614 nm, quite similar to that of monomeric PC. Upon trimer assembly, PC absorption red-shifts to 620 nm, while APC red-shifts to 650 nm (with a broad shoulder at 610–620 nm) (MacColl 2004). The source of this very significant red-shift in the absorption spectra could be caused by two different mechanisms: (i) the absorption maximum could be shifted as a direct result of the formation of a local protein/solvent environment which directly affects three chromophores (one per monomer), causing them to red shift. The three remaining chromophores would continue to absorb at a higher energy, explaining the presence of the shoulder. (ii) Formation of the trimer induces strong exciton coupling between two now adjacent chromophores, causing exciton splitting and the significant change in the absorption energy. The effect of monomer to trimer assembly on the absorption of PC and APC has been described in detail in a number of recent papers (MacColl 1998, 2003, 2004; Pizarro and Sauer 2001). Energy absorbed by the rods is transferred down into the APC core. There are three specialized core proteins bearing chromophores (besides the normal APC α and β subunits), called L_{CM}, β^{16} and α^{B}. The L_{CM} and α^{B} components serve as terminal energy acceptors from the bulk APC components and enable energy transfer to the RC chlorophyll. The function of the β^{16} core subunits is unclear since its absorption maxima are less red than L_{CM} or α^{B}.

4 Phycobilisome Assembly and Disassembly

As described in the previous sections, the PBS contains multiple copies of the different PBPs, which organize into the holo-complex. As in the case for all protein complexes, the instructions for assembly are primarily located within the protein subunits themselves. The enormous size and regular structure of the PBS dictates a process of self-assembly that is still not clear. It is, however, clear at this point that the equilibrium between free subunits, partially assembled substructures and the PBS is carefully tuned, in such a fashion that the absence of one component leads to the loss of other components. Anderson and Toole, reviewed in length many of the very early steps in PBS assembly, as understood 8–10 years ago (Anderson and Toole 1998). In this section we will describe the present state of understanding of this process. Specific residues are denoted using the sequences of the PBPs from *T. vulcanus*.

4.1 Monomer Assembly

Translation of the two PBP subunits of any type (APC, PC or PE) invariably leads to the formation of the ($\alpha\beta$) monomer (Figs. 2 and 3). This initial step may occur on folded or nearly folded polypeptides, which result from the significant α-helical nature of the polypeptides. The roles of chaperones in the folding of PBPs has been suggested (Anderson and Toole 1998), however, to date there is still no direct experimental evidence that polypeptide folding cannot occur spontaneously. Heterologous expression of the two PC subunits in *E. coli* resulted in the quantitative formation of ($\alpha\beta$) monomers (Arciero et al. 1988). This would indicate that any chaperone activity involved in this organizational step is of a general nature as it was performed by *E. coli* chaperones with high efficiency. Alternatively, chaperones are not required for this step. The initial steps of monomer formation are preceded by the lyase catalyzed ligation of the cofactors to the isolated subunits. This process apparently takes place mostly with the isolated subunits, but can also be performed at lower efficiency on higher-order oligomers. The possibility that the addition of cofactors at later stages could be a result of spontaneous ligation has been suggested, and also experimentally shown (Zhao et al. 2005).

The formation of the monomer is a critical assembly step in the formation of the PBS. If we look at the high-resolution PC structures available to us (Adir et al. 2002; Nield et al. 2003), we can see that the ($\alpha\beta$) monomer interface forms by the overlap of the two opposing *N*-terminal helix–turn–helix regions of both subunits (residues 1–42), with the greatest degree of overlap formed by the anti-parallel alignment of the Y helices (Fig. 2). The interface surface is quite flat, with almost no interpenetration between opposing polypeptide subunits, and it is thus not apparent how assembly occurs. Sequential translation of the two polypeptides from a single polycistronic mRNA could increase the local concentrations and thus drive complex formation. However, Glazer and co-workers (Cai et al. 2001) showed PC monomer formation when one of the PC subunits was expressed from a plasmid in transformed *Synecococcus* sp. PCC7120, while the other subunit was expressed from the endogenous gene. Therefore, while translation was not sequential, monomers were still able to efficiently form. It thus appears that the positions of the attracting species of each subunit are precise enough to afford the affinity necessary for complex formation. What are the forces that come into play during complex assembly? Most of the interactions in the final state are hydrophobic in nature, with three polar "locks" (Fig. 2B). The first of these is formed by residues Met1 (on both subunits) and αThr3-βAsp3 hydrogen bonds, the second is formed between αArg42 and βAsp25/βAsn21. The third polar anchor forms between αArg93 and βAsp13. The total surface area that is buried in the monomer interface is about 3000 Å^2, and yet the monomeric unit is extremely stable, at even extremely low concentrations. Unlike the higher assembly states of the PBPs, monomers are also stable in low ionic strength aqueous solutions.

Interestingly, the only side chains that significantly interpenetrate are the aromatic rings of Phe18 of both subunits, which align themselves with the long aliphatic chains of αArg93 and βArg13. In thermophilic species, an additional polar interaction comes into play, that is between αAsp21, βAsn35 and ring D of the β155 PCB, which shifts its orientation and latches onto the α subunit. The formation of higher-order oligomers $(\alpha\beta)_3$ trimers, $(\alpha\beta)_6$ hexamers, rods and core assemblies are most likely spontaneous, and result due to the mutual attraction of subunits. While it would be attractive to associate assembly with the presence of the various linker proteins, this may not be the case. PBP crystallization is usually initiated with purified $(\alpha\beta)_3$ trimeric protein, however, in almost all crystals of PC and PE, the organizational state of the PBP is that of extended rods, built up of $(\alpha\beta)_6$ hexameric units. Thus, at least to a close approximation these steps are independent of the linker proteins. Differences in hexamer association due to crystal packing effects can be identified, and the linkers may have a modifying effect on the final rod structure. It is quite clear that specific terminating linkers are required to stop the rod elongation process.

4.2 Assembly of the $(\alpha\beta)_3$ Trimer

The association of three monomeric units forms the trimer, an uneven closed circular disk with a radius of between 45–55 Å, a mean circumference of 314 Å and a width of ~ 30 Å (Adir et al. 2001). Formation of the trimeric ring creates an internal hole, which has the shape of an uneven circle with a diameter of between 15–50 Å. All PBP types associate into trimers, which are quite stable and do not disassociate at low concentrations [except for APC which can disassociate under very dilute conditions (MacColl 2004)], and are stable in relatively low ionic strength solutions. The overlap interface between the α subunit of one $(\alpha\beta)$ monomer and the β subunit of an adjacent monomer is considerably less than that found in the $(\alpha\beta)$ monomer interface (~ 1250 Å^2) or for $(\alpha\beta)_6$ hexamer interface (see below). The ionic/polar interactions were visualized for the first time in detail by Huber and co-workers in the structure of the *Synechococcus* sp. PCC 7002 PC (previously known as *Agmenellum quadruplicatum*) (Schirmer et al. 1986). In general, there are many contacts spread out over the entire overlap region. Additional contacts that strengthen this interaction surface in PC from thermophilic species include βGlu68-αArg86 and αSer72-βArg57. All of the hydrophobic interactions indicated previously for the mesophilic PC are conserved in the thermophilic species, and this may be the primary mode of trimer stability.

4.3 $(\alpha\beta)_6$ Hexamer and Rod Assembly

The $(\alpha\beta)_6$ hexamer is formed by the association of two trimers in a face-to face assembly step. While the hexamer interface zone is quite extensive (~ 2550 Å^2), with a buried surface area twice that of the trimer interface zone, the presence of high ionic strength phosphate buffer is required to preserve both hexamers and rods, which otherwise dissociate into trimers and monomers. This is perhaps indicative of the importance of PBS linker proteins in proper assembly in vivo. However, in most of the PC crystals described to date PC trimers readily reform into hexamers (Adir 2005). It is possible that crystal lattice packing dictates the formation of inter-trimer contacts lost when the linker proteins are removed, or that the hexamers found *in crystal* are not equivalent to those found in the PBS. It should be mentioned, however, that the structural complimentarity of the two assembled trimers is such that the rotation of one trimer relative to the other would lead to disrupting clashes of the polypeptides. To avoid such clashing, the hexamer would need to have a width that would be greater than the sum of the widths of two trimers. Measurements obtained from the different EM studies indicate that the hexamer width is indeed equal to that of two trimers. We can thus conclude with a high degree of confidence that hexamer assembly can be achieved without the presence of linker proteins. This may, however, not be the case under physiological conditions. Many ionic contacts were visualized and described by Huber and co-workers, and additional contacts are found in PC hexamer interface regions of thermophiles. A triad of polar interactions is created by each αAsp28 and the ring D of the β155 PCB on one trimeric ring with α'Arg33 of the monomer directly below on the second trimeric ring in *T. vulcanus* PC (Adir et al. 2001).

PC crystals typically pack the hexamers into "infinite" rods, and these have been suggested to mimic the rods found in the PBS in vivo. However, the relative rotational orientation of the hexamers is different in crystals of PC from different species. While in *F. diplosiphon, T. vulcanus, T. elongatus* and other PC crystals the hexamers pack directly one above the other, in the *C. caldarium* PC crystal, the hexamers are rotated allowing tighter packing (Stec et al. 1999). The rotation also brings PCB cofactors from adjacent hexamers into closer contact, providing an efficient inter-hexamer energy transfer pathway.

The mode of monomer $\rightarrow$ trimer $\rightarrow$ hexamer association creates a situation in which each of the PB species experiences a different protein/solvent environment (Adir 2005). The β84 PBs are directed into the hexamer internal space (which contains solvent and linker proteins), the α84 PBs are completely embedded between subunits in the trimer association domain, and the β155 PBs (in PC and PE) point out into the outer solvent. Each of the three cofactors thus experiences a different environment, which is one of the major factors in tuning the resulting PB absorption spectra which in turn determines its function. On the basis of experimental measurements, it has been suggested that the β84 PBs absorb at the lowest energy, and thus functions as fluorescing PBs (or f type). The α84 and β155 PBs absorb higher energy photons and are thus termed sensitizing pigments (or s type). S type PBs have been suggested to perform radiationless energy transfer to f type PBs (Teale and Dale 1970). In most PC crystals, the rod–rod interfaces suggest that the β155 PCB could perform energy transfer between rods, since there is only a separation of 36 Å between two β155 PCBs on adjacent rods (Padyana et al. 2001).

4.4 Rod-core Association and the Phycobilisome Structure

The assembly of the PBS requires the association of the core cylinders with the rods. In any multi-protein complex formation process, a certain degree of complimentarity is to be expected, and this has been born out by the great number of complex structures determined to date and deposited in the Protein Data Base. In many cases, the building blocks of the complex are of different shapes and sizes, with irregular surfaces such as in the case of ribosomal particles (Yonath 2005). In other cases, the interaction surfaces are more regular and extensive. This latter situation is certainly the case for the formation of the PBS rods, and core cylinders. The architectural question that arises is how to attach the rods to the uneven circumference created by the assembly of core cylinders. Visualization of PC crystal lattices also shows putative rod–rod interactions. Here the different crystals diverge further, lowering our degree of confidence that this is indeed the form of contacts in vivo. It is clear that in crystals, rods can only interact in either a parallel or perpendicular fashion. However, many PBS models show the rods radiating out from the central core cylinders at more-or-less even angles (Anderson and Toole 1998; Grossman et al. 2001; MacColl 1998). These models are based on the negatively stained EM studies performed on whole PBS complexes. However, in many of the published EM micrographs, it is clear that at least in some cases the PC rods are indeed parallel to one another (Glauser et al. 1992; Yu et al. 1981). On this basis other researchers have suggested that in the tricylindrical core-type PBS, the rods are arranged in three parallel doublets (Fig. 5) (Adir 2005; Glazer 1989; Huber 1989). If this is indeed the case, we may be able to learn more on PBS assembly by studying the interaction *in crystal*, and this may indicate again that the PBS indeed self assembles, without the need of either chaperones or linker proteins.

All PBP types form disks of similar dimensions (hexamers of PE, PC and double trimers APC). The PC disks (in rod assemblies) fit onto the APC cylinder circumference in perpendicular fashion. Electron micrographs in the literature have not so far had the resolution required to identify whether a linker protein (of the L_{RC} type) protrudes out of the terminal PC trimer and "latches" onto an APC disk. In any case, we should be able to use the molecular dimensions of the disks to build a geometrically competent model. As already indicated above, the two major models of the PBS show either three doublets of parallel rods attached to the core at right angles (with the membrane occupying the fourth position), or some version of six rods radiating out from the core. Any arrangement must take into account the dimension of the PC rods and APC core. In a tricylindrical core, the first layer of the core is composed of eight APC trimeric disks that form two cylinders, while the second layer has four disks making a single APC cylinder that fits into the "saddle" in the first layer interface. The total height of the core is thus less than twice the diameter of a single disk. Each disk is 110 Å in diameter and thus the total core height is about 190 Å. The lower layer of the core will be 220 Å in width, while the upper layer is 110 Å, leaving $\sim$ 55 Å on either side that must be filled in when a rod is assembled onto the core.

We will first examine the model which includes three rod doublets surrounding the core (Fig. 5). When the first two layers of rods attach to the first core layer (parallel to the membrane surface at 180° from each other) the circumference of the APC disk will exactly meet the aperture of the terminal PC disk, bringing the PC β84 cofactors (and possibly the L_{RC} linker) as close as possible with the APC β84 cofactors. However, the second layer of PC rods cannot now fit in the same fashion as the first layer because the total height of two rods is 220 Å, more than the height of the core. The second rod orifice will be mismatched with the circumference of the top APC cylinder. These two rods must contact the single top APC cylinder, which could create a situation where they are offset relative to the bottom rod by 55 Å (less than a complete hexamer). The third rod doublet juts out at 90° from the other two doublets and is in contact only with the top APC cylinder. Thus, this interaction interface must be very different, since both orifices cannot simultaneously meet the disk circumference (if they lie in a parallel fashion). As mentioned above in Sect. 3.3, we have suggested that the addition of extra PC trimers (such as the PC_{612} recently identified in our lab) could help in serving as flexible "washers" that help match the rods with the core (Adir 2005; Adir and Lerner 2003).

The second group of models proposes an arc of rods radiating out from the core. While visually pleasing, this arrangement is structurally problematic. The circumference of a hemidiscoidal core is 350 Å, which is not large enough to accommodate six 110 Å rods. These models usually try to compensate for this problem by either allowing the rods to interpenetrate, or by having the rods in a staggered formation (Bryant et al. 1979). The possibility of interpenetration of the rods is not likely when the crystal structures of PC hexamers are examined, since their outer surface is both rigid and without major gaps. Kuang and co-workers have recently performed EM studies on PBS from *N. flagelliforme*, in which they compared single particle images obtained from negatively stained samples with those obtained from samples prepared in vitreous water by flash-freezing techniques (Yi et al. 2005). The cryo-EM techniques allow imaging at angles not obtainable with the stained samples allowing reconstruction of multiple images, leading to a three-dimensional model of the PBS at about 30 Å resolution. One of the important observations in this study is that the PBS rods and core cylinders are all in the same plane, and not in a staggered formation. The rods also appear to form parallel doublets. Thus, from the available evidence it is more likely that the "three-rod doublets" model is closer to the real structure of the PBS, although it is clear that higher resolution studies of the entire complex are required.

4.5 Phycobilisome Disassembly

Cyanobacteria can be exposed to changes in their typical growth conditions which require modification in both the number and characteristics of the PBS. As already noted above, the expression of the genes encoding for PBS components can be altered by external environmental changes within a short period of time. However, more long-term changes also may occur that require variations in the PBS makeup. One of the most notable of these is the response of cyanobacteria to the lack of specific nutrients, especially nitrogen and sulfur. Under these conditions, the cyanobacterial cells change color from blue–green to yellow–green in a process known as chlorosis or bleaching (Allen and Smith 1969). Using spectroscopy and other techniques it was found that the bleaching occurs within a few hours of the onset of nutrient starvation and is a result of the loss of the PBS in an ordered fashion (Grossman et al. 1993). It has been proposed that the disassembly of the PBS can have two beneficial roles. During starvation the cellular metabolism decreases, while the rate of photosynthesis may be unchanged. This imbalance could lead to excessive absorption of excitation energy, which could then lead to the production of harmful radical species (Adir et al. 2003). Under extreme conditions this may cause to cell death and must be avoided. Since the rate of photosynthesis is directly coupled to the rate of energy absorption, uncoupling and degradation of the major antenna will be highly beneficial in avoiding overexcitation. A second role for PBS degradation would be to serve as an internal nutrient reservoir. The PBS can supply the cell with either amino acid residues for the synthesis of protein systems required for high affinity nutrient uptake, or by further degradation as building blocks for the synthesis of other metabolites. Collier and Grossman used a genetic screen to identify mutants unable to degrade PBSs during nutrient starvation in order to identify the molecular pathway of PBS degradation (Collier and Grossman 1994). The *nblA* (non-bleaching A) gene was thus identified and has since found to be present in all cyanobacteria. Nitrogen is the major nutrient whose lacking can lead to a 50-fold increase in the amount of nblA protein (Collier and Grossman 1994). In some cyanobacteria (such as *Synechococcus* sp. PCC 7942) sulfur and phosphorus limitation can also induce the *nblA* response (Collier and Grossman 1994) while in other species sulfur limitation does not influence the expression of *nblA* (Richaud et al. 2001). The size of proteins belonging to the nblA family is quite small, ranging from 54–65 residues corresponding to molecular masses of about 6.5–7.0 kDa. The sequence identity between nblA proteins is relatively poor ($\sim 30\%$), which is somewhat surprising since the PBS components with which the nblA interacts are highly homologous ($> 70\%$).

During starvation wild-type cells (but not *nblA* mutants) degrade two rod-linker proteins: L_R 33 and L_R 34.5, indicating a possible functional connection between linkers and the nblA protein (Richaud et al. 2001). This observation supports previous data showing that wild-type cells degrade the PBS by first decreasing the length of the rods (Anderson and Toole 1998; Sidler 1994). It was shown that in the cyanobacterium *Tolypothrix* PCC 7601, the nblA protein has affinity for both PC and PE subunits, but not for APC or for PBPs from other cyanobacterial species (Luque et al. 2003).

Beside *nblA*, there are four other genes that have been identified as being related to PBS degradation: *nblR, nblS, nblB and nblC.* The nblR protein controls *nblA* expression and is critical for survival under stress conditions (Schwarz and Grossman 1998). It has been suggested that *nblR* is controlled by *nblS*, a sensor histidine kinase that has a PAS domain (van Waasbergen et al. 2002). The nblB protein is necessary for PBS degradation during starvation, but unlike the nblA protein, the nblB levels are similar prior to or during nutrient starvation (Dolganov and Grossman 1999). This protein is partially homologous with PCB lyases, the enzyme involved in chromophore attachment to the PC apoprotein. The most recent addition to the arsenal of proteins involved in the disassembly of the PBS is the nblC protein, which is required for *nblA* expression (Sendersky et al. 2005).

Acknowledgements This work was supported by the Israel Science Foundation founded by the Israel Academy of Sciences and Humanities (438/02) and the Technion Fund for the Promotion of Research.

References

Adir N (2005) Elucidation of the molecular structures of components of the phycobilisome: reconstructing a giant. Photosynth Res 85:15–32

Adir N, Dobrovetsky Y, Lerner N (2001) Structure of C-phycocyanin from the thermophilic cyanobacterium *Synechococcus vulcanus* at 2.5 Å: Structural implications for thermal stability in phycobilisome assembly. J Mol Biol 313:71–81

Adir N, Lerner N (2003) The crystal structure of a novel unmethylated form of C-phycocyanin, a possible connector between cores and rods in pycobilisomes. J Biol Chem 278:25926–25932

Adir N, Vainer R, Lerner N (2002) Refined structure of C-phycocyanin from the cyanobacterium *Synechococcus vulcanus* at 1.6 Å: insights into the role of solvent molecules in thermal stability and co-factor structure. Biochim Biophys Acta 1556:168–174

Adir N, Zer H, Shochat S, Ohad I (2003) Photoinhibition—a historical perspective. Photosynth Research 76:343–370

Allen MM, Smith AJ (1969) Nitrogen chlorosis in blue-green algae. Arch Mikrobiol 69:114–120

Anderson LK, Toole CM (1998) A model for early events in the assembly pathway of cyanobacterial phycobilisomes. Mol Microbiol 30:467–474

Apt KE, Collier JL, Grossman AR (1995) Evolution of the phycobiliproteins. J Mol Biol 248:79–96

Arciero DM, Bryant DA, Glazer AN (1988) In vitro attachment of bilins to apophycocyanin. I. Specific covalent adduct formation at cysteinyl residues involved in phycocyanobilin binding in C-phycocyanin. J Biol Chem 263:18343–18349

Awramik SM (1992) The oldest records of photosynthesis. Photosynth Res 33:75–89

Barber J, Morris EP, da Fonseca PC (2003) Interaction of the allophycocyanin core complex with photosystem II. Photochem Photobiol Sci 2:536–541

Ben-Shem A, Frolow F, Nelson N (2003) Crystal structure of plant photosystem I. Nature 426:630–635

Blankenship RE, Hartman H (1998) The origin and evolution of oxygenic photosynthesis. Trends Biochem Sci 23:94–97

Blankenship RE, Olson JM, Miller M (1995) Antenna complexes from green photosynthetic bacteria. In: Blankenship RE, Madigan MT, Bauer CE (eds) Anoxygenic Photosynthetic Bacteria. Kluwer Academic Publishers, Dordrect, The Netherlands, pp 399–435

Brejc K, Ficner R, Huber R, Steinbacher S (1995) Isolation, crystallization, crystal structure analysis and refinement of allophycocyanin from the cyanobacterium Spirulina platensis at 2.3 Å resolution. J Mol Biol 249:424–440

Brown SB, Houghton JD, Vernon DI (1990) Biosynthesis of phycobilins. Formation of the chromophore of phytochrome, phycocyanin and phycoerythrin. J Photochem Photobiol B 5:3–23

Bryant DA, Guglielmi G, Tandeau de Marsac N, Castets AM, Cohen-Bazire G (1979) The structure of cyanobacterial phycobilisomes: A model. Arch Microbiol 123:113–127

Buick R (1992) The antiquity of oxygenic photosynthesis: evidence from stromatolites in sulphate-deficient Archaean lakes. Science 255:74–77

Cai YA, Murphy JT, Wedemayer GJ, Glazer AN (2001) Recombinant phycobiliproteins. Recombinant C-phycocyanins equipped with affinity tags, oligomerization, and biospecific recognition domains. Anal Biochem 290:186–204

Chang WR, Jiang T, Wan ZL, Zhang JP, Yang ZX, Liang DC (1996) Crystal structure of R-phycoerythrin from Polysiphonia urceolata at 2.8 Å resolution. J Mol Biol 262:721–731

Cogdell RJ, Gardiner AT, Roszak AW, Law CJ, Southall J, Isaacs NW (2004) Rings, ellipses and horseshoes: how purple bacteria harvest solar energy. Photosynth Res 81:207–214

Collier JL, Grossman AR (1994) A small polypeptide triggers complete degradation of light-harvesting phycobiliproteins in nutrient-deprived cyanobacteria. EMBO J 13:1039–1047

De Marais DJ (2000) Evolution. When did photosynthesis emerge on Earth? Science 289:1703–1705

Debreczeny MP, Sauer K, Zhou J, Bryant DA (1993) Monomeric C-phycocyanin at room temperature and 77K: Resolution of the absorption and fluorescence spectra of the individual chromophores and the energy-transfer rate constants. J Phys Chem 97:9852–9862

Dekker JP, Boekema EJ (2005) Supramolecular organization of thylakoid membrane proteins in green plants. Biochim Biophys Acta 1706:12–39

Dolganov N, Grossman AR (1999) A polypeptide with similarity to phycocyanin alpha-subunit phycocyanobilin lyase involved in degradation of phycobilisomes. J Bacteriol 181:610–617

Doust AB, Marai CN, Harrop SJ, Wilk KE, Curmi PM, Scholes GD (2004) Developing a structure-function model for the cryptophyte phycoerythrin 545 using ultrahigh resolution crystallography and ultrafast laser spectroscopy. J Mol Biol 344:135–153

Ducret A, Muller SA, Goldie KN, Hefti A, Sidler WA, Zuber H, Engel A (1998) Reconstitution, characterization and mass analysis of the pentacylindrical allophycocyanin core complex from the cyanobacterium Anabaena sp. PCC 7120. J Mol Biol 278:369–388

Duerring M, Huber R, Bode W (1988) The structure of gamma-N-methylasparagine in C-phycocyanin from *Mastigocladus laminosus* and *Agmenellum quadriplicatum*. FEBS Lett 236:167–170

Duerring M, Schmidt GB, Huber R (1991) Isolation, crystallization, crystal structure analysis and refinement of constitutive C-phycocyanin from the chromatically adapting cyanobacterium *Fremyella diplosiphon* at 1.66 Å resolution. J Mol Biol 217:577–592

Fairchild CD, Jones IK, Glazer AN (1991) Absence of glycosylation on cyanobacterial phycobilisome linker polypeptides and rhodophytan phycoerythrins. J Bacteriol 173:2985–2992

Ferreira KN, Iverson TM, Maghlaoui K, Barber J, Iwata S (2004) Architecture of the photosynthetic oxygen-evolving center. Science 303:1831–1838

Forster T (1948) Zwischenmolekulare Energiewanderung und Fluoreszenz. Ann Physik 2:55–75

Forster T (1965) Delocalized excitation and excitation transfer. In: Sinanoglu O (ed) Modern Quantum Chemistry. Academic Press, New York, pp 93–137

Frankenberg N, Mukougawa K, Kohchi T, Lagarias JC (2001) Functional genomic analysis of the HY2 family of ferredoxin-dependent bilin reductases from oxygenic photosynthetic organisms. Plant Cell 13:965–978

Frigaard NU, Chew AG, Li H, Maresca JA, Bryant DA (2003) Chlorobium tepidum: insights into the structure, physiology, and metabolism of a green sulfur bacterium derived from the complete genome sequence. Photosynth Res 78:93–117

Frigaard N-U, Vassilieva EV, Li H, Milks KJ, Zhao J, Bryant DA (eds) (2001) The Remarkable Chlorosome. CSIRO Publishing, Melbourne, Australia

Fuglistaller P, Mimuro M, Suter F, Zuber H (1987) Allophycocyanin complexes of the phycobilisome from *Mastigocladus laminosus*. Influence of the linker polypeptide L8.9C on the spectral properties of the phycobiliprotein subunits. Biol Chem Hoppe Seyler 368:353–367

Gantt E, Conti SF (1966a) Granules associated with the chloroplast lamellae of *Porphyridium cruentum*. J Cell Biol 29:423–434

Gantt E, Conti SF (1966b) Phycobiliprotein localization in algae. Brookhaven Symp Biol 19:393–405

Gantt E, Lipschultz CA (1972) Phycobilisomes of *Porphyridium cruentum*. I. Isolation. J Cell Biol 54:313–324

Glauser M, Bryant DA, Frank G, Wehrli E, Rusconi SS, Sidler W, Zuber H (1992) Phycobilisome structure in the cyanobacteria *Mastigocladus laminosus* and *Anabaena* sp. PCC 7120. Eur J Biochem 205:907–915

Glazer AN (1985) Light harvesting by phycobilisomes. Annu Rev Biophys Biophys Chem 14:47–77

Glazer AN (1989) Light guides. Directional energy transfer in a photosynthetic antenna. J Biol Chem 264:1–4

Glazer AN, Bryant DA (1975) Allophycocyanin B (λ_{max} 671, 618 nm): a new cyanobacterial phycobiliprotein. Arch Microbiol 104:15–22

Gomez-Lojero C, Perez-Gomez B, Shen G, Schluchter WM, Bryant DA (2003) Interaction of ferredoxin:NADP+ oxidoreductase with phycobilisomes and phycobilisome substructures of the cyanobacterium *Synechococcus* sp. strain PCC 7002. Biochemistry 42:13800–13811

Grossman AR, Bhaya D, He Q (2001) Tracking the light environment by cyanobacteria and the dynamic nature of light harvesting. J Biol Chem 276:11449–11452

Grossman AR, Schaefer MR, Chiang GG, Collier JL (1993) The phycobilisome, a light-harvesting complex responsive to environmental conditions. Microbiol Rev 57:725–749

Holm L, Sander C (1993) Structural alignment of globins, phycocyanins and colicin A. FEBS Lett 315:301–306

Huang L, McCluskey MP, Ni H, LaRossa RA (2002) Global gene expression profiles of the cyanobacterium *Synechocystis* sp. strain PCC 6803 in response to irradiation with UV-B and white light. J Bacteriol 184:6845–6858

Huber R (1989) Nobel lecture. A structural basis of light energy and electron transfer in biology. Embo J 8:2125–2147

Klotz AV, Leary JA, Glazer AN (1986) Post-translational methylation of asparaginyl residues. Identification of beta-71 γ-N-methylasparagine in allophycocyanin. J Biol Chem 261:15891–15894

Kondo K, Geng XX, Katayama M, Ikeuchi M (2005) Distinct roles of CpcG1, CpcG2 in phycobilisome assembly in the cyanobacterium *Synechocystis* sp. PCC 6803. Photosynth Res 84:269–273

Liu JY, Jiang T, Zhang JP, Liang DC (1999) Crystal structure of allophycocyanin from red algae *Porphyra yezoensis* at 2.2 Å resolution. J Biol Chem 274:16945–16952

Liu LN, Chen XL, Zhang YZ, Zhou BC (2005) Characterization, structure and function of linker polypeptides in phycobilisomes of cyanobacteria and red algae: an overview. Biochim Biophys Acta 1708:133–142

Loll B, Kern J, Zouni A, Saenger W, Biesiadka J, Irrgang KD (2005) The Antenna System of Photosystem II From *Thermosynechococcus elongatus* at 3.2 Angstrom Resolution. Photosynth Res 86:175–184

Lundell DJ, Glazer AN (1983) Molecular architecture of a light-harvesting antenna. Structure of the 18 S core-rod subassembly of the *Synechococcus* 6301 phycobilisome. J Biol Chem 258:894–901

Luque I, Ochoa De Alda JA, Richaud C, Zabulon G, Thomas JC, Houmard J (2003) The NblAI protein from the filamentous cyanobacterium *Tolypothrix* PCC 7601: regulation of its expression and interactions with phycobilisome components. Mol Microbiol 50:1043–1054

MacColl R (1998) Cyanobacterial phycobilisomes. J Struct Biol 124:311–334

MacColl R (2004) Allophycocyanin and energy transfer. Biochim Biophys Acta 1657:73–81

MacColl R, Eisele LE, Menikh A (2003) Allophycocyanin: trimers, monomers, subunits, and homodimers. Biopolymers 72:352–365

MacDonald TM, Dubois L, Smith LC, Campbell DA (2003) Sensitivity of cyanobacterial antenna, reaction center, CO2 assimilation transcripts and proteins to moderate UVB: light acclimation potentiates resistance to UVB. Photochem Photobiol 77:405–412

Melkozernov AN, Blankenship RE (2005) Structural and functional organization of the peripheral light-harvesting system in Photosystem I. Photosynth Res 85:33–50

Migita CT, Zhang X, Yoshida T (2003) Expression and characterization of cyanobacterium heme oxygenase, a key enzyme in the phycobilin synthesis. Properties of the heme complex of recombinant active enzyme. Eur J Biochem 270:687–698

Mimuro M, Kikuchi H, Murakami A (1999) Structure, function of Phycobiliosmes. In: Singhal GS, Renger G, Sopory SK, Irrgang KD, Govindjee (eds) Concepts in Photobiology. Kluwer Academic Publishers, Dordrecht, pp 104–135

Montgomery BL, Casey ES, Grossman AR, Kehoe DM (2004) Apl A, a member of a new class of phycobiliproteins lacking a traditional role in photosynthetic light harvesting. J Bacteriol 186:7420–7428

Nield J, Rizkallah PJ, Barber J, Chayen NE (2003) The 1.45 Å three-dimensional structure of C-phycocyanin from the thermophilic cyanobacterium *Synechococcus elongatus*. J Struct Biol 141:149–155

Padyana AK, Bhat VB, Madyastha KM, Rajashankar KR, Ramakumar S (2001) Crystal structure of a light-harvesting protein C-phycocyanin from *Spirulina platensis*. Biochem Biophys Res Commun 282:893–898

Piven I, Ajlani G, Sokolenko A (2005) Phycobilisome linker proteins are phosphorylated in *Synechocystis* sp. PCC 6803. J Biol Chem 280:21667–21672

Pizarro SA, Sauer K (2001) Spectroscopic study of the light-harvesting protein C-phycocyanin associated with colorless linker peptides. Photochem Photobiol 73:556–563

Reuter W, Wiegand G, Huber R, Than ME (1999) Structural analysis at 2.2 Å of orthorhombic crystals presents the asymmetry of the allophycocyanin-linker complex, AP.LC7.8, from phycobilisomes of *Mastigocladus laminosus*. Proc Natl Acad Sci USA 96:1363–1368

Richaud C, Zabulon G, Joder A, Thomas JC (2001) Nitrogen or sulfur starvation differentially affects phycobilisome degradation and expression of the nblA gene in *Synechocystis* strain PCC 6803. J Bacteriol 183:2989–2994

Sauer K, Scheer H (1988) Exitation transfer in C-phycocyanin. Forster transfer rate and exciton calculations based on new crystal structure data for C-phycocyanins from *Agmenellum quadruplaticum* and *Mastigocladus laminosus*. Biochim Biophys Acta 936:157–170

Schirmer T, Bode W, Huber R, Sidler W, Zuber H (1985) X-ray crystallographic structure of the light-harvesting biliprotein C-phycocyanin from the thermophilic cyanobacterium *Mastigocladus laminosus* and its resemblance to globin structures. J Mol Biol 184:257–277

Schirmer T, Huber R, Schneider M, Bode W, Miller M, Hackert ML (1986) Crystal structure analysis and refinement at 2.5 Å of hexameric C-phycocyanin from the cyanobacterium *Agmenellum quadruplicatum*. The molecular model and its implications for light-harvesting. J Mol Biol 188:651–676

Schwarz R, Grossman AR (1998) A response regulator of cyanobacteria integrates diverse environmental signals and is critical for survival under extreme conditions. Proc Natl Acad Sci USA 95:11008–11013

Sendersky E, Lahmi R, Shaltiel J, Perelman A, Schwarz R (2005) Nbl C, a novel component required for pigment degradation during starvation in *Synechococcus* PCC 7942. Mol Microbiol 58:659–668

Sidler WA (1994) Phycobilisome and phycobiliprotein structures. In: Bryant DA (ed) The Molecular Biology of Cyanobacteria. Kluwer Academic Publishers, Dordrect, pp 139–216

Stec B, Troxler RF, Teeter MM (1999) Crystal structure of C-phycocyanin from *Cyanidium caldarium* provides a new perspective on phycobilisome assembly. Biophys J 76:2912–2921

Steglich C, Frankenberg-Dinkel N, Penno S, Hess WR (2005) A green light-absorbing phycoerythrin is present in the high-light-adapted marine cyanobacterium *Prochlorococcus* sp. MED4. Environ Microbiol 7:1611–1618

Sugishima M, Hagiwara Y, Zhang X, Yoshida T, Migita CT, Fukuyama K (2005) Crystal structure of dimeric heme oxygenase-2 from *Synechocystis* sp. PCC 6803 in complex with heme. Biochemistry 44:4257–4266

Sun L, Wang S (2003) Allophycocyanin complexes from the phycobilisome of a thermophilic blue-green alga *Myxosarcina concinna Printz*. J Photochem Photobiol B 72:45–53

Swanson RV, Glazer AN (1990) Phycobiliprotein methylation. Effect of the γ-N-methylasparagine residue on energy transfer in phycocyanin and the phycobilisome. J Mol Biol 214:787–796

Tandeau de Marsac N, Cohen-Bazire G (1977) Molecular composition of cyanobacterial phycobilisomes. Proc Natl Acad Sci USA 74:1635–1639

Teale FW, Dale RE (1970) Isolation and spectral characterization of phycobiliproteins. Biochem J 116:161–169

Ting CS, Rocap G, King J, Chisholm SW (2002) Cyanobacterial photosynthesis in the oceans: the origins and significance of divergent light-harvesting strategies. Trends Microbiol 10:134–142

van Waasbergen LG, Dolganov N, Grossman AR (2002) *nblS*, a gene involved in controlling photosynthesis-related gene expression during high light and nutrient stress in *Synechococcus elongatus* PCC 7942. J Bacteriol 184:2481–2490

Wilk KE, Harrop SJ, Jankova L, Edler D, Keenan G, Sharples F, Hiller RG, Curmi PM (1999) Evolution of a light-harvesting protein by addition of new subunits and rearrangement of conserved elements: crystal structure of a cryptophyte phycoerythrin at 1.63 Å resolution. Proc Natl Acad Sci USA 96:8901–8906

Xiong J, Bauer CE (2002) Complex evolution of photosynthesis. Annu Rev Plant Biol 53:503–521

Yamanaka G, Glazer AN, Williams RC (1978) Cyanobacterial phycobilisomes. Characterization of the phycobilisomes of *Synechococcus* sp. 6301. J Biol Chem 253:8303–8310

Yi ZW, Huang H, Kuang TY, Sui SF (2005) Three-dimensional architecture of phycobilisomes from *Nostoc flagelliforme* revealed by single particle electron microscopy. FEBS Lett 579:3569–3573

Yonath A (2005) Antibiotics targeting ribosomes: resistance, selectivity, synergism and cellular regulation. Annu Rev Biochem 74:649–679

Yu MH, Glazer AN (1982) Cyanobacterial phycobilisomes. Role of the linker polypeptides in the assembly of phycocyanin. J Biol Chem 257:3429–3433

Yu MH, Glazer AN, Williams RC (1981) Cyanobacterial phycobilisomes. Phycocyanin assembly in the rod substructures of anabaena variabilis phycobilisomes. J Biol Chem 256:13130–13136

Zhao KH, Su P, Bohm S, Song B, Zhou M, Bubenzer C, Scheer H (2005) Reconstitution of phycobilisome core-membrane linker, LCM, by autocatalytic chromophore binding to ApcE. Biochim Biophys Acta 1706:81–87

Microbiol Monogr (2)
J.M. Shively: Complex Intracellular Structures in Prokaryotes
DOI 10.1007/7171_021/Published online: 13 May 2006

Chlorosomes: Antenna Organelles in Photosynthetic Green Bacteria

Niels-Ulrik Frigaard[1] (✉) · Donald A. Bryant[2]

[1]Institute of Molecular Biology and Physiology, University of Copenhagen, Sølvgade 83H, 1307 Copenhagen, Denmark
nuf@mermaid.molbio.ku.dk

[2]Department of Biochemistry and Molecular Biology, The Pennsylvania State University, University Park, Pennsylvania, 16802, USA

Abstract Chlorosomes are the light-harvesting antenna organelles found in two groups of bacteria, the green sulfur bacteria and the green filamentous bacteria, collectively known as photosynthetic green bacteria. Chlorosomes consist mostly of aggregated bacteriochlorophyll (BChl) *c*, *d*, or *e* and are the largest antenna structures known. Unlike other light-harvesting antenna structures, the major antenna pigments (BChl *c*, *d*, or *e*) form aggregates that do not require a protein scaffold. This is possible because these BChls possess structural modifications that do not occur in other naturally occurring chlorophyll derivatives. These properties allow the formation of very large and efficient antennae that permit phototrophic growth at remarkably low light intensities. The BChl aggregates are enveloped in a monolayer protein–lipid membrane with a high content of glycolipids. Chlorosomes also contain small amounts of BChl *a* bound to the CsmA protein, other protein species of mostly unknown function, carotenoids, and isoprenoid quinones. Chlorosomes from thermophilic green bacteria also contain wax esters. Although very little is known about how chlorosomes are synthesized by the cells, significant efforts have been devoted to understanding the structural organization and energy transfer characteristics of the BChl aggregates in chlorosomes. This is in part because these aggregates serve as models of self-assembling systems and their potential use as light-harvesting nanostructures in artificial photosystems. Recent progress in genetic manipulation of BChl *c*, carotenoids, and chlorosome envelope proteins in green sulfur bacteria has allowed a better understanding of how these components function in chlorosomes.

1 Introduction

All photosynthetic organisms have membrane-embedded photosynthetic reaction centers that directly generate reducing power and indirectly generate an energy-conserving transmembrane proton gradient (Blankenship 2002). These photosynthetic reaction centers have similar basic structures and mechanisms, and thus they appear to share a common evolutionary ancestor (Blankenship 2002; Olson and Blankenship 2004). The reaction centers receive excitation energy from peripheral antenna pigments within the reaction center complex, and typically also receive excitation energy from separate pigment–protein complexes, or antennae, which are either embedded in the membrane or attached to the cytoplasmic side of the membrane. In contrast to photosynthetic reaction centers, nature has devised many different light-harvesting antenna structures for photosynthetic organisms, whose protein components do not share a common evolutionary ancestor (Blankenship 2002; Green 2003; Green et al. 2003; Olson and Blankenship 2004).

Green sulfur bacteria were first described a century ago by Nadson (1906) and later in pure culture by van Niel (1932). Nadson realized that the absorption spectrum of the major chlorophyll species of these bacteria (BChl *c*) resembles that of chlorophyll *a* in solution, but that its absorption spectrum in vivo and its chemical properties are different (Nadson 1912). Krasnovsky and coworkers found in the late 1950s that the unique chlorophyll of green sulfur bacteria forms aggregates in vitro with red-shifted absorption, and they proposed this as a model of pigment organization in the antenna of these organisms (Krasnovsky and Pakshina 1959; reviewed in Blankenship et al. 1995). This ability of the chlorosome BChls to self-organize into antenna structures has generated much interest in them as models of self-assembling antenna systems and as light-harvesting nanostructures for use in artificial photosystems (reviewed in Balaban 2004, 2005). The structure of BChl *c* and *d* was determined by Holt and other researchers in the 1960s (reviewed in Smith 2003). However, it was not until the early 1960s that researchers realized that BChl *c* is located in large organelles observable by thin section electron microscopy and originally denoted "*Chlorobium* vesicles" (Fig. 1; Cohen-Bazire et al. 1964). Later, in 1974 when chlorosomes were discovered in the green filamentous bacterium *Chloroflexus aurantiacus* and it was realized that this bacterium belongs to a different taxonomic group (the Chloroflexi) than the green sulfur bacteria (the Chlorobi), "*Chlorobium* vesicles" were renamed chlorosomes (Pierson and Castenholz 1974; Staehelin et al. 1978).

The genomes of 15 strains of chlorosome-containing bacteria have been or are currently being sequenced (Table 1). The first genome to be fully se-

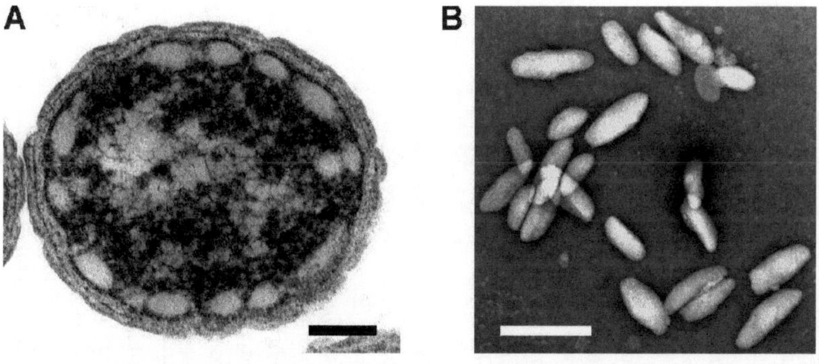

Fig. 1 **a** Transmission electron micrograph of a thin-section of a *Chl. tepidum* cell showing the chlorosomes as large *white bodies* attached to the cytoplasmic membrane. The view is parallel to the long *axis* of the cell and the chlorosomes. The *bar* represents 100 nm. (Reproduced from Frigaard et al. 2002 with permission from the American Society for Microbiology.) **b** Transmission electron micrograph of negatively stained chlorosomes isolated from *Chl. tepidum*. The view is perpendicular to the long *axis* of the chlorosomes. The *bar* represents 200 nm. (Reproduced from Frigaard et al. 2004a with permission from the American Society for Microbiology.)

Table 1 Strains of Chlorobi and Chloroflexi whose genomes have been or will soon be sequenced[1]

Strain	Name	Phylum	Phototrophic	Chlorosomes	BChls
BS1	*Chlorobium* sp. BS1[2]	Chlorobi	Yes	Yes	*a, e*
CaD3	*Chlorobium chlorochromatii*[3]	Chlorobi	Yes	Yes	*a, c*
DSMZ 13031^T	*"Chlorobium ferrooxidans"*	Chlorobi	Yes	Yes	*a, c*
DSMZ 245^T	*Chlorobium limicola*	Chlorobi	Yes	Yes	*a, c*
DSMZ 266^T	*Chlorobium phaeobacteroides*	Chlorobi	Yes	Yes	*a, e*
DSMZ 265	*Chlorobium vibrioforme* f. sp. *thiosulfatophilum*[4]	Chlorobi	Yes	Yes	*a, c, d*
ATCC 49652^T	*Chlorobium tepidum* TLS[5]	Chlorobi	Yes	Yes	*a, c*
ATCC 35110^T	*Chloroherpeton thalassium*	Chlorobi	Yes	Yes	*a, c*
DSMZ 273^T	*Pelodictyon luteolum*[6]	Chlorobi	Yes	Yes	*a, c*
DSMZ 5477^T	*Pelodictyon phaeoclathratiforme* BU-1[7]	Chlorobi	Yes	Yes	*a, e*
DSMZ 271^T	*Prosthecochloris aestuarii* SK413	Chlorobi	Yes	Yes	*a, c*
DSMZ 9485^T	*Chloroflexus aggregans* MD-66	Chloroflexi	Yes	Yes	*a, c*
ATCC 29366^T	*Chloroflexus aurantiacus* J-10-fl	Chloroflexi	Yes	Yes	*a, c*
UdG9001	*Chloronema giganteum*	Chloroflexi	Yes	Yes	*a, c, d*
	Chlorothrix halophila	Chloroflexi	Yes	Yes	*a, c*
	Heliothrix oregonensis	Chloroflexi	Yes	No	*a*
RS-1	*Roseiflexus* sp. RS-1	Chloroflexi	Yes	No	*a*
DSMZ 13941^T	*Roseiflexus castenholzii*	Chloroflexi	Yes	No	*a*
DSMZ 785^T	*Herpetosiphon aurantiacus*	Chloroflexi	No	No	None

[1] For *Chl. tepidum* see Eisen et al. (2002), for all others see http://www.jgi.doe.gov
[2] Also called *Chlorobium phaeobacteroides* BS1 or *Prosthecochloris* sp. BS1
[3] Previously called *"Chlorochromatium aggregatum"* epibiont
[4] Suggested new name: *Chlorobium phaeovibrioides* (Imhoff 2003)
[5] Suggested new name: *Chlorobaculum tepidum* (Imhoff 2003)
[6] Suggested new name: *Chlorobium luteolum* (Imhoff 2003)
[7] Suggested new name: *Chlorobium clathratiforme* (Imhoff 2003)

quenced and annotated was that of *Chlorobium tepidum* (Eisen et al. 2002). This genome sequence enabled significant progress in understanding the biosynthesis of BChl *c* and *d* and carotenoids and their functions in chlorosomes (Frigaard et al. 2003, 2006; Frigaard and Bryant 2004). Comparative genomic analyses within the Chlorobi, within the Chloroflexi, between Chlorobi and Chloroflexi, and between chlorosome-containing and non-chlorosome-containing Chloroflexi, will certainly produce much new information on how chlorosomes function, on how their components are formed, and how chlorosomes and their components have evolved. At the time of writing, genome sequence information is only available for *Cfx. aurantiacus* and nine of the green sulfur bacteria in Table 1.

This review gives an overview of chlorosome structure and function with special emphasis on the recent progress that has been made in understanding the biosynthesis, function, and evolution of chlorosome BChls, carotenoids, and proteins. Other excellent recent reviews describing chlorosomes and photosynthesis in green bacteria are also available (Blankenship et al. 1995; Blankenship and Matsuura 2003).

2 Occurrence and Significance of Chlorosomes

Chlorosomes only occur in two bacterial phyla, the Chlorobi and the Chloroflexi. The unique structure of the chlorosome allows accumulation of large amounts of antenna pigments in the cell and thus allows efficient light harvesting and phototrophic growth at low light intensities.

2.1 The Chlorobi and the Chloroflexi

The Chlorobi, or green sulfur bacteria, are obligately photolithoautotrophic and strictly anaerobic organisms (Overmann 2000; Garrity and Holt 2001a). They comprise a coherent group of organisms that are phylogenetically and physiologically distinctive. These organisms are commonly found in anoxic and sulfide-rich freshwater and estuarine environments including lakes, sediments, microbial mats, and sulfide-rich hot springs. They contain a type I (also called iron–sulfur-type) photosynthetic reaction center similar to the reaction center in heliobacteria and photosystem I in cyanobacteria and plants (Hauska et al. 2001). Most green sulfur bacteria are dependent on oxidation of sulfide or elemental sulfur for growth and are poor at using organic compounds except acetate. In addition to chlorosomes, green sulfur bacteria have a unique BChl *a*-containing antenna complex, the Fenna–Matthews–Olson (FMO) protein, which is situated between the chlorosomes and the cytoplasmic membrane (Blankenship et al. 1995).

Members of the phylum Chloroflexi are physiologically more diverse and many are capable of both chemotrophic and phototrophic growth (Garrity and Holt 2001b; Hanada and Pierson 2002). The family Chloroflexaceae within the Chloroflexi comprises multicellular filamentous anoxygenic photosynthetic bacteria (FAPs), and are also called phototrophic filamentous bacteria. The Chloroflexaceae frequently form microbial mats and have been observed in marine and hypersaline environments as well as hot springs and freshwater habitats (Hanada and Pierson 2002). They are obligately or facultatively phototrophic. All members of the Chloroflexaceae characterized to date contain a type II (also called quinone-type) photosynthetic reaction center similar to the reaction center in purple bacteria and photosystem II in cyanobacteria and plants (Feick et al. 1996; Yamada et al. 2005). In addition to chlorosomes, Chloroflexaceae have a cytoplasmic membrane-embedded, light-harvesting complex similar to the light-harvesting LH1 complex of purple bacteria, denoted B808–866 (Xin et al. 2005). Many Chloroflexaceae are capable of aerobic respiration in the dark, under which conditions expression of the photosynthetic apparatus is repressed. Some Chloroflexi are not photosynthetic and only grow aerobically as chemoheterotrophs (e.g., *Herpetosiphon aurantiacus*; Holt and Lewin 1968; Garrity and Holt 2001b). Whereas all green sulfur bacteria characterized to date are obligately photosynthetic and contain chlorosomes, chlorosomes in Chloroflexi are only found within a subgroup of the Chloroflexaceae often referred to as "green filamentous bacteria", "green filamentous anoxygenic phototrophs", or simply as "green FAPs", which include the characterized genera *Chloroflexus*, *Oscillochloris*, *Chloronema*, and *Chlorothrix* (Hanada and Pierson 2002; Klappenbach and Pierson 2004). Phototrophic Chloroflexaceae without chlorosomes include the characterized genera *Heliothrix* and *Roseiflexus*; these organisms are sometimes referred to as "red FAPs" (Hanada and Pierson 2002; Klappenbach and Pierson 2004).

2.2 Significance of Chlorosomes

The light intensities are exceedingly low in the anoxic layers of stratified lakes and microbial mats in which chlorosome-containing organisms typically occur. For example, a layer of BChl *e*-containing green sulfur bacteria occurs at the chemocline of the Black Sea at a depth of approx. 100 m (Overmann et al. 1992). The light intensity at this depth is nearly 10^6 times less than at the surface and has been determined to be about 3 nmol photons $m^{-2}\,s^{-1}$. Under these conditions, a single chlorophyll molecule will absorb one photon in a period of about 6 h. Another remarkable example is the recent recovery of a green sulfur bacterium from a black smoker at 2391 m beneath the surface of the Pacific Ocean (Beatty et al. 2005). These cells are proposed to be able to extract sufficient excitation energy from infrared geothermal radiation from the hot smoker to allow phototrophic growth.

Although the photon fluxes may be very small in the environments of green sulfur bacteria, a single cell can contain 50 million BChl molecules, and thus the number of absorption events per unit time can still be significant. The protein content of chlorosomes is relatively low because the BChl *c* aggregates do not require a pigment-binding protein for their organization and assembly. This enables cells to accumulate large amounts of BChl *c* – representing up to 30% of the carbon content of the cells (Borrego et al. 1999) – without the need for synthesizing what would be energetically unfavorable amounts of protein as well. This in turn enables the cells to grow at very low light intensities.

Green sulfur bacteria usually live in low light environments and chlorosome formation does not appear to be extensively regulated in these organisms. In contrast, green filamentous bacteria at the surface of microbial mats may experience high light conditions. Under such conditions BChl *c* biosynthesis is repressed or missing altogether in green filamentous bacteria, and the organisms harvest sufficient light energy using their smaller BChl *a*-based antennae.

2.3
Types of Chlorosomes

Although chlorosomes from green sulfur bacteria and green filamentous bacteria are similar, there are some important differences, especially in the molecular structure of the pigments and in their protein composition. Most research has been performed on chlorosomes from various strains of green sulfur bacteria (recently especially *Chl. tepidum*) and the green filamentous bacterium *Cfx. aurantiacus*. On the basis of this work, researchers have often designated chlorosomes as either "*Chlorobium*-type" or "*Chloroflexus*-type". However, chlorosomes from a newly characterized member of the green filamentous bacteria, *Chloronema* sp. UdG9001, have properties resembling both "*Chlorobium*-type" and "*Chloroflexus*-type" chlorosomes (Gich et al. 2003).

3
Techniques for Studying Chlorosomes

Chlorosomes are relatively easy to isolate, but it should be noted that the isolation method of choice might affect subsequent measurements. A large variety of biophysical techniques have yielded much information on the pigment organization in chlorosomes and recently genetic techniques have added to the repertoire of approaches. However, new techniques and approaches are still necessary to reveal more about the structure of the chlorosome, which still is not satisfactorily resolved.

3.1
Isolation of Chlorosomes

Chlorosomes are typically isolated after disrupting cells with a French pressure-cell treatment; chlorosomes are then isolated by ultracentrifugation on sucrose density gradients (Oelze and Golecki 1995). Inclusion of 2 M NaSCN (Gerola and Olson 1986) or detergents (Feick and Fuller 1984) during the cell disruption and isolation procedure greatly enhances the separation of cytoplasmic membranes and chlorosomes. However, differences in the isolation method have been found to affect subsequent measurements on the isolated chlorosomes (Oelze and Golecki 1995). For example, detergents remove lipids and most of the proteins from the chlorosome envelope and, for this reason, the use of detergents to release chlorosomes from membranes is not recommended (Vassilieva et al. 2002b).

3.2
Measurements on Chlorosomes

The structure of chlorosomes has been investigated by transmission, freeze-fracture, and cryo- electron microscopy (Fig. 1; Staehelin et al. 1978, 1980; Oelze and Golecki 1995; Psencik et al. 2004; Hohmann-Marriott 2005) and atomic force microscopy (Zhu et al. 1995; Martinez-Planells et al. 2002; Frigaard et al. 2005). The pigment composition is usually determined using high-pressure liquid chromatography (HPLC) combined with absorption or mass spectrometric detection (Borrego and Garcia-Gil 1994; Airs and Keely 2002). The protein composition has been studied by polyacrylamide gel electrophoresis (Feick and Fuller 1984; Chung and Bryant 1996a,b; Vassilieva et al. 2002a,b), immunoblotting (Chung 1996a,b; Vassilieva et al. 2002a,b), and mass spectrometry (Persson et al. 2000; Milks et al. 2005). Steady-state and time-resolved optical and magnetic resonance techniques including absorption spectroscopy, fluorescence spectroscopy, circular and linear dichroism, NMR, and X-ray scattering have provided a wealth of information on the organization and energy transfer properties of the BChls and carotenoids (reviewed in Blankenship et al. 1995; Blankenship and Matsuura 2003; Psencik et al. 2004).

3.3
Studies of BChl Aggregation In Vitro

Chlorosome BChls spontaneously form aggregates in vitro with a red-shifted absorption similar to that observed in chlorosomes (Krasnovsky and Pakshina 1959). By varying the exact solvent composition, aggregate formation can be controlled. In nonpolar solvents such as chlorinated alkanes, dimers and higher aggregates are readily obtained (Smith et al. 1983; Olson and Pedersen 1990). BChl aggregates also form in aqueous suspension in the presence of lipids, either from purified components (BChl *c* and MGDG; Uehara et al. 1994) or from a crude pigment–lipid extract of chlorosomes (Hirota et al. 1992; Miller et al. 1993). These studies have been extended to include both synthetic and semisynthetic tetrapyrroles (reviewed in Balaban 2004, 2005).

3.4 Studies of Chlorosomes Using Mutagenesis

Inactivation of a gene is a powerful approach for studying the function of its product. A gene inactivation method based on natural transformation and homologous recombination has been established in the BChl *c*-containing green sulfur bacterium *Chl. tepidum* (Frigaard and Bryant 2001; Frigaard et al. 2004c). Targeted gene inactivation has not yet been reported for any green filamentous bacterium or any BChl *e*-containing green sulfur bacterium. Null-mutant strains of *Chl. tepidum* have also been generated that lack the enzymes for specific steps in the biosynthetic pathways for BChl *c* and carotenoids, and thus these strains have modified pigment contents. Characterization of cell cultures and isolated chlorosomes of such mutant strains has allowed many conclusions to be drawn about the function of these pigments (Sects. 5.1.2 and 5.3.2). Null-mutant strains of *Chl. tepidum* have also been generated that lack chlorosome proteins (Sect. 5.4.1). Most mutants lacking only a single chlorosome protein have little or no phenotype. However, mutants lacking more than one chlorosome protein appear to have significantly affected chlorosomes. Some components are essential for photosynthesis in *Chl. tepidum* (e.g., BChl *a* and possibly CsmA) and cannot be completely eliminated by mutagenesis because *Chl. tepidum* is obligately photoautotrophic.

4 Structure of the Chlorosome

Figure 2 shows a model of the chlorosome and photosynthetic apparatus in the green sulfur bacterium *Chl. tepidum*. Green filamentous bacteria do not contain an FMO protein. Instead, the chlorosome is thought to interact directly with the LH1-like antenna complex (denoted B808–866) embedded in the cytoplasmic membrane, which functions as an intermediate in the transfer of excitation energy from the chlorosome to the reaction center (Xin et al. 2005). Both types of chlorosomes contain a baseplate structure that presum-

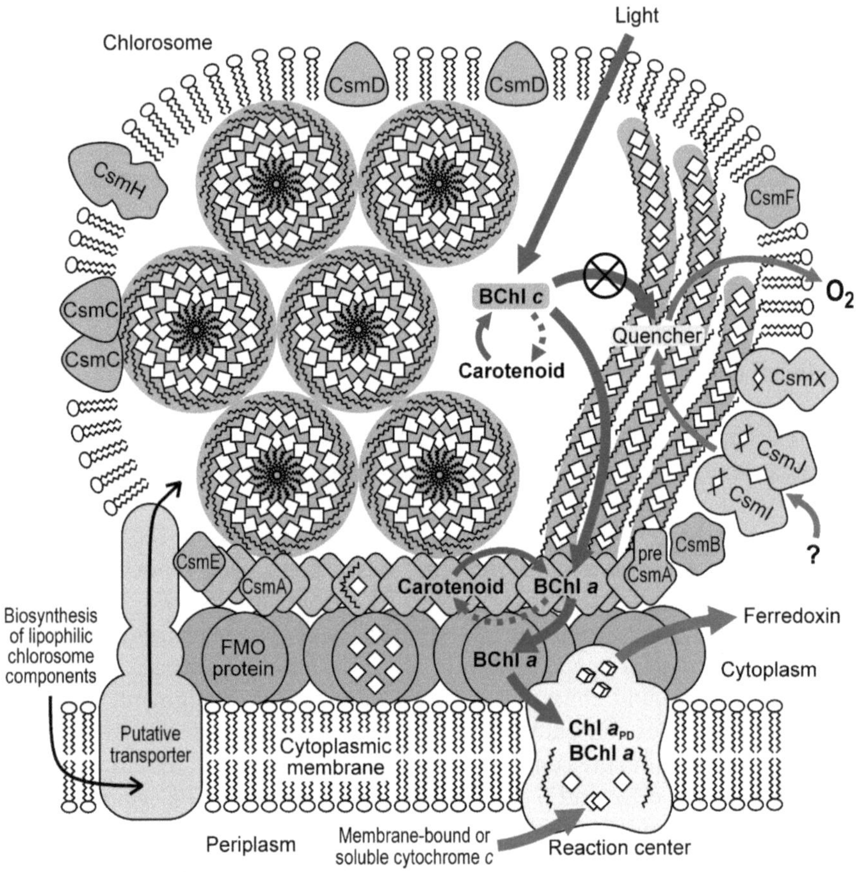

Fig. 2 Model of the chlorosome and photosynthetic membrane in *Chl. tepidum*. The commonly favored rod model of BChl *c* aggregation is shown on the *left side* of the chlorosome interior (Nozawa et al. 1994) and the recently proposed alternative lamellar model of BChl *c* aggregation is shown on the *right side* of the chlorosome interior (Pšenčík et al. 2004). Singlet excitation energy transfer is shown by *red lines*, quenching of excited BChl triplets by carotenoids is shown by *dotted red lines*, and electron transfer is shown by *blue lines*. In the presence of O_2 (which is detrimental to the bacteria), the quencher in the chlorosome is activated and prevents excitation transfer from BChl *c* to the reaction center and thus prevents photosynthetic electron transfer. The quencher is activated by oxidation, with the electrons probably being delivered directly or indirectly to O_2 and it is inactivated by reduction probably directly or indirectly by the chlorosome proteins CsmI and CsmJ. See text for details. (Modified from Frigaard et al. 2003, 2005 and Frigaard and Bryant 2004.)

ably forms the attachment site of the chlorosome to the FMO protein or the B808-866 complex.

4.1
Dimensions

In transmission electron microscopy, isolated chlorosomes from *Chl. tepidum* appear about 110–180 nm long and 40–60 nm in diameter (Frigaard et al. 2004a). In atomic force microscopy, isolated chlorosome from *Chl. tepidum* appear a little larger, about 170–260 nm long, 90–160 nm wide, and 30–40 nm high (Martinez-Planells et al. 2002; Frigaard et al. 2005). Chlorosomes from *Cfx. aurantiacus* are in general smaller than chlorosomes from *Chl. tepidum* and vary greatly in their dimensions and composition, depending on growth conditions (Golecki and Oelze 1987; Oelze and Golecki 1995; Martinez-Planells et al. 2002).

Rod-like structures have been observed in electron microscopy of disintegrated chlorosomes (Staehelin et al. 1978, 1980; Bryant et al. 2002). It was proposed that these rods represent the natural BChl aggregate structure in chlorosomes. These rods have a diameter of 10 nm in *Chlorobium limicola* and a diameter of 5 nm in *Cfx. aurantiacus*. In *Chlorobium limicola*, the presence of 10–30 rod elements per chlorosome was described.

4.2
Enumeration of Components

A chlorosome from *Chl. tepidum* contains about 150 000–300 000 BChl *c* molecules (Martinez-Planells et al. 2002; Montaño et al. 2003a). Based on chemical measurements of the ratio of one chlorosome component to another, it can be calculated that a chlorosome from *Chl. tepidum* containing about 200 000 BChl *c* molecules also contains about 2500 BChl *a* molecules, 20 000 carotenoid molecules, 15 000 chlorobiumquinone molecules, 3000 menaquinone-7 molecules, 5000 protein molecules (of which about half are CsmA), and 20 000 lipid molecules (glycolipids, phospholipids, and wax esters) (Bryant et al. 2002; Martinez-Planells et al. 2002; Montaño et al. 2003a; Frigaard et al. 2004; M. Miller, P.G. Sørensen, R.P. Cox, manuscript in preparation).

Although the size varies significantly with growth conditions, a typical chlorosome from *Cfx. aurantiacus* contains about 50 000 BChl *c* molecules, in addition to about 2000 BChl *a* molecules, 15 000 carotenoid molecules, and 6000 menaquinone-10 molecules (Frigaard et al. 1997).

4.3
The Protein–Lipid Chlorosome Envelope

Given the highly hydrophobic nature of the chlorosomal BChl aggregates and results from freeze-fracture electron microscopy (Staehelin et al. 1978, 1980), it is thought that the polar lipids form a monolayer membrane envelope around the chlorosome. Based on measurements on highly purified chlorosomes from *Chl. tepidum* (0.25 g protein per g BChl *c*, Vassilieva et al. 2002b and 0.05 g polar lipid per g BChl *c*, M. Miller, P.G. Sørensen, R.P. Cox, manuscript in preparation), this envelope has a protein-to-lipid ratio of about 5 w/w, (not counting wax esters), which is comparable to the ratio of roughly 3 w/w in bacterial cytoplasmic membranes. In agreement with this model, agglutination experiments with *Ricinus* lectin or antibodies directed towards galactose have shown that the galactose moieties of the glycolipids are exposed on the chlorosome surface (Holo et al. 1985; Vassilieva et al. 2002b). Similarly, experiments with agglutination with antibodies, protease treatments, and chemical cross-linking have shown that each of the chlorosome proteins is also exposed on the chlorosome surface (Chung and Bryant 1996a,b; Vassilieva et al. 2002b; Li et al. 2005).

The chlorosome protein contents of cells do not vary much in *Chl. vibrioforme* (Vassilieva et al. 2002a) or phototrophically grown *Cfx. aurantiacus* (Foidl et al. 1998), even when the BChl *c* content varies fivefold or more. These observations support the idea that the protein–lipid chlorosome envelope constitutes a "bag" that is constitutively synthesized under phototrophic conditions and can be filled to a greater or lesser extent with BChl *c* depending on the light intensity (Oelze and Golecki 1995; Foidl et al. 1998). Chlorosomes completely devoid of BChl *c*, denoted "carotenosomes", have been isolated from a BChl *c*-less *bchK* mutant of *Chl. tepidum* (Frigaard et al. 2005). The physical shape of these carotenosomes studied by atomic force microscopy is reminiscent of a flattened chlorosome and may represent a "bag" devoid of BChl *c*.

4.4
The Chlorosome Baseplate

The BChl *a*-CsmA pigment–protein complex is located in a distinct region of the chlorosome envelope called the baseplate. Cross-linking studies show that CsmA forms a paracrystalline array (Bryant et al. 2002; Frigaard et al. 2005), and that CsmA can also be cross-linked to the FMO protein (Li et al. 2005). This paracrystalline array is presumably what can be observed as striations in freeze-fracture electron micrographs of chlorosomes and is nearly as large as the basal surface of a complete chlorosome (Staehelin et al. 1980; Golecki and Oelze 1987). The BChl *a* of the baseplate absorbs maximally at about 795 nm, and its optical properties indicate that it is not as highly aggregated as BChl *c*. Stoichiometric measurements strongly suggest that each CsmA monomer binds a single BChl *a* molecule (Sect. 5.4.4).

5
Chlorosome Components and Their Function

Some chlorosome components are unique, or nearly so, to the chlorosomes (BChl *c*/*d*/*e*, chlorosome proteins, chlorobiumquinone), whereas other components also are present elsewhere in the cell (e.g., BChl *a*, carotenoids, menaquinone, various lipids). Gene products currently known to be specifically involved in chlorosome formation are listed in Table 2. The molecular structures of selected chlorosome components are shown in Fig. 3.

Table 2 Gene products in *Chl. tepidum* specific for chlorosome formation [1]

Gene product	Function
Chlorosome proteins [2]	
CsmA	Binds BChl *a* and probably also carotenoid(s) Forms the chlorosome baseplate
CsmB	Possibly involved in organization of BChl *c* aggregates
CsmC	Possibly involved in organization of BChl *c* aggregates
CsmD	Possibly involved in organization of BChl *c* aggregates
CsmE	Possibly binds BChl *a* (based on sequence similarity with CsmA)
CsmF	Possibly involved in organization of BChl *c* aggregates
CsmH	Possibly involved in organization of BChl *c* aggregates
CsmI	[2Fe – 2S] protein. Involved in redox control of energy transfer
CsmJ	[2Fe – 2S] protein. Involved in redox control of energy transfer
CsmX	Probable [2Fe – 2S] protein. May be involved in redox control of energy transfer
BChl c biosynthesis [3]	
BchK	BChl *c* synthase
BchQ	$C\text{-}8^2$ Methyltransferase
BchR	$C\text{-}12^1$ Methyltransferase
BchS	Magnesium chelatase subunit S
BchU	C-20 Methyltransferase
BchV	$C3^1$-Vinyl hydratase, S-stereochemistry

[1] See text for details. All genes encoding the products shown, with the exception of *csmA*, have been inactivated
[2] All chlorosome proteins are located in the chlorosome envelope
[3] Only enzymes specific for BChl *c* biosynthesis are listed

Substituents in BChls and Chl *a*

	R_3	R_7	R_8	7,8-bond	R_{12}	R_{13^2}	R_{20}	R_{17^3}
BChl *a*	$-CO-CH_3$	M	E	single	M	$-CO-OCH_3$	H	(P)
BChl *c*	$-CHOH-CH_3$	M	E,Pr,I,N	double	M,E	H	M	(F,S)
BChl *d*	$-CHOH-CH_3$	M	E,Pr,I,N	double	M,E	H	H	(F)
BChl *e*	$-CHOH-CH_3$	-CHO	E,Pr,I,N	double	E	H	M	(F)
BChl *f*	$-CHOH-CH_3$	-CHO	E,Pr,I,N	double	M,E	H	H	(F)
Chl *a*	$-CH=CH_2$	M	E	double	M	$-CO-OCH_3$	H	(P)

Parenthesis indicates that other substituents are also found naturally.

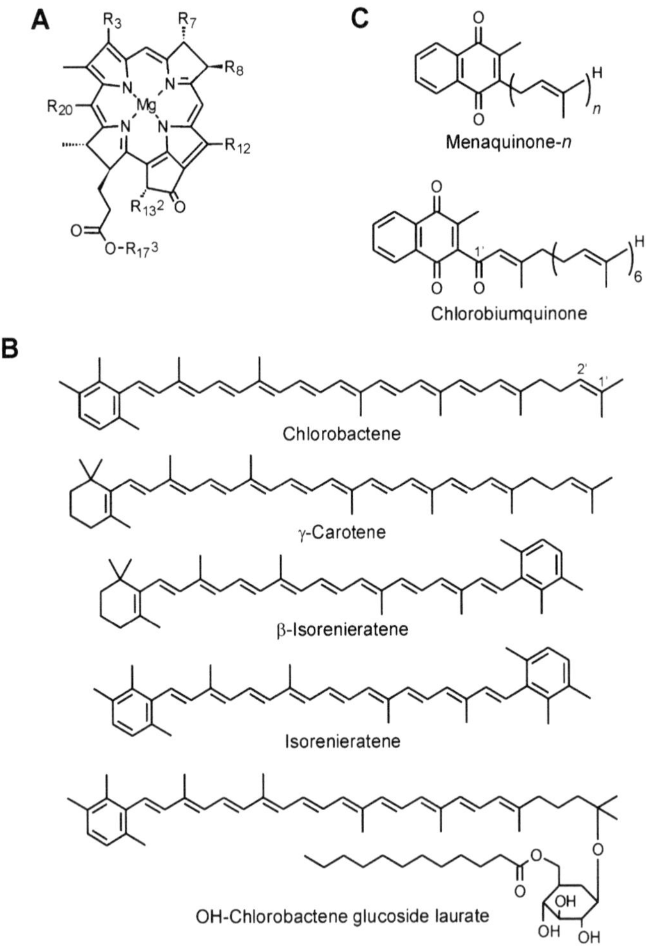

Fig. 3 Structures of chlorosome components (not exhaustive list). **a** Bacteriochlorophylls, (although not known to be components of chlorosomes, BChl *f* and Chl *a* are also shown for comparison), **b** carotenoids, and **c** isoprenoid quinones and other lipids. Abbreviations: *BChl* bacteriochlorophyll, *Chl* chlorophyll, *E* ethyl, *F* farnesyl, *I iso*-butyl, *M* methyl, *P* phytyl, *Pr n*-propyl, *S* stearyl

5.1 Bacteriochlorophylls *c*, *d*, and *e*

Chlorosomes always contain BChl *c*, *d*, or *e*, and these BChls are only found in chlorosomes. Cell cultures of green sulfur bacteria containing BChl *c* or *d* are dark green in color, whereas cultures of green sulfur bacteria containing BChl *e* appear brown in color. This is not due to differences in the carotenoid content or composition but is due to the special optical properties of aggregated BChl *e*, which has a strong absorption maximum in the region 500 to 550 nm (Steensgaard et al. 2000).

5.1.1 Organization in Aggregates

The chlorophyll species that constitute the major antenna pigment of chlorosomes are BChl *c*, *d*, or *e*, sometimes called "chlorosome BChls" or "*Chlorobium* BChls". Usually only one of these chlorophyll types is found in the chlorosome, although BChl *c* and *d* sometimes occur in chlorosomes from a single organism (Table 1). Chlorosome BChls have a C-3^1 hydroxyl group but lack a C-13^2 methylcarboxyl group (Fig. 3a). The C-3^1 hydroxyl group allows formation of a BChl dimer in which the hydroxyl group in one molecule binds to the central magnesium in the other molecule. It is thought that larger oligomeric structures form by aggregation of a large number of BChl dimers. Rod-like structures have been observed in electron microscopy of disintegrated chlorosomes (Sect. 4.1). In *Chlorobium*-type chlorosomes these rods are about 10 nm in diameter and in *Chloroflexus*-type chlorosomes they are about 5 nm in diameter. In both types of chlorosomes, the rods are 100–200 nm long and probably span the entire length of the chlorosome. It is thought that these rods predominantly consist of aggregated chlorosome BChls. The aggregated BChls interact strongly and are highly coupled excitonically. In methanol solution, monomeric BChl *c* absorbs at 671 nm – in comparison, aggregated multimeric BChl *c* absorbs maximally around 740 and 750 nm. Recent studies have challenged the prevailing rod–element models for BChl organization in chlorosomes. Psencik et al. (2004) have suggested that the BChl dimers form lamellar structures in which the hydrophobic tails of the BChls from adjacent layers interact to form bilayer-like structures (Fig. 2).

1-Hexanol disrupts BChl *c*/*d*/*e* aggregation and causes monomerization of these BChls in chlorosomes; this is observable as a dramatic reversible change in absorption maximum from $\sim$ 750 nm to $\sim$ 670 nm (Matsuura and Olson 1990; Wang et al. 1995). Hexanol, which is only slightly soluble in water, presumably enters the chlorosome interior due to its hydrophobic hexyl moiety, and causes disruption of the dimers by binding of its polar hydroxyl group to the central magnesium of the BChls. Upon removal of the hexanol, the aggregation and optical properties of the BChl are restored. It has proven difficult to isolate the BChl *a*–CsmA complex. However, a fraction from chlorosomes that is highly enriched in this complex has recently been isolated by the use of hexanol in combination with detergent (Sakuragi et al. 1999; Montaño et al. 2003b).

5.1.2 Methylations

The BChls of green sulfur bacterial chlorosomes are methylated in the C-8^2 and C-12^1 positions (Fig. 3a). Although the absorption properties of the monomeric BChl are unaffected by these methylations, aggregates formed in vivo or in vitro by chlorosome BChls with these methylations absorb at a red-shifted wavelength compared to the unmethylated BChls. The genes encoding the enzymes responsible for the C-8^2 and C-12^1 methylation, *bchQ* and *bchR*, have been identified in *Chl. tepidum* (Frigaard et al. 2003, 2006; Gomez Maqueo Chew et al., manuscript in preparation). *Chl. tepidum* mutants that lack either the C-8^2 or C-12^1 methyltransferases, or both, have been characterized biochemically and physiologically (Gomez Maqueo Chew et al., manuscript in preparation; Frigaard et al. 2003, 2006). These studies have shown that increased methylation causes a red shift in the absorption maximum of the chlorosomes, a broadening of the absorption near-infrared absorption band, and an increased extinction coefficient of the aggregated BChl *c* (Gomez Maqueo Chew et al., manuscript in preparation). Moreover, strains lacking the ability to methylate BChl *c*, especially at the C-12 position, are unable to increase their cellular content of BChl *c* in response to low light intensity but retain a BChl *c* level per cell that is approximately threefold lower than that contained in the wild type when grown at low light intensity. The C-8^2 and C-12^1 methylations thus appear to be an extremely important component of the mechanism of adaptation to low light conditions. Although methylations at C-8^2 and C-12^1 are not observed in *Cfx. aurantiacus*, which interestingly was isolated from a very high light intensity environment from a mat at Yellowstone National Park (Pierson and Castenholz 1974), such methylations have been observed in *Chloronema* sp. UdG9001 (Gich et al. 2003), an organism that typically grows at very low light intensity. Similar results were obtained with a *bchU* mutant of *Chl. tepidum* that was unable to synthesize BChl *c* and only produced BChl *d* (Maresca et al. 2004). This mutant contained a reduced amount of chlorosome BChl relative to the wild type, had a lower in vivo molar extinction coefficient for its chlorosomes, and grew slower than the wild type at low light intensity.

5.2 Bacteriochlorophyll *a*

Chlorosomes contain a small amount of BChl *a*, which unlike BChl *c*, is bound to a protein, CsmA (Sect. 5.4.4). The ratio of BChl *a* to BChl *c* in whole cells of *Chl. tepidum* is typically about 0.03; most ($\sim 60\%$) of this BChl *a* is associated with the FMO protein, and the remainder is distributed between the chlorosomes ($\sim 30\%$) and the reaction centers ($\sim 10\%$) (Frigaard et al. 2003). Because the ratio of BChl *a* to BChl *c* in chlorosomes isolated from *Chl. tepidum* typically is about 0.01 (Frigaard et al. 2003), the BChl *a* component is hardly detectable in the absorption spectrum due to masking by BChl *c*. However, it is clearly detectable in fluorescence emission spectra, and many optical kinetic studies and other spectroscopic evidence show that this BChl *a* is an important intermediate species that participates in transferring excitation energy from the BChl antenna of the chlorosome to the reaction center (Blankenship et al. 1995; Blankenship and Matsuura 2003).

5.3 Carotenoids

In photosynthetic organisms in general, carotenoids function in light harvesting, photoprotection and, in some systems, structure stabilization (Frank and Cogdell 1998). In green bacteria, carotenoids are present both in the chlorosomes and in the cytoplasmic membrane.

5.3.1 Types of Carotenoids

Green sulfur bacteria that synthesize BChl *c* or *d* contain only monocyclic carotenoids, primarily chlorobactene (φ,ψ-carotene), whereas BChl *e*-containing green sulfur bacteria contain almost entirely dicyclic carotenoids, primarily isorenieratene (φ,φ-carotene) and β-isorenieratene (β,φ-carotene) (Takaichi 1999). The significance of this correlation is not yet known. The carotenoids of green sulfur bacteria are unusual in having an aromatic φ end group (Fig. 3b). The only other known organisms synthesizing carotenoids with φ end groups are certain actinomycetes (Britton 1998). The phylogenetically divergent, BChl *c*-containing green sulfur bacterium *Chloroherpeton thalassium* synthesizes γ-carotene as its principal carotenoid (Garrity and Holt 2001a).

Most of the carotenoids in green sulfur bacteria are located in the chlorosomes (about 90% in *Chl. tepidum*; Frigaard et al. 1997; Takaichi and Oh-oka et al. 1999). These carotenoids are not easily removed by detergent treatment of isolated chlorosomes (Frigaard et al. 1998). This observation suggests that the majority of the carotenoids are located in the chlorosome interior. The major carotenoid in cell extracts of the BChl *c*-containing *Chl. tepidum* is chlorobactene; minor carotenoids include γ-carotene, 1′,2′-dihydrochlorobactene, 1′,2′-dihydro-γ-carotene, OH-chlorobactene, and the laurate esters of both OH-chlorobactene glucoside and OH-γ-carotene glucoside (Takaichi et al. 1997). The laurate esters are presumably predominantly located in the cytoplasmic membrane and the reaction center (Takaichi and Oh-oka et al. 1999). The major carotenoids in cell extracts of the BChl *e*-containing *Chl. phaeobacteroides* strain 1549 are isorenieratene and β-isorenieratene; minor carotenoids are β-carotene, 7,8-dihydro-β-carotene, β-zeacarotene, and chlorobactene (Hirabayashi et al. 2004).

The major carotenoids in *Cfx. aurantiacus* grown under photosynthetic and anoxic conditions are γ-carotene, β-carotene, and fatty acid esters of OH-γ-carotene glucoside (Halfen et al. 1972; Takaichi 1999). Although all these carotenoid species are found in the chlorosomes, most of the total carotenoid content in the cells are found in the cytoplasmic membrane fraction (Schmidt 1980).

5.3.2 Functions of Carotenoids

In green bacteria, the light-harvesting function involves absorption of blue light with wavelengths from 400 to 550 nm and transfer of excitation energy to BChl *a* and the chlorosome BChls via singlet states. Steady-state fluorescence measurements on isolated chlorosomes indicate that the efficiency of energy transfer from carotenoids to BChl *c* is 50–80% (van Dorssen et al. 1986; Melø et al. 2000). Singlet excitation transfer from carotenoids to BChl *a* also occurs in the chlorosome baseplate (Montaño et al. 2003b; Frigaard et al. 2005). However, a more important role of carotenoids in chlorosomes may be quenching of both BChl *a* and BChl *c* excited triplet states, which occur with an efficiency of close to 100% (Melø et al. 2000; Carbonera et al. 2001; Arellano et al. 2000a, 2002).

The functions of carotenoids in green bacteria have also been investigated by chemical inhibition of carotenoid biosynthesis and by genetic manipulation of carotenoid biosynthesis. Foidl et al. (1997) isolated chlorosomes from *Cfx. aurantiacus* in which the content of colored carotenoids was reduced by 90% by growing the cells in the presence of 2-hydroxybiphenyl. The content of BChl *a* and CsmA in these chlorosomes was reduced to about half, whereas the content of two other chlorosome proteins, CsmM and CsmN, was essentially unchanged. A similar decrease in both BChl *a* and CsmA was found in chlorosomes isolated from carotenoid-deficient cells of *Chl. phaeobacteroides* strain CL1401 treated with 2-hydroxybiphenyl (Arellano et al. 2000b, 2001). In *crtB* and *crtP* mutants of *Chl. tepidum*, which are completely devoid of colored carotenoids, the cellular content of BChl *c* is similar to wild type, whereas the BChl *a* content is reduced by about one-third (Frigaard et al. 2004b). In addition, the growth of these mutants is significantly inhibited at both low and high light intensities. These results indicate that carotenoids have important structural, and possibly photoprotection, roles in the BChl *a*–CsmA complex in the chlorosome baseplate.

5.4 Proteins

Studies of the chlorosome proteins have lagged behind studies of the chlorosome pigments. Only in the past few years has a clear function been ascribed to a chlorosome protein, the BChl *a*-binding CsmA protein (Sect. 5.4.4).

5.4.1 Chlorosome Proteins in *Chl. tepidum*

A total of ten chlorosome proteins have been detected in *Chl. tepidum*: CsmA (CT1942), CsmB (CT2054), (CT2054), CsmC (CT1943), CsmD (CT2064), CsmE (CT2062), CsmF (CT1046), CsmH (CT1417), CsmI (CT1382), CsmJ (CT0651), and CsmX (CT0652), (Chung et al. 1994; Chung and Bryant 1996a,b; Vassilieva et al. 2001, 2002b). CsmA is a BChl *a*-binding protein (see Sect. 5.4.4), CsmI, CsmJ, and CsmX are iron-sulfur proteins (see section Sect. 5.4.5), but very little information is available on the role of the other chlorosome proteins.

Protease susceptibility mapping and chlorosome agglutination experiments have shown that all chlorosome proteins in *Chl. tepidum* are exposed on the surface of isolated chlorosomes (Chung and Bryant 1996a,b; Vassilieva et al. 2002b; Milks et al. 2005). Chemical cross-linking studies of the proteins in isolated chlorosomes have revealed that all proteins interact with at least one other protein (Li et al. 2005, 2006). CsmA, CsmC, and CsmD form homodimers and higher homomultimers. CsmA also interacts with CsmB, CsmD, and CsmF. CsmB, which is the second-most abundant chlorosome protein, can interact with all other chlorosome proteins, with the possible exception of CsmA and CsmF. Pre-CsmA has a different pattern of cross-linking than CsmA. Proteolytic processing at the *C*-terminus might modify the structure of this protein in such a way that it can be incorporated into the paracrystalline baseplate structure.

The ten chlorosome proteins in *Chl. tepidum* represent only four structural motifs: CsmA/CsmE, CsmB/CsmF, CsmC/CsmD, and the adrenodoxin-type [2Fe – 2S] ferredoxin domain of CsmI, CsmJ, and CsmX (Vassilieva et al. 2001, 2002b). Interestingly, CsmH seems to be a fusion of the CsmB/CsmF and CsmC/CsmD motifs, and the iron-sulfur proteins CsmI, CsmJ, and CsmX seem to be a fusion of the CsmA/CsmE and ferredoxin motifs. This suggests that some of the proteins have evolved by gene shuffling and duplication. An examination of the sequences reveals that these proteins are not particularly hydrophobic, and that they do not contain sequences that are predicted to form transmembrane alpha-helices. However, the proteins "behave" as though they are hydrophobic membrane proteins. Their structure is expected to be adapted to a monolayer lipid membrane rather than the bilayer cytoplasmic membrane. The chlorosome envelope may have some resemblance to the outer membrane of Gram negative bacteria that has a highly asymmetric bilayer structure. This could suggest that the chlorosome proteins will contain beta-sheet motifs, which would lead to a different membrane targeting inside the cell.

The only chlorosome protein that has not been eliminated by gene inactivation is CsmA, presumably because its function as BChl *a*-binding protein and chlorosome baseplate component is essential (Frigaard et al. 2004a). The BChl *c* absorption of chlorosomes from a *csmC* mutant of *Chl. tepidum* is slightly blue-shifted, and the chlorosomes are smaller than those of wild type. This suggests that CsmC may interact in some way with the BChl *c* aggregates (Frigaard et al. 2004a). CsmE is about 50% identical in sequence to CsmA and has a conserved histidine residue that could in principle also serve as a ligand to BChl *a*, but there is at present no experimental evidence to support this (Chung and Bryant 1996b). No obvious phenotype was detected in *Chl. tepidum* single-locus mutants lacking CsmB, CsmD, CsmE, CsmF, or CsmH (Frigaard et al. 2004a).

In order to study the roles of some of these proteins further, mutants lacking two, three or even four chlorosome proteins have been constructed (H. Li and D.A. Bryant, manuscript in preparation; Li 2006). It is possible to delete all three Fe-S proteins with little obvious effect under normal growth conditions; however, the absence of these proteins clearly reduces the ability to reverse energy transfer quenching after an exposure to oxygen (see Sect. 6.2). Mutants completely lacking the CsmC/CsmD or the CsmB/CsmF motifs have reduced BChl *c* contents per cell; for example, a mutant lacking CsmC, CsmD, CsmE, and CsmH has only about half the BChl *c* of wild type. When CsmC and CsmD are missing, the absorption maximum of the BChl *c* in chlorosomes is blue-shifted to 737 nm, but when CsmB and CsmF are missing, the absorption is slightly red-shifted to 756 nm relative to the wild type (750 nm). Electron microscopic analyses of these chlorosomes also revealed striking differences. Wild-type chlorosomes are 165 ± 60 nm long, are 60 ± 25 nm in length and have a length-to-width ratio of 2.7 ± 0.6. When CsmC and CsmD are missing, the chlorosomes have significantly reduced dimensions (length: 100 ± 30 nm; width: 45 ± 15 nm; ratio 2.1 ± 0.6). Further deletion of CsmH returned the ratio (2.4 ± 0.7) to nearly the wild-type value, however. Mutants lacking CsmB and CsmF were about 25% longer and 15% wider than wild-type chlorosomes (length 210 ± 80 nm; width 70 ± 20 nm; ratio 3.1 ± 0.6). Further deletion of CsmH caused the dimensions to again be more similar to wild-type chlorosomes. These studies indicate that the shape and absorption properties of chlorosomes depend in a complex way on the CsmB, CsmC, CsmD, CsmF, and CsmH proteins (H. Li and D.A. Bryant, manuscript in preparation; Li 2006).

The multitude of apparently redundant chlorosome proteins may be related to the unusual structure of the chlorosome. Unlike other pigment–protein complexes, the chlorosome is a flexible structure in the sense that there is no fixed stoichiometry between BChl *c* and the protein components. Besides the BChl *a*-binding CsmA protein and the three iron-sulfur proteins, there are six other chlorosome proteins in *Chl. tepidum*, which are conserved with approximately the same number of copies in all green sulfur bacteria examined. Based on cross-linking studies, we know that these proteins interact and that they are not organized in a paracrystalline manner like CsmA. None of these proteins has a demonstrated prosthetic group, and single-locus mutants show little if any phenotype. However, multiple-loci mutants have significant alterations in the size and organization of the BChl *c* aggregates. We propose that the multitude of these proteins ensures that the protein–lipid envelope is a flexible structure by "breaking order" and thus preventing a rigid paracrystalline organization like that of CsmA in the baseplate.

5.4.2 Chlorosome Proteins from Other Green Sulfur Bacteria

Gel electrophoresis and immunoblotting experiments have shown that the same suite of ten chlorosome proteins are present in *Chl. tepidum*, *Chl. vibrioforme* strain 8327, and *Chl. phaeobacteroides* strain 1549 (Frigaard et al. 2001; Vassilieva et al. 2002a). Genome sequence analyses of the nine strains of green sulfur bacteria currently available (Table 1; http://www.jgi.doe.gov/) show a similar trend. Both *csmA* and *csmC*, which encode the only chlorosome proteins with a demonstrated role in affecting pigment organization, are present as single-copy genes in all strains. The *csmD* gene is only found in two strains. The *csmB*, *csmE*, *csmF*, and *csmH* genes are present in all strains; the *csmB* and *csmE* genes are even present in multiple copies in some strains. At least two of the three iron-sulfur proteins (CsmI, CsmJ, CsmX) are also produced by all strains. It appears that the structural constraints for the CsmB/CsmF proteins are more relaxed or that these proteins have fewer interactions with other molecules, since these proteins are the least conserved of the protein motif families. This is somewhat surprising because CsmB, which is the second-most abundant chlorosome protein, appears from chemical cross-linking experiments to interact with most other chlorosome proteins (Li et al. 2005). It is also somewhat surprising that CsmB, CsmE, CsmF, and CsmH are also conserved in all strains, since mutants lacking one of these proteins have no detectable defects in chlorosome formation and growth rate (Frigaard et al. 2004a), although mutants lacking several of these chlorosome proteins have obvious defects in chlorosome formation and growth rate (Li et al., manuscript in preparation). One possible conclusion is that, because of its critical importance to the survival of the organism, the chlorosome has evolved to be a highly robust structure that can withstand the loss of any single protein component while retaining nearly all of its functionality.

5.4.3
Chlorosome Proteins in *Cfx. aurantiacus*

It is interesting that the same four structural motif families found in chlorosomes of *Chl. tepidum* describe the proteins of the chlorosomes of *Cfx. aurantiacus*. In addition to a 5.7-kDa CsmA (P09928), chlorosomes from *Cfx. aurantiacus* contain two proteins of unknown function, 11-kDa CsmM (AAG15234) and 18-kDa CsmN (AAG15235) (Feick and Fuller 1984; Niedermeier et al. 1994). CsmM and CsmN are distantly related to the CsmC and CsmD proteins in *Chl. tepidum* (Vassilieva et al. 2002b), and these proteins are also located in the chlorosome envelope (Wullink et al. 1991). Analysis of isolated chlorosomes by SDS-PAGE and *N*-terminal sequencing of distinct protein bands, in combination with genome sequence information, has potentially identified three additional chlorosome proteins: CsmO (AAN85641), CsmP (AAG15237), and CsmY (ZP_00017757) (Frigaard et al. 2001; Vassilieva et al. 2002b). CsmO is clearly similar to CsmB and CsmF of *Chl. tepidum*. CsmY has a putative iron-sulfur cluster binding motif and strong sequence similarity to CsmX from *Chl. tepidum*. Based on its sequence similarity to CsmM, another potential chlorosome protein, CsmQ (ZP_00018889), has also been tentatively identified in the *Cfx. aurantiacus* genome.

5.4.4
The BChl *a*-Binding CsmA Protein

The most abundant protein species in chlorosomes from both green sulfur bacteria and green filamentous bacteria is the CsmA protein (Chung et al. 1994). In *Chl. tepidum*, the 6.2-kDa CsmA accounts for about half of the protein present in the chlorosome (Frigaard et al. 2004a). CsmA was initially thought to be a BChl *c*-binding protein (Feick and Fuller 1984; Wechsler et al. 1985; Wagner-Huber et al. 1988). However, recent evidence clearly shows that CsmA is a BChl *a*-binding protein in both *Cfx. aurantiacus* (Sakuragi et al. 1999; Montaño et al. 2003b) and *Chl. tepidum* (Bryant et al. 2002; Frigaard et al. 2004a, 2005). CsmA binds one BChl *a* molecule and probably one or two carotenoid molecules per monomer (Sakuragi et al. 1999; Bryant et al. 2002; Montaño et al. 2003b; Frigaard et al. 2005). CsmA is highly conserved within the green sulfur bacteria (more than 85% amino acid sequence identity within the genome-sequenced strains); the CsmA protein of the green filamentous bacteria is more distantly related to that of green sulfur bacteria (about 30% sequence identity). The protein has a single conserved histidine residue that most likely serves as the ligand to BChl *a*.

The *csmA* gene of *Chl. tepidum* encodes a 79-residue polypeptide denoted pre-CsmA, which is detected in small amounts in isolated chlorosomes. Most of this protein retains its *N*-terminal methionine but 20 amino acid residues are removed by *C*-terminal processing in the mature form of the protein (Chung et al. 1994; Persson et al. 2000). Chemical cross-linking studies of isolated *Chl. tepidum* chlorosomes show that the CsmA protein forms homooligomers up to at least octamers (Bryant et al. 2002; Li et al. 2005). This oligomerization of CsmA is independent of BChl *c*, because it is also observed in carotenosomes, which are chlorosomes devoid of BChl *c* (Frigaard et al. 2005). The CsmA–CsmA interactions are very strong – CsmA dimers are observed even on denaturing SDS polyacrylamide gels. Cross-linking studies also suggests that CsmA interacts with the FMO protein in *Chl. tepidum* (Li et al. 2005), consistent with the baseplate localization of CsmA in the model shown in Fig. 2. Protease mapping experiments have shown that the *N*-terminal region of CsmA is exposed on the surface of isolated chlorosomes (Milks et al. 2005), and thus it is probably this region of CsmA that interacts with the FMO protein.

5.4.5 The Iron-Sulfur Cluster Proteins CsmI, CsmJ, and CsmX

Three of the chlorosomes proteins in *Chl. tepidum*, CsmI, CsmJ, and CsmX are iron-sulfur proteins (Vassilieva et al. 2001, 2002b). Cross-linking studies show that CsmI and CsmJ form homodimers and heterodimers, and thus probably form $CsmI_2CsmJ_2$ heterotetramers (Li et al. 2005). The content of CsmX is about tenfold lower than that of either CsmI or CsmJ (Frigaard et al. 2004a). The redox midpoint potentials of each cluster have been determined in chlorosomes from wild type and mutants lacking one or two of these proteins (T.W. Johnson, H. Li, N.-U. Frigaard, D.A. Bryant, J.H. Golbeck, manuscript in preparation). In the wild type, CsmI and CsmJ have midpoint potentials of about – 346 mV and + 90 mV, respectively. However, in mutants lacking CsmJ, the CsmI cluster has a midpoint potential of about – 200 mV, and in mutants lacking CsmI, the CsmJ cluster has a midpoint potential of about 0 mV. Thus, it appears that the interaction of CsmI and CsmJ affects the midpoint potential of their iron-sulfur clusters.

A *csmI csmJ csmX* triple mutant is clearly defective in the inactivation of the redox-dependent BChl *c* fluorescence quencher, which is produced in chlorosomes of cells that have been exposed to oxygen (H. Li, N.-U. Frigaard, D.A. Bryant, manuscript in preparation; Li 2006). The iron-sulfur proteins of chlorosomes thus apparently function in restoring energy transfer from the chlorosome to the reaction center after an oxic-to-anoxic transition by reducing the chlorosome-associated quencher, most likely chlorobiumquinone (Frigaard et al. 1997; Blankenship and Matsuura 2003; Fig. 2).

The genome of *Cfx. aurantiacus* encodes a homolog of these iron-sulfur proteins, denoted CsmY, which appears to be present in the chlorosomes (Sect. 5.4.3; Frigaard et al. 2001).

5.5 Isoprenoid Quinones

The only isoprenoid quinone found in green filamentous bacteria is menaquinone, mostly menaquinone-10 (Hale et al. 1986; Hanada and Pierson 2002). Green sulfur bacteria contain menaquinone-7 and two additional quinones that are unique to green sulfur bacteria, chlorobiumquinone (1′-oxomenaquinone-7) and 1′-hydroxymenaquinone-7 (Powls and Redfearn 1969; Frigaard et al. 2003). 1′-Hydroxymenaquinone-7 is usually found in small amounts and probably is an intermediate in the biosynthesis of chlorobiumquinone from menaquinone. In one experiment, 23 mmol of menaquinone-7 and 63 mmol of chlorobiumquinone per mole of BChl *c* were found in *Chl. tepidum* chlorosomes (46% and 90% of the total cellular content, respectively), whereas 126 mmol of menaquinone-10 per mole of BChl *c* was found in *Cfx. aurantiacus* chlorosomes (33% of the total cellular content) (Frigaard et al. 1997). These isoprenoid quinones are not easily removed by detergent treatment of isolated chlorosomes, suggesting that the majority of the quinones are located in the chlorosome interior (Frigaard et al. 1998). The function of menaquinone in chlorosomes is unknown. In *Chlorobium*-type chlorosomes chlorobiumquinone appears to function as a redox-dependent quencher of BChl *c* excitation (Frigaard et al. 1997; Sect. 6.2).

5.6 Lipids

The major polar lipids in chlorosomes are glycolipids and phospholipids (Schmidt 1980; Holo et al. 1985). Schmidt (1980) found a content of 0.30 g of glycolipids and 0.18 g of phospholipids per g of BChl *c* in *Cfx. aurantiacus* chlorosomes and 0.24 g of glycolipids and 0.07 g of phospholipids per g of BChl *c* in chlorosomes from *Chlorobium* sp. strain 6230 (= strain DSMZ 249). In contrast, Miller et al. found only a total of 47 mmol of polar lipids per mole of BChl *c* (about 0.05 w/w) in *Chl. tepidum* chlorosomes with a distribution of 55% glycolipids, 16% phosphatidylglycerol, 12% aminoglycosphingolipid, 8% phosphatidylethanolamine, and 3% diphosphatidylglycerol (M. Miller, P.G. Sørensen, R.P. Cox, manuscript in preparation). The large difference in total polar lipids between the two mentioned *Chlorobium* strains is probably in part due to differences in chlorosome preparation methodology. It has been reported that the major glycolipid in green sulfur bacteria is monogalactosyldiacylglycerol (MGDG). However, Miller et al. found that the major glycolipid in *Chl. tepidum* chlorosomes is a 1-diglycosyl-2,3-diacylglycerol tentatively identified as a 1-rhamnosylgalactosyl-2,3-diacylglycerol (RGDG). *Cfx. aurantiacus* contains both MGDG and digalactosyldiacylglycerol (DGDG) (Knudsen et al. 1982; Holo et al. 1985). Glycolipids are also found in cyanobacteria but are rare in other phototrophic bacteria. (It should be noted that Jensen et al. (1991), who identified the unusual aminoglycosphingolipid in *Chlorobium* sp. strain 6230, concluded that this lipid is located exclusively in the cytoplasmic membrane of this organism. These results differ from those of Miller et al. (manuscript in preparation), who found the aminoglycosphingolipid both in the cytoplasmic membrane and the chlorosomes.)

In addition to polar lipids, the two thermophilic chlorosome-containing green bacteria, *Chl. tepidum* and *Cfx. aurantiacus*, contain wax esters (Knudsen et al. 1982; Beyer et al. 1983; Wahlund et al. 1991). Wax esters are esters of long-chain fatty acids and long-chain fatty alcohols and are rarely found in bacteria (Knudsen et al. 1982; see also Vol. 1 in this series, Chap. 6 by M. Wältermann and A. Steinbüchel). The content of wax esters in *Chl. tepidum* chlorosomes is 59 mmol per mole of BChl *c* (about 0.033 w/w), most of which is tetradecanoic acid esterified with hexadecanol (M. Miller, P.G. Sørensen, R.P. Cox, manuscript in preparation). Chlorosomes contain most (80%) of the wax esters in the *Chl. tepidum* cells. Wax esters are not found in mesophilic green sulfur bacteria. It is therefore possible that wax esters play a specialized role in stabilizing membranes and hydrophobic domains in the chlorosomes under high temperature conditions.

5.7 Other Chlorosome Components

A hydrophobic compound, denoted component F, of unknown structure and function has been identified in *Chl. tepidum* (Frigaard et al. 2004b). At least some of it appears to be located in the chlorosomes, and its content is roughly ten times less than that of BChl *a*. Component F has absorption peaks at 285, 325, and 453 nm in a methanol-based HPLC solvent mixture. The chromophore may be related to a carotenoid, but the molecule itself is not a carotenoid because it is present in a phytoene synthase-deficient *crtB* mutant. This compound may be related to the outer-membrane flexirubin-type pigments found in some *Cytophaga*, *Flavobacterium*, and *Flexibacter* species.

6 Energy Transfer in Chlorosomes

The excitation energy of the antenna pigments in the cell is transferred through successive pigment species to the reaction center. Each successive pigment species has progressively red-shifted absorption and fluorescence spectra and thus decreasing energy levels, which causes a "funneling" of the excitations into the reaction center (Blankenship 2002). The transfer of excitation energy between the pigment species takes place on a picosecond timescale and is described in detail elsewhere (Blankenship et al. 1995; Blankenship and Matsuura 2003).

6.1 From Chlorosome to Reaction Center

BChl *c*, *d*, or *e* in the chlorosome is the primary antenna in green bacteria. Excitation energy from BChl *c* (maximum absorption in the range 740–760 nm) is transferred to BChl *a* in the chlorosome baseplate (maximum absorption around 795 nm). In green sulfur bacteria the energy is transferred via the BChl *a*-containing FMO protein (maximum absorption at 809 nm) to the BChl *a* antenna in the reaction center. In green filamentous bacteria the energy is transferred via the membrane-integral B808–866 complex (maximum absorption at 808 and 866 nm) to the BChl *a* antenna in the reaction center.

6.2 Redox-Dependent Regulation in Green Sulfur Bacteria

Although green sulfur bacteria are strictly anaerobic, they may transiently encounter O_2 in their natural environments. O_2 is probably toxic to green sulfur bacteria because of formation of reactive oxygen species derived from non-specific reactions between O_2 and strong cellular reductants like those generated by the photosynthetic reaction center. The *Chl. tepidum* genome encodes many enzymes that appear to protect against such reactive oxygen species, including superoxide dismutase (Eisen et al. 2002). Another protective measure appears to prevent generation of reduced cellular reductants by shutting down the excitation energy transfer in the chlorosome under aerobic conditions. This mechanism is not found in the green filamentous bacterium *Cfx. aurantiacus*. *Cfx. aurantiacus* may not benefit from such a mechanism because its reaction centers do not generate strong cellular reductants and because the organism can easily adapt to aerobic conditions.

The quenching mechanism of chlorosome BChl *c* excitations is observed both in intact cells and in isolated chlorosomes (van Dorssen et al. 1986; Wang et al. 1990). It is active under oxidizing conditions and inactive under reducing conditions. Although BChl *c* radicals appear to cause a similar redox-dependent quenching in isolated chlorosomes (van Noort et al. 1997), it has been shown that the quenching mechanism is mostly due to the chlorobiumquinone (Frigaard et al. 1997, 1998). The midpoint potential of the quenching in isolated chlorosomes is about – 150 mV at pH 7 and appears to be a one-electron, one-proton reaction (Blankenship et al. 1990,

1993). The midpoint potential of the two-electron, two-protein reduction of chlorobiumquinone is + 39 mV at pH 7 whereas that of menaquinone is – 81 mV at pH 7 (Redfearn and Powls 1968). It is therefore proposed that the quenching mechanism in the chlorosomes involves a chlorobiumquinone–BChl *c* complex, which is inactivated by reduction of the quinone to the semiquinone (Frigaard et al. 1998). Chlorobiumquinone is probably more suitable than menaquinone as a quencher because the midpoint potential of the chlorobiumquinone–BChl *c* complex may be in a physiologically more suitable range.

The physiological effect of the chlorosome quenching mechanism has been demonstrated by flash-induced oxidation of cytochrome *c* in intact cells (Frigaard and Matsuura 1999; Frigaard et al. 1999). The primary electron donors to the oxidized photosynthetic reaction center are various cytochrome *c*, whose oxidation can be measured by time-resolved absorption spectroscopy. The measurements showed that excitation of BChl *c* in *Chl. tepidum* causes cytochrome *c* oxidation under anaerobic conditions, but not under aerobic conditions. Since the quenching mechanism only acts on chlorosome BChl *c*, flash-induced cytochrome *c* oxidation by excitation of the much smaller BChl *a* antennae is identical under aerobic and anaerobic conditions. In *Cfx. aurantiacus*, flash-induced cytochrome *c* oxidation is identical under aerobic and anaerobic conditions for both BChl *c* excitation and BChl *a* excitation. However, if certain soluble, low-molecular synthetic quinones are added to the cells, cytochrome *c* oxidation by BChl *c* excitation becomes sensitive to redox conditions like in *Chl. tepidum* (Frigaard et al. 1999).

When green sulfur bacterial cells return to anoxic conditions after being exposed to O_2, the chlorosome quencher is slowly reduced by cellular reductants. *Chl. tepidum* mutants lacking the iron-sulfur chlorosome proteins CsmI or CsmJ appear to be deficient in this restoration mechanism, and it thus appears that the CsmI CsmJ proteins reduce the quencher inside the chlorosome (H. Li, N.-U. Frigaard, D.A. Bryant, manuscript in preparation; Li 2006). It still remains to be demonstrated that the redox-dependent quenching mechanism increases the survival rate of green sulfur bacteria under aerobic conditions.

7
Evolution of Chlorosomes

The core components of photosynthesis (reaction centers, antennae, and pigments) observed in extant phototrophs clearly are not derived from a single ancestral organism (Blankenship 2002; Green 2003). For example, the chlorosome proteins and the BChl *c* biosynthetic enzymes are clearly closely related in green sulfur bacteria and green filamentous bacteria, although the reaction centers, carotenoid biosynthetic enzymes, and peripheral antennae

(B808–866 in green filamentous bacteria, FMO protein in green sulfur bacteria) are not. This strongly indicates that chlorosomes were laterally transferred between photosynthetic members of these two lineages. Alternatively, if chlorosomes and reaction centers were laterally transferred simultaneously, the reaction centers of one lineage could have been subsequently exchanged during evolution. It is not clear in which lineage chlorosomes first evolved. However, it seems very likely that BChl *c* biosynthesis evolved from the ancestral BChl *a*/Chl *a* biosynthetic pathway by gene duplication and divergence and the recruitment of a few new genes from other pathways. The acquisition of BChl *c*/*d*/*e* provided these organisms with an immense competitive advantage in low light environments. It has been proposed that the FMO protein may have evolved from the reaction center core subunit in green sulfur bacteria (Olson and Raymond 2003).

The BChl *c* biosynthetic pathway has been studied intensively in *Chl. tepidum* since the genome sequence became available (Frigaard et al. 2003, 2006; Frigaard and Bryant 2004). All enzymes identified so far, except BchU, appear to have been generated by duplication of genes involved in BChl *a*/Chl *a* biosynthesis. It seems obvious that BChl *a* biosynthesis must have preceded BChl *c* biosynthesis because BChl *c* only functions as an antenna pigment in chlorosomes, whereas Chl *a* and BChl *a* and its derivatives function both as antenna pigments and as essential electron transfer cofactors in the reaction center.

Chlorophyllide *a* is probably the last common intermediate in the biosynthetic pathway of BChl *c*, BChl *a*, and Chl a_{PD} in *Chl. tepidum*. Presumably the C-13^2 methylcarboxyl group of chlorophyllide *a* is removed by one or two unidentified enzymes to yield 3-vinyl-bacteriochlorphyllide *d*. The 3-vinyl-bacteriochlorophyllide intermediate is converted by a C-8^2 methyltransferase (BchQ), a C-12^1 methyltransferase (BchR), a hydratase (BchV), and a BChl *c*/*d* synthase (BchK) to yield BChl *c* or *d*. All these enzymes are BChl *c*/*d*-specific and are related to other chlorophyll biosynthetic enzymes: BchQ and BchR to the BchE oxidative cyclase, BchV to the BchF hydratase, and BchK to the BchG BChl *a* synthase. BChl *c*/*d* biosynthesis is also dependent on a BChl *c*/*d*-specific isozyme (BchS) of the large subunit of magnesium chelatase (BchH). The only difference between BChl *c* and BChl *d* biosynthesis is the action of a C-20 methyltransferase (BchU), which probably acts on the porphyrin intermediates prior to esterification (Maresca et al. 2004). BchU is related to the CrtF methyltransferase in carotenoid biosynthesis in purple bacteria.

BChl *e* is probably formed by the action of an unidentified enzyme that converts the C-7 methyl group into a formyl group. This enzyme could also be a radical SAM enzyme related to the BchE oxidative cyclase.

This scenario for the evolution of the biosynthetic pathway for BChl *c*/*d*/*e* follows a typical model for the evolution of new functions in biology: gene duplication followed by divergence. Removal of the methylcarboxyl group at C-13^2 is likely to follow a similar pattern. A methylesterase alone might be sufficient to effect the removal of this substituent, since spontaneous decarboxylation of the resulting product is known to occur in similar molecules during chlorophyll degradation and turnover in plants (Frigaard et al. 2006).

8
Concluding Remarks

It is obvious that chlorosomes are the largest and most efficient light-harvesting antenna structures identified so far and that these antennae allow the organisms that harbor them to grow phototrophically in near darkness. It will be interesting to obtain more information on the green sulfur bacteria identified on the black smoker at the bottom of the ocean, where these organisms supposedly grow phototrophically using their chlorosomes to convert geothermal radiation to excitation energy (Beatty et al. 2005). If this is true, it is probably the first example of photosynthesis in a natural environment that is not dependent on solar radiation. We only know of two groups of organisms that contain chlorosomes, the green sulfur bacteria and the green filamentous bacteria, which have quite different evolutionary histories. It therefore seems obvious that chlorosomes have been laterally transferred between these two groups. It remains to be determined if chlorosomes are present in other yet-uncharacterized phototrophic organisms. It is possible that the recent approach of metagenomics, when applied to DNA isolated from photosynthetic microbial mats, will reveal novel organisms that contain chlorosomes (Tringe and Rubin 2005).

We know very little about how chlorosomes are synthesized in cells. Even though significant effort has been devoted to understanding the structure and mechanism of energy transfer in the chlorosome BChl aggregates, still much remains to be understood. Are the BChl *c* molecules arranged in rods or lamellae? How are the other chlorosome components – carotenoids, BChl *a*, quinones, lipids, and proteins – organized, and what are their precise functions? The mechanism and precise function of the redox-dependent energy transfer in chlorosome is also still not clear. Structural information on the relatively simple BChl *a*-containing CsmA protein at the atomic level may give biophysicists new insights into how photosynthesis works. It is also possible that the idea of mimicking the self-assembling properties of chlorosome BChl in artificial antennas for use in nanotechnological applications may succeed.

Acknowledgements This work was supported by grant DE-FG02-94ER20137 from the US Department of Energy to D.A.B. and grant 21-04-0463 from The Danish Natural Science Research Council to N.-U. F.

References

Airs RL, Keely BJ (2002) Atmospheric pressure chemical ionisation liquid chromatography/mass spectrometry of bacteriochlorophylls from Chlorobiaceae: characteristic fragmentations. Rapid Comm Mass Spectrom 16:453–461

Arellano JB, Melø TB, Borrego CM, Garcia-Gil J, Naqvi KR (2000a) Nanosecond laser photolysis studies of chlorosomes and artificial aggregates containing bacteriochlorophyll *e*: evidence for the proximity of carotenoids and bacteriochlorophyll *a* in chlorosomes from *Chlorobium phaeobacteroides* strain CL1401. Photochem Photobiol 72:669–675

Arellano JB, Psencik J, Borrego CM, Ma Y-Z, Guyoneaud R, Garcia-Gil J, Gillbro T (2000b) Effect of carotenoid biosynthesis inhibition on the chlorosome organization in *Chlorobium phaeobacteroides* strain CL1401. Photochem Photobiol 71:715–723

Arellano JB, Borrego CM, Martínez-Planells A, Garcia-Gil LJ (2001) Effect of carotenoid deficiency on cells and chlorosomes of *Chlorobium phaeobacteroides*. Arch Microbiol 175:226–233

Arellano JB, Melø TB, Borrego CM, Naqvi KR (2002) Bacteriochlorophyll e monomers, but not aggregates, sensitize singlet oxygen: implications for a self-photoprotection mechanism in chlorosomes. Photochem Photobiol 76:373–380

Balaban ST (2004) Light-harvesting nanostructures. In: Nalwa HS (ed) Encyclopedia of nanoscience and nanotechnology, vol 4. American Scientific, p 505–559

Balaban ST (2005) Tailoring porphyrins and chlorins for self-assembly in biomimetic artificial antenna systems. Acc Chem Res 38:612–623

Beatty JT, Overmann J, Lince MT, Manske AK, Lang AS, Blankenship RE, Van Dover CL, Martinson TA, Plumley FG (2005) An obligately photosynthetic bacterial anaerobe from a deep-sea hydrothermal vent. PNAS 102:9306–9310

Beyer P, Falk H, Kleining H (1983) Particulate fractions from *Chloroflexus aurantiacus* and distribution of lipids and polyprenoid forming activities. Arch Microbiol 134:60–63

Blankenship RE (2002) Molecular mechanisms of photosynthesis. Blackwell Science, Oxford

Blankenship RE, Matsuura K (2003) Antenna complexes from green photosynthetic bacteria. In: Green BR, Parson WW (eds) Light-harvesting antennas in photosynthesis. Kluwer Academic, Dordrecht, p 195–217

Blankenship RE, Olson JM, Miller M (1995) Antenna complexes from green photosynthetic bacteria. In: Blankenship RE, Madigan MT, Bauer CE (eds) Anoxygenic photosynthetic bacteria. Kluwer Academic, Dordrecht, p 399–435

Borrego CM, Garcia-Gil LJ (1994) Separation of bacteriochlorophyll homologs from green photosynthetic sulfur bacteria by reversed-phase HPLC. Photosynth Res 41:157–164

Borrego CM, Gerola PD, Miller M, Cox RP (1999) Light intensity effects on pigment composition and organisation in the green sulfur bacterium *Chlorobium tepidum*. Photosynth Res 59:159–166

Britton G (1998) Overview of carotenoid biosynthesis. In: Britton G, Liaaen-Jensen S, Pfander H (eds) Carotenoids: biosynthesis and metabolism, vol 3. Birkhäuser, Basel, p 13–147

Bryant DA, Vassilieva EV, Frigaard N-U, Li H (2002) Selective protein extraction from *Chlorobium tepidum* chlorosomes using detergents. Evidence that CsmA forms multimers and binds bacteriochlorophyll *a*. Biochemistry 41:14403–14411

Carbonera D, Bordignon E, Giacometti G, Agostini G, Vianelli A, Vannini C (2001) Fluorescence and absorption detected magnetic resonance of chlorosomes from green bacteria *Chlorobium tepidum* and *Chloroflexus aurantiacus* – a comparative study. J Phys Chem B 105:246–255

Chung S, Frank G, Zuber H, Bryant DA (1994) Genes encoding two chlorosome proteins from the green sulfur bacteria *Chlorobium vibrioforme* strain 8327D and *Chlorobium tepidum*. Photosynth Res 41:261–275

Chung S, Bryant DA (1996a) Characterization of *csmB* genes from *Chlorobium vibrioforme* 8327D and *Chlorobium tepidum* and overproduction of the *Chlorobium tepidum* CsmB protein in *Escherichia coli*. Arch Microbiol 166:234–244

Chung S, Bryant DA (1996b) Characterization of the *csmD* and *csmE* genes from *Chlorobium tepidum*. The CsmA, CsmC, CsmD, and CsmE proteins are components of the chlorosome envelope. Photosynth Res 50:41–59

Cohen-Bazire G, Pfennig N, Kunisawa R (1964) The fine structure of green bacteria. J Cell Biol 22:207–225

Eisen JA, Nelson KE, Paulsen IT, Heidelberg JF, Wu M, Dodson RJ, Deboy R, Gwinn ML, Nelson WC, Haft DH, Hickey EK, Peterson JD, Durkin AS, Kolonay JL, Yang F, Holt I, Umayam LA, Mason T, Brenner M, Shea TP, Parksey D, Nierman WC, Feldblyum TV, Hansen CL, Craven MB, Radune D, Vamathevan J, Khouri H, White O, Gruber TM, Ketchum KA, Venter JC, Tettelin H, Bryant DA, Fraser CM (2002) The complete genome sequence of *Chlorobium tepidum* TLS, a photosynthetic, anaerobic, green-sulfur bacterium. PNAS 99:9509–9514

Feick RG, Fuller RC (1984) Topography of the photosynthetic apparatus of *Chloroflexus aurantiacus*. Biochemistry 23:3693–3700

Feick R, Ertlmaier A, Ermler U (1996) Crystallization and X-ray analysis of the reaction center from the thermophilic green bacterium *Chloroflexus aurantiacus*. FEBS Lett 396:161–164

Foidl M, Golecki JR, Oelze J (1997) Phototrophic growth and chlorosome formation in *Chloroflexus aurantiacus* under conditions of carotenoid deficiency. Photosynth Res 54:219–226

Foidl M, Golecki JR, Oelze J (1998) Chlorosome development in *Chloroflexus aurantiacus*. Photosynth Res 55:109–114

Frank HA, Cogdell RJ (1996) Carotenoids in photosynthesis. Photochem Photobiol 63:257–264

Frigaard N-U, Bryant DA (2001) Chromosomal gene inactivation in the green sulfur bacterium *Chlorobium tepidum* by natural transformation. Appl Environ Microbiol 67:2538–2544

Frigaard N-U, Bryant DA (2004) Seeing green bacteria in a new light: genomics-enabled studies of the photosynthetic apparatus in green sulfur bacteria and filamentous anoxygenic phototrophic bacteria. Arch Microbiol 182:265–276

Frigaard NU, Matsuura K (1999) Oxygen uncouples light absorption by the chlorosome antenna and photosynthetic electron transfer in the green sulfur bacterium *Chlorobium tepidum*. Biochim Biophys Acta 1412:108–117

Frigaard N-U, Takaichi S, Hirota M, Shimada K, Matsuura K (1997) Quinones in chlorosomes of green sulfur bacteria and their role in the redox-dependent fluorescence studied in chlorosome-like bacteriochlorophyll *c* aggregates. Arch Microbiol 167:343–349

Frigaard N-U, Matsuura K, Hirota M, Miller M, Cox RP (1998) Studies of the location and function of isoprenoid quinones in chlorosomes from green sulfur bacteria. Photosynth Res 58:81–90

Frigaard NU, Tokita S, Matsuura K (1999) Exogenous quinones inhibit photosynthetic electron transfer in *Chloroflexus aurantiacus* by specific quenching of the excited bacteriochlorophyll *c* antenna. Biochim Biophys Acta 1413:108–116

Frigaard N-U, Vassilieva EV, Li H, Milks KJ, Zhao J, Bryant DA (2001) The remarkable chlorosome. In: Proceedings of the 12th international congress on photosynthesis (PS2001). CSIRO, Melbourne, article S1–S3

Frigaard N-U, Voigt GD, Bryant DA (2002) *Chlorobium tepidum* mutant lacking bacteriochlorophyll *c* made by inactivation of the *bchK* gene, encoding bacteriochlorophyll *c* synthase. J Bacteriol 184:3368–3376

Frigaard N-U, Gomez Maqueo Chew A, Li H, Maresca JA, Bryant DA (2003) *Chlorobium tepidum*: insights into the structure, physiology, and metabolism of a green sulfur bacterium derived from the complete genome sequence. Photosynth Res 78: 93–117

Frigaard N-U, Li H, Milks KJ, Bryant DA (2004a) Nine mutants of *Chlorobium tepidum* each unable to synthesize a different chlorosome protein still assemble functional chlorosomes. J Bacteriol 186:646–653

Frigaard N-U, Maresca JA, Yunker CE, Jones AD, Bryant DA (2004b) Genetic manipulation of carotenoid biosynthesis in the green sulfur bacterium *Chlorobium tepidum*. J Bacteriol 186:5210–5220

Frigaard N-U, Sakuragi Y, Bryant DA (2004c) Gene inactivation in the cyanobacterium *Synechococcus* sp. PCC 7002 and the green sulfur bacterium *Chlorobium tepidum* using *in vitro*-made DNA constructs and natural transformation. In: Carpentier R (ed) Photosynthesis research protocols. Methods in molecular biology, vol 274. Humana, Totowa, p 325–340

Frigaard N-U, Li H, Martinsson P, Das SK, Frank HA, Aartsma TJ, Bryant DA (2005) Isolation and characterization of carotenosomes from a bacteriochlorophyll *c*-less mutant of *Chlorobium tepidum*. Photosynth Res 86:101–111

Frigaard N-U, Gomez Maqueo Chew A, Maresca JA, Bryant DA (2006) Bacteriochlorophyll biosynthesis in green bacteria. In: Grimm B, Porra R, Rüdiger W, Scheer H (eds) Chlorophylls and bacteriochlorophylls: biochemistry, biophysics, functions and applications. Advances in photosynthesis and respiration, vol 25. Springer, Dordrecht, The Netherlands, p 201–221

Garrity GM, Holt JG (2001a) Phylum BXI. Chlorobi phy. nov. In: Boone DR, Castenholz RW (eds) Bergey's manual of systematic bacteriology, vol 1, 2nd edn. Springer, Heidelberg Berlin New York, p 601–623

Garrity GM, Holt JG (2001b) Phylum BVI. Chloroflexi phy. nov. In: Boone DR and Castenholz RW (eds) Bergey's manual of systematic bacteriology, vol 1, 2nd edn. Springer, Heidelberg Berlin New York, p 427–446

Gerola PD, Olson JM (1986) A new bacteriochlorophyll a-protein complex associated with chlorosomes of green sulfur bacteria. Biochim Biophys Acta 848:69–76

Gibson J, Pfennig N, Waterbury JB (1984) *Chloroherpeton thalassium* gen nov et spec nov, a non-filamentous, flexing and gliding green sulfur bacterium. Arch Microbiol 138: 96–101

Gich F, Airs RL, Danielsen M, Keely BJ, Abella CA, Garcia-Gil J, Miller M, Borrego CM (2003) Characterization of the chlorosome antenna of the filamentous anoxygenic phototrophic bacterium *Chloronema* sp strain UdG9001. Arch Microbiol 180: 417–426

Golecki JR, Oelze J (1987) Quantitative relationship between bacteriochlorophyll content, cytoplasmic membrane structure and chlorosome size in *Chloroflexus aurantiacus*. Arch Microbiol 148:236–241

Green BR (2003) The evolution of light-harvesting antennas. In: Green BR, Parson WW (eds) Light-harvesting antennas in photosynthesis. Kluwer Academic, Dordrecht, p 129–168

Green BR, Anderson JM, Parson WW (2003) Photosynthetic membranes and their light-harvesting antennas. In: Green BR, Parson WW (eds) Light-harvesting antennas in photosynthesis. Kluwer Academic, Dordrecht, p 1–28

Hale MB, Blankenship RE, Fuller RC (1983) Menaquinone is the sole quinone in the facultatively aerobic green photosynthetic bacterium *Chloroflexus aurantiacus*. Biochim Biophys Acta 723:376–382

Halfen LN, Pierson BK, Francis GW (1972) Carotenoids of a gliding organism containing bacteriochlorophylls. Arch Mikrobiol 82:240–246

Hanada S, Pierson BK (2002) The family Chloroflexaceae. In: The prokaryotes: an evolving electronic resource for the microbiological community, 3rd edn. Release 3.11. Springer, Heidelberg Berlin New York, http://link.springer-ny.com/link/service/books/10125/

Hanada S, Hiraishi A, Shimada K, Matsuura K (1995) *Chloroflexus aggregans* sp nov, a filamentous phototrophic bacterium which forms dense cell aggregates by active gliding movement. Intl J Syst Bacteriol 45:676–681

Hanada S, Takaichi S, Matsuura K, Nakamura K (2002) *Roseiflexus castenholzii* gen nov, sp nov, a thermophilic, filamentous, photosynthetic bacterium that lacks chlorosomes. Intl J Syst Evol Microbiol 52:187–193

Hauska G, Schoedl T, Remigy H, Tsiotis G (2001) The reaction center of green sulfur bacteria. Biochim Biophys Acta 1507:260–277

Heising S, Richter L, Ludwig W, Schink B (1999) *Chlorobium ferrooxidans* sp nov, a phototrophic green sulfur bacterium that oxidizes ferrous iron in coculture with a *Geospirillum* sp. strain. Arch Microbiol 172:116–124

Hirabayashi H, Ishii T, Takaichi S, Inoue K, Uehara K (2004) The role of carotenoids in the photoadaptation of the brown-colored sulfur bacterium *Chlorobium phaeobacteroides*. Photochem Photobiol 79:280–285

Hirota M, Moriyama T, Shimada K, Miller M, Olson JM, Matsuura K (1992) High degree of organization of bacteriochlorophyll *c* in chlorosome-like aggregates spontaneously assembled in aqueous-solution. Biochim Biophys Acta 1099:271–274

Hohmann-Marriott MF, Blankenship RE, Roberson RW (2005) The ultrastructure of *Chlorobium tepidum* chlorosomes revealed by electron microscopy. Photosynth Res 86:145–154

Holo H, Broch-Due M, Ormerod JG (1985) Glycolipids and the structure of chlorosomes in green bacteria. Arch Microbiol 143:94–99

Holt JG, Lewin RA (1968) *Herpetosiphon aurantiacus* gen et sp n, a new filamentous gliding organism. J Bacteriol 95:2407–2408

Imhoff JF (2003) Phylogenetic taxonomy of the family Chlorobiaceae on the basis of 16S rRNA and fmo (Fenna-Matthews-Olson protein) gene sequences. Intl J Syst Evol Microbiol 53:941–951

Jensen MT, Knudsen J, Olson JM (1991) A novel aminoglycosphingolipid found in *Chlorobium limicola* f. *thiosulfatophilum* 6230. Arch Microbiol 156:248–254

Klappenbach JA, Pierson BK (2004) Phylogenetic and physiological characterization of a filamentous anoxygenic photoautotrophic bacterium *Candidatus* Chlorothrix halophila gen nov, sp nov, recovered from hypersaline microbial mats. Arch Microbiol 181:17–25

Knudsen E, Jantzen E, Bryn K, Ormerod JG, Sirevåg R (1982) Quantitative and structural characteristics of lipids in *Chlorobium* and *Chloroflexus*. Arch Microbiol 135:149–154

Krasnovsky AA, Pakshina EV (1959) The photochemical and spectral properties of bacterioviridin of green sulfur bacteria. Doklady Acad Nauk SSSR 127:215–218

Li H (2006) Organization and function of chlorosome proteins in the green sulfur bacterium *Chlorobium tepidium*. PhD thesis, The Pensilvania State University, p 205

Li H, Frigaard N-U, Bryant DA (2005) Locations and interactions of chlorosome proteins on the chlorosome envelope in *Chlorobium tepidum*: insights from cross-linking experiments. In: van der Est A, Bruce D (eds) Photosynthesis: fundamental aspects to global perspectives. Allen, Lawrence, p 116–119

Maresca JA, Gomez Maqueo Chew A, Ros Ponsatí M, Frigaard N-U, Bryant DA (2004) The *bchU* gene of *Chlorobium tepidum* encodes the C-20 methyltransferase in bacteriochlorophyll *c* biosynthesis. J Bacteriol 186:2558–2566

Martinez-Planells A, Arellano JB, Borrego CA, López-Iglesias C, Gich F, Garcia-Gil J (2002) Determination of the topography and biometry of chlorosomes by atomic force microscopy. Photosynth Res 71:83–90

Matsuura K, Olson JM (1990) Reversible conversion of aggregated bacteriochlorophyll *c* to the monomeric form by 1-hexanol in chlorosomes from *Chlorobium* and *Chloroflexus*. Biochim Biophys Acta 1019:233–238

Melø TB, Frigaard N-U, Matsuura K, Naqvi KR (2000) Electronic energy transfer involving carotenoid pigments in chlorosomes of two green bacteria: *Chlorobium tepidum* and *Chloroflexus aurantiacus*. Spectrochim Acta Part A 56:2001–2010

Milks KJ, Danielsen M, Persson S, Jensen ON, Cox RP, Miller M (2005) Chlorosome proteins studied by MALDI-TOF-MS: Topology of CsmA in *Chlorobium tepidum*. Photosynth Res 86:113–121

Miller M, Gillbro T, Olson JM (1993) Aqueous aggregates of bacteriochlorophyll *c* as a model for pigment organization in chlorosomes. Photochem Photobiol 57:98–102

Montaño GA, Bowen BP, LaBelle JT, Woodbury NW, Pizziconi VB, Blankenship RE (2003a) Characterization of *Chlorobium tepidum* chlorosomes: A calculation of bacteriochlorophyll c per chlorosome and oligomer modeling. Biophys J 85:2560–2565

Montaño GA, Wu HM, Lin S, Brune DC, Blankenship RE (2003b) Isolation and characterization of the B798 light-harvesting baseplate from the chlorosomes of *Chloroflexus aurantiacus*. Biochemistry 42:10246–10251

Nadson GA (1906) The morphology of inferior Algae II: *Chlorobium limicola* Nads, the green chlorophyll bearing microbe. Bull Jard Bot St Petersb 6:190

Nadson GA (1912) *Chlorobium limicola* Nads, the green microorganism with nonfunctioning chlorophyll. Izv St Peterburg Bot Sada 12:55–73

Niedermeier G, Shiozawa JA, Lottspeich F, Feick R (1994) The primary structure of two chlorosome proteins from *Chloroflexus aurantiacus*. FEBS Lett 342:61–65

Nozawa T, Ohtomo K, Suzuki M, Nakagawa H, Shikama Y, Konami H, Wang Z-Y (1994) Structures of chlorosomes and aggregates BChl *c* in *Chlorobium tepidum* from solid state high resolution CP/MAS ^{13}C NMR. Photosynth Res 41:211–223

Oelze J, Golecki JR (1995) Membranes and chlorosomes of green bacteria: structure, composition and development. In: Blankenship RE, Madigan MT, Bauer CE (eds) Anoxygenic photosynthetic bacteria. Kluwer Academic, Dordrecht, p 259–278

Olson JM, Pedersen JP (1990) Bacteriochlorophyll *c* monomers, dimers, and higher aggregates in dichloromethane, chloroform, and carbon-tetrachloride. Photosynth Res 25:25–37

Olson JM, Blankenship RE (2004) Thinking about the evolution of photosynthesis. Photosynth Res 80:373–386

Olson JM, Raymond J (2003) The FMO protein is related to PscA in the reaction center of green sulfur bacteria. Photosynth Res 75:277–285

Overmann J (2000) The family Chlorobiaceae. In: The prokaryotes: an evolving electronic resource for the microbiological community, 3rd edn. Release 3.1. Springer, Heidelberg Berlin New York, http://link.springer-ny.com/link/service/books/10125/

Overmann J, Cypionka H, Pfennig N (1992) An extremely low-light-adapted phototrophic sulfur bacterium from the Black Sea. Limnol Oceanogr 37:150–155

Persson S, Sönksen CP, Frigaard N-U, Cox RP, Roepstorff P, Miller M (2000) Pigments and proteins in green bacterial chlorosomes studied by matrix-assisted laser desorption ionization mass spectrometry. Eur J Biochem 267:450–456

Pierson BK, Castenholz RW (1974) A phototrophic gliding filamentous bacterium of hot springs, *Chloroflexus aurantiacus*, gen. and sp. nov. Arch Microbiol 100:5–24

Pšenčík J, Ikonen TP, Laurinmäki P, Merckel MC, Butcher SJ, Serimaa RE, Tuma R (2004) Lamellar organization of pigments in chlorosomes, the light harvesting complexes of green photosynthetic bacteria. Biophys J 87:1165–1172

Powls R, Redfearn ER (1969) Quinones of the Chlorobacteriaceae – properties and possible function. Biochim Biophys Acta 172:429–437

Redfearn ER, Powls R (1968) The quinones of green photosynthetic bacteria. Biochem J 106:50P

Sakuragi Y, Frigaard N-U, Shimada K, Matsuura K (1999) Association of bacteriochlorophyll *a* with the CsmA protein in chlorosomes of the photosynthetic green filamentous bacterium *Chloroflexus aurantiacus*. Biochim Biophys Acta 1413:172–180

Schmidt K (1980) A comparative study on the composition of chlorosomes (*Chlorobium* vesicles) and cytoplasmic membranes from *Chloroflexus aurantiacus* strain Ok-70-fl and *Chlorobium limicola* f. *thiosulfatophilum* strain 6230. Arch Microbiol 124:21–31

Smith KM, Kehres LA, Fajer J (1983) Aggregation of the bacteriochlorophylls *c*, *d*, and *e*. Models for the antenna chlorophylls of green and brown photosynthetic bacteria. J Am Chem Soc 105:1387–1389

Smith KM (2003) Chlorosome chlorophylls (bacteriochlorophylls *c*, *d*, and *e*): structures, partial syntheses, and biosynthetic proposals. In: Kadish KM, Smith KM, Guilard R (eds) Chlorophylls and bilins: biosynthesis, structure, and degradation. Elsevier Science (USA), p 157–182

Staehelin LA, Golecki JR, Drews G (1978) Visualization of the supramolecular architecture of chlorosomes (*Chlorobium* type vesicles) in freeze-fractured cells of *Chloroflexus aurantiacus*. Arch Microbiol 119:269–277

Staehelin LA, Golecki JR, Drews G (1980) Supramolecular organization of chlorosomes (*Chlorobium* vesicles) and of their membrane attachment sites in *Chlorobium limicola*. Biochim Biophys Acta 589:30–45

Steensgaard DB, Wackerbarth H, Hildebrandt P, Holzwarth AR (2000) Diastereoselective control of bacteriochlorophyll *e* aggregation. 3^1-S-BChl e is essential for the formation of chlorosome-like aggregates. J Phys Chem B 104:10379–10386

Takaichi S (1999) Carotenoids and carotenogenesis in anoxygenic photosynthetic bacteria. In: Frank HA, Young AJ, Britton G, Cogdell RJ (eds) The photochemistry of carotenoids. Kluwer Academic, Dordrecht, The Netherlands, p 39–69

Takaichi S, Oh-oka H (1999) Pigment composition in the reaction center complex from the thermophilic green sulfur bacterium, *Chlorobium tepidum*: Carotenoid glucoside esters, menaquinone and chlorophylls. Plant Cell Physiol 40:691–694

Takaichi S, Wang ZY, Umetsu M, Nozawa T, Shimada K, Madigan MT (1997) New carotenoids from the thermophilic green sulfur bacterium Chlorobium tepidum: 1′,2′-dihydro-γ-carotene, 1′,2′-dihydrochlorobactene, and OH-chlorobactene glucoside ester, and the carotenoids composition of different strains. Arch Microbiol 168:270–276

Tringe SG, Rubin EM (2005) Metagenomics: DNA sequencing of environmental samples. Nat Rev Genet 6:805–814

Uehara K, Mimuro M, Ozaki Y, Olson JM (1994) The formation and characterization of the *in vitro* polymeric aggregates of bacteriochlorophyll c homologs from *Chloro-*

bium limicola in aqueous suspension in the presence of monogalactosyl diglyceride. Photosynth Res 41:235–243

van Dorssen RJ, Gerola PD, Olson JM, Amesz J (1986) Optical and structural properties of chlorosomes of the photosynthetic green sulfur bacterium *Chlorobium limicola*. Biochim Biophys Acta 848:77–82

van Niel CB (1932) On the morphology and physiology of the purple and green sulphur bacteria. Arch Microbiol 3:1–112

van Noort PI, Zhu Y, LoBrutto R, Blankenship RE (1997) Redox effects on the excited-state lifetime in chlorosomes and bacteriochlorophyll *c* oligomers. Biophys J 72:316–325

Vassilieva EV, Antonkine ML, Zybailov BL, Yang F, Jakobs CU, Golbeck JH, Bryant DA (2001) Electron transfer may occur in the chlorosome envelope: The CsmI and CsmJ proteins of chlorosomes are 2Fe – 2S ferredoxins. Biochemistry 40:464–473

Vassilieva EV, Ormerod JG, Bryant DA (2002a) Biosynthesis of chlorosome proteins is not inhibited in acetylene-treated cultures of *Chlorobium vibrioforme*. Photosynth Res 71:69–81

Vassilieva EV, Stirewalt VL, Jakobs CU, Frigaard N-U, Inoue-Sakamoto K, Baker MA, Sotak A, Bryant DA (2002b) Subcellular localization of chlorosome proteins in *Chlorobium tepidum* and characterization of three new chlorosome proteins: CsmF, CsmH, and CsmX. Biochemistry 41:4358–4370

Wagner-Huber R, Brunisholz R, Frank G, Zuber H (1988) The bacteriochlorophyll *c*/*e*-binding polypeptides from chlorosomes of green photosynthetic bacteria, FEBS Lett 239:8–12

Wahlund TM, Woese CR, Castenholz RW, Madigan MT (1991) A thermophilic green sulfur bacterium from New Zealand hot springs, *Chlorobium tepidum* sp. nov. Arch Microbiol 156:81–90

Wang J, Brune DC, Blankenship RE (1990) Effects of oxidants and reductants on the efficiency of excitation transfer in green photosynthetic bacteria. Biochim Biophys Acta 1015:457–463

Wang ZY, Marx G, Umetsu M, Kobayashi M, Mimuro M, Nozawa T (1995) Morphology and spectroscopy of chlorosomes from *Chlorobium tepidum* by alcohol treatments. Biochim Biophys Acta 1232:187–196

Wechsler T, Suter F, Fuller RC, Zuber H (1985) The complete amino-acid sequence of the bacteriochlorophyll *c* binding polypeptide from chlorosomes of the green photosynthetic bacterium *Chloroflexus aurantiacus*. FEBS Lett 181:173–178

Wullink W, Knudsen J, Olson JM, Redlinger TE, van Bruggen EFJ (1991) Localization of polypeptides in isolated chlorosomes from green phototrophic bacteria by immunogold labeling electron microscopy. Biochim Biophys Acta 1060:97–105

Xin YY, Lin S, Montano GA, Blankenship RE (2005) Purification and characterization of the B808–866 light-harvesting complex from green filamentous bacterium *Chloroflexus aurantiacus*. Photosynth Res 86:155–163

Yamada M, Zhang H, Hanada S, Nagashima KVP, Shimada K, Matsuura K (2005) Structural and spectroscopic properties of a reaction center complex from the chlorosome-lacking filamentous anoxygenic phototrophic bacterium *Roseiflexus castenholzii*. J Bacteriol 187:1702–1709

Zhu YW, Ramakrishna BL, van Noort PI, Blankenship RE (1995) Microscopic and spectroscopic studies of untreated and hexanol-treated chlorosomes from *Chloroflexus aurantiacus*. Biochim Biophys Acta 1232:197–207

Microbiol Monogr (2)
J.M. Shively: Complex Intracellular Structures in Prokaryotes
DOI 10.1007/7171_022/Published online: 13 May 2006

Gas Vesicles of Archaea and Bacteria

Felicitas Pfeifer

Institute for Microbiology and Genetics, University of Technology Darmstadt,
Schnittspahnstrasse 10, 64287 Darmstadt, Germany
pfeifer@bio.tu-darmstadt.de

Abstract Gas vesicles are hollow proteinaceous structures of spindle or cylinder shape produced by many cyanobacteria, heterotrophic bacteria and Archaea. Because of their gas content, gas vesicles decrease the cell density and provide neutral or even positive buoyancy to cells. The rigid wall is formed solely from protein and is freely permeable to gas molecules. A major constituent of the wall is the small hydrophobic protein GvpA that is arranged along 4.5 nm wide ribs running perpendicular to the long axis of the gas vesicle. The surface tension at the hydrophobic inner surface excludes the formation of water droplets inside. A second protein, GvpC, is attached to the outside surface strengthening the structure, and five additional Gvp proteins are present in minor amounts. Gas vesicle formation involves the expression of 10 to 14 gas vesicle protein (*gvp*) genes. Ten *gvp* genes have been identified in the cyanobacterium *Microcystis aeruginosa*, whereas 14 *gvp* genes are found in halophilic Archaea. Gas-vesiculate microorganisms occur abundantly in aqueous environments, but recently, homologues of the *gvp* genes have been also detected in sporulating soil bacteria such as *Bacillus megaterium* and *Streptomyces coelicolor*, raising the question of additional functions of Gvp proteins.

1
Introduction

Gas vesicles are found exclusively in prokaryotic microorganisms and occur in bacteria as well as in Archaea. They are produced by many cyanobacteria and heterotrophic bacteria from the aerobic surface epilimnia of lakes, but appear most abundantly in anoxygenic photosynthetic bacteria of the anaerobic, cold and deep hypolimnia of stratified lakes. Gas vesicles provide a gas-filled space within the cells, allowing the microorganism to avoid sinking and to seek and maintain a position in the water column where the conditions are optimal for growth (Walsby 1994). In the domain of Archaea, gas vesicles are found in various haloarchaea and anaerobic methane producers such as *Methanosarcina barkeri* (Petter 1931; Houwink 1956; Archer and King 1984). Piles of gas vesicles constitute the gas vacuoles, which are easily seen as light refractile bodies in cells by phase-contrast microscopy (Fig. 1).

Gas vacuoles were discovered in the late 19th century by H. Klebahn, who investigated waterbloom-forming cyanobacteria and demonstrated that the light refractile, intracellular bodies indeed contain gas and provide buoyancy (Klebahn 1895). Similar results have been obtained with gas vacuoles of the extremely halophilic archaeon *Halobacterium (halobium) salinarum* (Petter 1931; Houwink 1956). Bowen and Jensen (1965) showed that piles of small, cylindrical gas vesicles measuring 75 nm in diameter and up to 1 μm in length constitute the gas vacuoles of cyanobacteria. Gas vesicles develop from hollow biconical structures that grow until they reach a certain width, and further growth occurs through the formation of a cylindrical mid section (Waaland and Branton 1969; Lehmann and Jost 1971). Upon treatment of cells with pressure the gas vesicles collapse and paired "membranes" are found in their place (Bowen and Jensen 1965; Stoeckenius and Kunau 1968). The wall of the gas vesicle is formed by a single layer of protein and is only 2 nm thick. Neither lipids nor carbohydrates are present. The main constituent is the 7 to

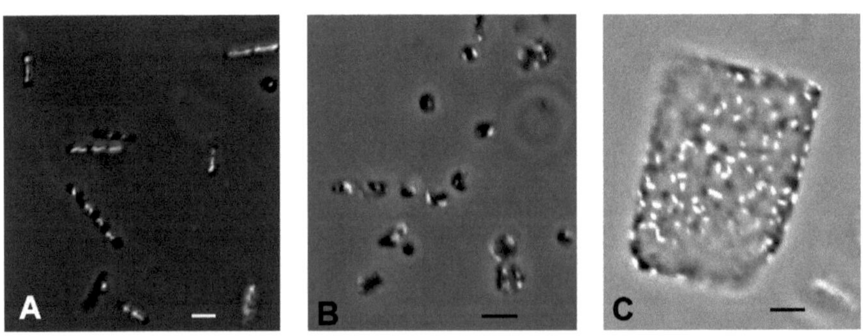

Fig. 1 Phase contrast microscopic appearance of *Hbt. salinarum* (**A**), *Hrr. vacuolatum* (**B**) and *Hqu. walsbyi* (**C**). The *bar* indicates 1 μm in each case

8 kDa protein GvpA, one of the most hydrophobic proteins known. GvpA is arranged along ribs running perpendicular to the long axis, and a second protein, GvpC, attaches at the outside surface and strengthens the gas vesicle structure (Walsby and Hayes 1988).

The following review summarizes some ecological aspects of gas-vacuolate microorganisms and describes the properties of the gas vesicles found in cyanobacteria and halophilic Archaea. Gas vesicles have been studied extensively by Anthony Walsby over the past 40 years (an excellent review: Walsby 1994). Further insights into the gas vesicle formation came from the investigation of *gvp* genes and their regulation. Up to ten *gvp* genes were identified in cyanobacteria, whereas 12 to 14 *gvp* genes are found in haloarchaea. Recently, *gvp* genes have been also detected in bacteria living in soil, such as *Bacillus megaterium* and several *Actinomyces* species. The putative function of gas vesicles in these microorganisms is not yet clear, but vesicular caps on spores have been described for certain *Clostridium* species (Krasil'nikov et al. 1971).

2 Ecology and Gas Vesicle Producing Microorganisms

Gas vesicles occur in 53 genera from five phyla of bacteria and in two phyla of Archaea (Walsby 1994). Gas-vacuolate aquatic microorganisms are able to maintain a certain position in the water body, or even float towards the surface when sufficient amounts of gas vesicles are present. The major function of gas vesicles is to provide buoyancy and enable the cells to counteract sinking. Gas vesicles allow the photosynthetic bacteria and the halophilic archaeon *Halobacterium salinarum* to reach regions of higher light irradiance for photosynthetic growth. With a few exceptions, gas vesicles are restricted to non-flagellated species, but one exception is *Hbt. salinarum*. Both swimming with flagella and vertical movements by gas vesicles require a fine-tuned coordination to avoid conflicting reactions of the cell in response to environmental factors.

2.1 Gas Vesicles of Halophilic Archaea

Halophilic Archaea thrive in solar salt evaporation ponds, salt lakes like the Great Salt Lake in Utah, the Dead Sea and also in shallow brine pools and brine lakes of arid regions. Oxygen is less soluble in brine than in water, and steep diffusion gradients develop. Many haloarchaea are able to live under anaerobic conditions utilizing nitrate or dimethyl sulphoxide (DMSO) as external electron acceptors, and some are also able to ferment amino acids. Gas vesicles were first described for *Halobacterium salinarum* (formerly designated *Hbt. halobium* or *Hbt. salinarium*) (Fig. 1A). This rod-shaped strain contains up to 70 spindle-shaped gas vesicles per cell that are present throughout growth (Fig. 2A) (Stoeckenius and Kunau 1968). Gas vesicles of haloarchaea exhibit a larger width compared to the bacterial counterparts (diameters of 200 to 250 nm versus 33 to 75 nm in bacteria), but can reach a similar length (Larsen et al. 1967; Englert et al. 1990). A reason for the larger diameter is that they are not exposed to high pressure. The negligible turgor pressure of the haloarchaeal cell is due to the accumulation of similar KCl concentrations inside as NaCl found outside. In addition, the depth of a brine pool is usually less than 1 meter.

Besides being gas-vesiculate, *Hbt. salinarum* produces bacteriorhodopsin (BR) that acts as a light-driven proton pump under low oxygen tension and light (Oesterhelt 1998). The cells prefer regions of long-wavelength light (the absorption maximum of BR is 560 nm), whereas UV light is avoided. Two retinal-based photosensor proteins enable *Halobacterium* to monitor the quality of light, and the movement of their polar flagella is regulated by a signal transduction pathway (Yao and Spudich 1992; Zhang et al. 1996). The vertical positioning by gas vesicles and swimming due to the movements of flagella could counteract each other. To ensure that the cells are able to flee UV light and move deeper in the brine, gas vesicle formation is suppressed by UV light. Consequently most haloarchaeal isolates position themselves at a certain depth in the saltwater column rather than floating at the surface. In this way they also avoid exposure to rainfall that could lead to a sudden reduction in the outside salt concentration and cause lysis of the cells. In laboratory cultures, *Hbt. salinarum* wild type forms a pellicle at the surface when a liquid culture is left standing at the lab bench for a few days (Fig. 3A).

Another gas-vesiculate haloarchaeon is the flat and triangular *Haloferax mediterranei* that produces longer and cylinder-shaped gas vesicles, similar to *Hbt. salinarum* PHH4 (Fig. 2B). In both species less than 20 gas vesicles

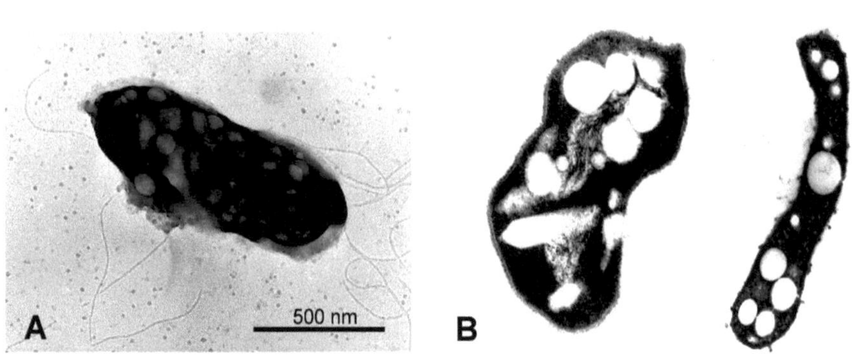

Fig. 2 Electron micrographs of *Hbt. salinarum* (**A**) and thin sections of *Hfx. mediterranei* (**B**). The gas vesicles are visible as white bodies inside the cells

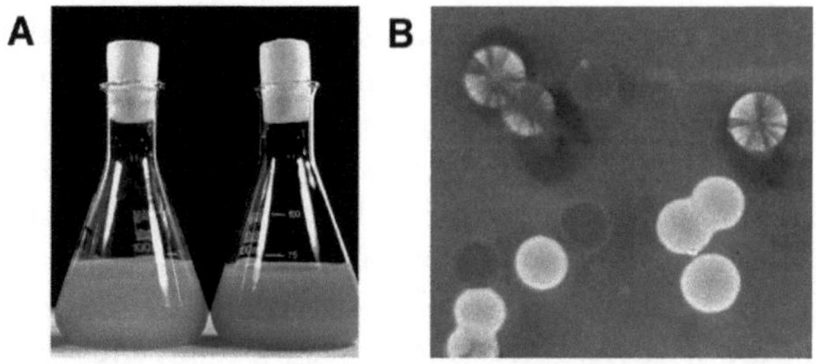

Fig. 3 Flotation of *Hbt. salinarum* PHH1 in liquid cultures (**A**) and phenotype of colonies on solid medium (**B**). **A** Gas vesiculate cells remain swimming or floating in the culture and form a pellicle at the surface after two days standing on the lab bench. **B** The colonies of gas-vesiculate (Vac$^+$) cells are pink, whereas gas vesicle minus (Vac$^-$) cells form red and translucent colonies

are found per cell, and they occur in the stationary growth phase only. In addition, *Hfx. mediterranei* produces gas vesicles only when the salt concentration of the medium exceeds 17% (Englert et al. 1990). Cylindrical gas vesicles are also found in *Halorubrum (Natronobacterium) vacuolatum* isolated from the Kenyan alkaline soda lake Magadi, but they are only produced in media containing more than 15% salt (Fig. 1B) (Mwatha and Grant 1993; Mayr and Pfeifer 1997). The most unusual haloarchaeon is the large and square, but extremely thin gas-vesiculate *Haloquadratum walsbyi* (Fig. 1C) (Parkes and Walsby 1981; Burns et al. 2004; Bolhuis et al. 2004). For all these haloarchaea it might be important to escape the bottom of the brine where the oxygen concentration is low and salt crystals form at saturating salt concentrations.

2.2 Buoyancy Regulation in Cyanobacteria

Cyanobacteria need light to gain energy and these organisms have to move to positions where the light intensity is sufficient for phototrophic growth. Light decreases exponentially with depth due to the scattering or the absorption of light by the lake water (Walsby 1994). The filamentous species *Oscillatoria, Planktothrix, Anabaena* and *Calothrix*, but also the unicellular *Microcystis aeruginosa* and *Dactylococcopsis salina* are gas vesicle producers that have been investigated in some detail. Cyanobacteria appear to regulate their buoyancy in response to light, producing large amounts of gas vesicles at low irradiances and less gas vesicles at high irradiances. This leads to an upward movement of the cells from deeper regions of the lake, but avoids floating near the surface.

As already mentioned in the case of haloarchaea, flotation at the surface exposes the cells to UV light, and sinking might be more favorable. *Anabaena flos-aquae* regulate buoyancy by the counteracting effects of gas vesicles and the accumulation of carbohydrate ballast. Carbohydrates are considerably denser than water and an accumulation during photosynthesis results in sinking, whereas the larger amounts of gas vesicles produced at low light intensity enable flotation. Both features allow *Anabaena* to perform vertical migrations during the day, and cyanobacterial waterblooms that appear at the lake surface in the morning and disappear by the afternoon are most likely due to this phenomenon (Walsby 1994). Calculations on the extent of the vertical movements of the cells yield that single cells move up and down only a few centimeters per day, whereas large colonies of cyanobacteria may move tens of meters (Walsby 1994). Faster moving colonies even migrate below the metalimnion during the day and then return to the water surface at night using gas vesicles (Walsby and McAllister 1987). About 3–10% of the cell volume must be occupied by gas vesicles to provide buoyancy (Oliver and Walsby 1984; Thomas and Walsby 1985).

2.3 Gas-vesiculate Anoxygenic Phototrophs and Heterotrophic Bacteria

A large variety of gas-vesiculate bacteria occur abundantly in the anaerobic hypolimnia of stratified lakes, such as iron bacteria (*Siderocapsa, Ochrobium*), purple sulphur bacteria (*Thiodictyon, Thiopedia, Ectothiorhodospira*), and green sulphur bacteria (*Pelodictyon, Ancalochloris, Chlorochromatium, Chloronema*) (Clark and Walsby 1978; Overmann et al. 1991; Walsby 1994). These anoxigenic photosynthetic bacteria harvest the longer-wavelength far-red light and are often found in discrete layers in the anaerobic hypolimnia. Gradients of light, iron and sulphide may influence their vertical distribution. Similar to cyanobacteria, the green sulphur bacterium *Pelodictyon phaeoclathratiforme* regulates cell density by the accumulation of carbohydrate ballast and a decrease in gas vesicle production at high light intensity (Overmann et al. 1991).

Gas-vesiculate bacteria are found abundantly in cold marine and cold freshwater habitats (Staley et al. 1989; Gosink et al. 1997; Jung et al. 2004). A study performed on bacteria from sea ice of Antarctica yields many filamentous gas-vesiculate cells that compose up to 83 to 91% of the total viable bacteria in certain sites. A remarkable diversity of phototrophic purple bacteria is also present in the permanently ice covered, meromictic Lake Fryxell, Antarctica, where the filamentous *Rhodoferax antarcticus* is found in high abundance at certain locations in the near-freezing water column (Karr et al. 2003; Jung et al. 2004). Also the psychrophilic sulphate-reducer *Desulforhopalus vacuolatus* synthesizes gas vesicles (Isaksen and Teske 1996). Gas vesicles are also present in the aerobic or facultative anaerobic bacteria *Ancylobacter (Microcyclus) aquaticus, Aquabacter spiritensis* and in appendaged bacteria (*Prosthecomicrobium, Stella*), or the sheathed bacterium *Leptothrix* (Staley 1968; van Ert and Staley 1971; Konopka et al. 1977; Staley et al. 1987; Irgens et al. 1989). In all these cases, gas vesicles are used to position the microorganism at a favorable depth in the water body.

3 Structure of the Gas Vesicle Wall and Physical Properties

The properties of the gas vesicles depend on the structural features of the proteins constituting the gas vesicle wall. The predominant constituent of gas vesicles is the GvpA protein, and the minor gas vesicle protein GvpC is attached to the outside. The function and the structural features of these two proteins and also of isolated gas vesicles have been investigated in detail to gain further insights into this peculiar structure.

3.1 Structure of the Gas Vesicle Wall Formed by GvpA and GvpC

Electron micrographs of collapsed, and also of intact, frozen hydrated gas vesicles show 4.5-nm wide ribs running perpendicular to the axis of the gas vesicle (Fig. 4) (Stoeckenius and Kunau 1968; Walsby 1972). The ribs are formed by a helix of low pitch rather than by a stack of hoops, and consist of a single layer of GvpA (Offner et al. 1998). The amino acid sequence of the GvpA protein of cyanobacteria (Hayes et al. 1986; Damerval et al. 1987) and haloarchaea (Jones et al. 1991; Pfeifer and Englert 1992) are highly conserved, especially in a central region of 50 amino acids (Fig. 5). This region is very hydrophobic, and hydrophobic interactions are most likely responsible for the tight contacts between GvpA molecules. Gas vesicles are insoluble in detergents and aqueous solutions, and during SDS-polyacrylamide gel-electrophoresis the structures remain as aggregates in the separating gel. However, in cell lysates, the monomer of the 8 kDa GvpA is easily detectable by Western analyses indicating that the monomer is not too hydrophobic to enter the SDS-polyacrylamide gel (Englert et al. 1992; Englert and Pfeifer 1993; Li and Cannon 1998).

MALDI-TOF mass spectrometry studies with gas vesicles of *Anabaena flos-aquae* and *Hbt. salinarum* indicate no post-translational modifications of GvpA (Belenky et al. 2004). Proteases and peptidases have been used to determine which peptide bonds of GvpA are exposed at the outer surface. The proteinase trypsin (cleaves R-X and K-X) is able to cleave K59-I60 in some GvpA molecules of *Hbt. salinarum*, whereas all lysines found in the conserved central portion are inaccessible (Fig. 5). The endoproteinase Glu-C from *Staphylococcus aureus* (cleaves D-X and E-X) degrades the charged C-terminal portion of the haloarchaeal GvpA and leads to an immediate collapse of the gas vesicles. Again, the central D-X or E-X bonds are not affected (Belenky et al. 2004). Thus, it appears that the C-terminal portion of GvpA is

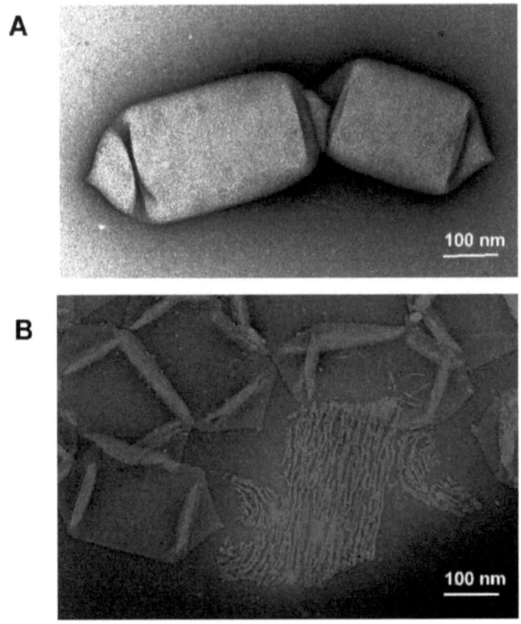

Fig. 4 Electron micrographs of isolated gas vesicles of *Hfx. mediterranei*. **A** Intact, cylinder-shaped gas vesicles, **B** collapsed gas vesicles

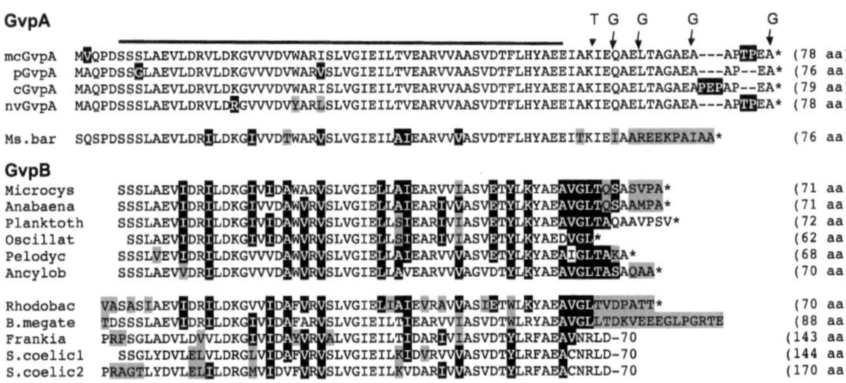

Fig. 5 Alignment of GvpA sequences derived from Archaea and Bacteria. **A** The pGvpA, cGvpA, mcGvpA and nvGvpA are from haloarchaea; Ms.bar from *Methanosarcina barkeri*. Amino acid mutations are *shaded in black*. **B** The GvpA sequences of aquatic bacteria derive from *Microcystis, Anabaena, Planktothrix, Oscillatoria, Pelodyction* and *Ancylobacter*, and from the soil bacteria *Rhodobacter, Bacillus megaterium*, Frankia and *Streptomyces coelicolor*. Amino acids typical for bacteria that are distinct from Archaea are *shaded in black*. *Numbers in parentheses* describe the length of the particular GvpA protein. The *bar* on *top* denotes the conserved part of the proteins. *Letters* and *arrows* above the haloarchaeal GvpA proteins indicate sites that are accessible to protease trypsin (T; cleaves R-X and K-X) and the endoproteinase Glu-C (G; cleaves D-X and E-X)

exposed at the exterior surface and is important for an intact gas vesicle structure. Similar analyses have also been done with cyanobacterial gas vesicles.

No crystal structure of GvpA is yet available, since GvpA forms aggregates that cannot be dissolved without denaturing. X-ray analysis has been done with layers of collapsed gas vesicles from cyanobacteria and haloarchaea (Blaurock and Walsby 1976; Blaurock and Wober 1976). A stacking periodicity with multiples of 2 nm (*Halobacterium*) or 1.95 nm (*Anabaena flos-aquae*) has been found that relates to the thickness of the wall. The subunits along the ribs appear to be regularly spaced at intervals of 1.15 nm (Fig. 6) (Blaurock and Walsby 1976). Assuming two adjacent antiparallel β-chains of GvpA, the 1.15 nm could be the distance to the next parallel chain. The regularly spaced subunits along the ribs have been confirmed by atomic force microscopy (AFM) and slope in an angle of 54° to the rib axis (McMaster et al. 1996). The angle of 54° is very close to the "magic" angle at which the transverse and longitudinal stresses are equal in the wall of a cylindrical structure (Walsby 1994). Peptide chains of 14 to 16 amino acids in a β-sheet structure could be accommodated here. However, secondary structure programs predict that only part of GvpA is able to form β-sheets, and also the analysis of gas vesicles by circular dichroism (CD) spectra implies the presence of random coil and α-helices besides β-sheets. It is possible that GvpA could form an amyloid structure, but the crystal structures of GvpA in the monomer and crystalline form in the gas vesicle are required to solve this question.

GvpC strengthens the wall formed by GvpA as a second structural protein. GvpC can be removed from the gas vesicle by SDS or urea treatment, and gas

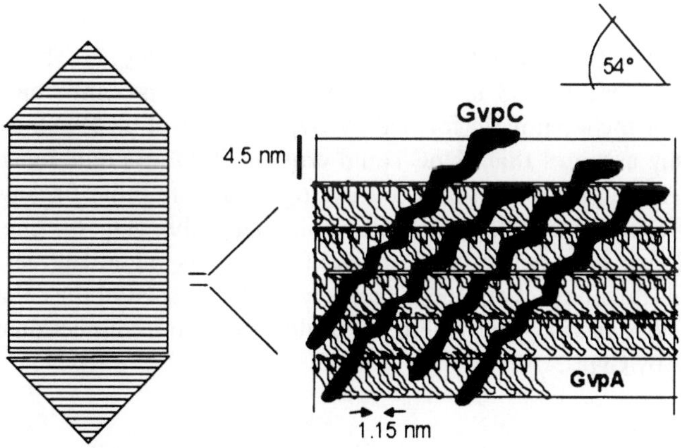

Fig. 6 Schematic representation of the gas vesicle structure. The ribs formed by GvpA are perpendicular to the *long axis*. GvpA is arranged as a regular structure oriented at an angle of 54° in each rib. GvpC (*black structures*) strengthens the gas vesicles by attaching with the 33-aa or 38-aa repeats to the outside surface and connecting 5 GvpA molecules from adjacent ribs each (Walsby 1994)

pGvpC, *Hbt. salinarum*

```
MSVTDKRDEMSTARDKFAESQQEFESYADEFAADITA-KQDDVSDLVDAITDFQAEMT
                         NTTDAFHTYGDEFAAEVDH-LRADIDAQRDVIR
                         EMQDAFEAYADIFATDIAD-KQ-DIGNLLAAIEALRTEMN
                         STHGAFEAYADDFAADVAA-LR-DISDLVAAIDDFQEEFI
                         AVQDAFDNYAGDFDAEIDQ-LHAAIADQHDSFD
                         ATADAFAEYRDEFYRIEVEALLEAINDFQQDIGDFRAEFE
                         TTEDAFVAFARDFYGHEITAEEGAAEAEAEPVEADAD
VEAEAEVSPDEAGGESAGTEEEETEPAEVETAAPEVEGSPADTADEAEDTEAEEETEEEA
PEDMVQCRVCGEYYQAITEPHLQTHDMTIQEYRDEYGEDVPLRPDDKT
```

GvpC, *Anabaena flos-aquae*

```
MISLMAKIRQEHQSIAEKVAELSLETREFLSVTTAKRQEQAEKQAQELQAF
                         YKDLQETSQQFLSETAQARIAQAEKQAQELLAF
                         HKELQETSQQFLSATAQARIAQAEKQAQELLAF
                         YQEVRETSQQFLSATAQARIAQAEKQAQELLAF
                         HKELQETSQQFLSATADARTAQAKEQKESLLKFRQDLFVSIFG
```

Fig. 7 Amino acid sequence of GvpC's derived from *Halobacterium salinarum* and *Anabaena flos-aquae*, indicating the amino acid repeats of 38-aa (haloarchaea) or 33-aa (cyanobacteria). Identical amino acids are marked in *bold*

vesicles remain intact but become three-fold weaker (Hayes et al. 1992). Immunological studies show that GvpC attaches to both the end-cones and the cylinder of cyanobacterial gas vesicles (Buchholz et al. 1993). The sequence of GvpC exhibits several 33 to 34 amino acid (aa) repeats in varying number (4 to 5 in cyanobacteria and 6 to 7 in haloarchaea) that are according to secondary structure predictions in α-helical structure (Fig. 7). The length calculated for a repeat of 33 aa in the α-helical structure is 5 nm which exceeds the width of the 4.5 nm rib formed by GvpA (Hayes et al. 1988). Because of this, Walsby assumes that GvpC could cross a rib at an angle by contacting five adjacent GvpA proteins and connecting them with other packages of five GvpA molecules in adjacent ribs like a stabilizing fibre (Fig. 6). The GvpA to GvpC ratio has been determined to be 25 : 1, which is in good agreement with this model (Buchholz et al. 1993). The very thick ribs seen in the collapsed gas vesicle of *Hfx. mediterranei* in Fig. 3B could be due to a connection of several GvpA ribs by GvpC.

3.2 Gas Permeability and Critical Collapse Pressure

Gas vesicles are freely permeable by gases and cannot be inflated by gas. The gas space rapidly fills with gas by diffusion, and the gas inside is in equilibrium with the gas dissolved in the surrounding (Walsby 1994). No storage of gas occurs. Gas vesicles are permeable to O_2, N_2, H_2, CO_2, CO, CH_4, Ar and even to the large compound C_4F_8 (perfluorocyclobutane), implying that the wall must have pores (Walsby 1982). Walsby calculated that a pore diameter of at least 0.63 nm is required to allow the penetration of C_4F_8. Since water molecules are much smaller than C_4F_8, water vapour is possibly able to enter the gas vesicle. However, the surface tension at the hydrophobic inner surface excludes liquid water. Water vapour cannot form a droplet inside, because the vapour pressure in a droplet is higher over the droplet than over a planar water surface (Walsby 1994). Water vapour will always condense outside the gas vesicle, and any droplet formed inside (or remaining here during the formation of the particle) will evaporate. For gas vesicles of *Microcystis*, the exponential filling time has been calculated as 0.4 μs, and the permeability coefficient of the wall as $k = 32\ \text{mm s}^{-1}$ (Walsby et al. 1992). The latter value is more than 1000-fold higher than the gas permeability coefficient of a C_{16} lipid monolayer.

Gas vesicles irreversibly collapse at higher pressures, and the critical pressure at which the collapse occurs has been determined for different microorganisms. The critical collapse pressure varies from a mean value of 0.09 MPa determined for *Hbt. salinarium* to more than 1 MPa for gas vesicles of *Oscillatoria* (Walsby and Bleything 1988), and correlates inversely with the width of the gas vesicles. The largest gas vesicles (radius $r = 100$ nm) found in haloarchaea collapse at the lowest critical pressure, whereas the gas vesicles of *Dactylococcopsis salina, Microcystis*, and *Oscillatoria* are smaller ($r = 55$ to 31 nm) and collapse at pressures of 0.3, 0.6 and 1 MPa, respectively (Walsby 1994). This correlation has been confirmed by environmental studies with *Planktothrix rubescens*. Isolates derived from deeper regions of the lake indeed contain narrower gas vesicles than isolates collected closer to the surface (Bright and Walsby 1999; Beard et al. 2000a).

The differences in the width of the gas vesicles appear to correlate with the size of GvpC, and especially with the number of 33-aa repeats found here. Shorter GvpC variants (i.e. containing a smaller number of the 33-aa repeats) are found with isolates from deeper water, whereas larger GvpC proteins are present in cells collected near the surface water that contain the wider gas vesicles (Beard et al. 1999, 2000a). Similar variations in GvpC are also observed with two *Microcystis aeruginosa* isolates living at different depths in a lake (Dunton and Walsby 2005). The GvpC proteins found with the wider gas vesicles in cells collected near the surface exhibit an increase in the number of the 33-aa repeats. In addition, this increase is the only difference observed between the two isolates, since none of the nine remaining *gvp* genes shows a significant difference. *In vitro* experiments demonstrate that the complete removal of GvpC causes a three-fold decrease in the critical pressure (Buchholz et al. 1993). The addition of GvpC proteins with a complete number of 33-aa repeats restores the strength, whereas reduced numbers of 33-aa repeats result in a lower critical pressure (Kinsman et al. 1995).

4
Genes Involved in Gas Vesicle Formation in Haloarchaea

The gene encoding GvpA was first identified in the cyanobacterium *Calothrix* (Damerval et al. 1987) and enabled the isolation of the respective genes in the two *Halobacterium salinarum* (formerly *Hbt. halobium*) strains NRC-1 and PHH1 (DasSarma et al. 1987; Horne et al. 1988). The *gvpA* genes of the two haloarchaea are located on large plasmids (pNRC100 or pHH1), and both genes are affected by a mutation rate of 10^{-2} caused by frequent integrations of insertion elements (ISH-elements) (Pfeifer et al. 1981; DasSarma et al. 1988; Horne et al. 1991). This high frequency of transpositions is typical for these two strains, and the mutations due to the action of ISH-elements occur especially in the more AT-rich regions in the plasmids and the chromosome of *Hbt. salinarum* (Pfeifer and Betlach 1985; Pfeifer and Blaseio 1989). With respect to gas vesicle formation the high frequency of mutation can be easily observed with the phenotype of colonies on agar plates (Fig. 3B). The colonies of gas-vesiculate (Vac^+) cells are pink, whereas gas vesicle negative (Vac^-) cells are red and translucent.

A second gene, c-*gvpA*, is found in *Hbt. salinarum* PHH1 and *Hbt. salinarum* PHH4 (or *gvpA2* in the case of *Hbt. salinarum* NRC-1), and a third one, mc-*gvpA*, has been identified in the chromosome of the moderately halophilic, gas-vesiculate *Haloferax mediterranei*. These *gvpA* genes are all surrounded by additional *gvp* genes identified through mutation analyses. In each case, the entire *gvp* gene cluster comprises 14 *gvp* genes arranged as oppositely oriented units *gvpACNO* and *gvpDEFGHIJKLM* (Fig. 8) (Englert et al. 1992a; DasSarma et al. 1994). A fourth *gvp* gene cluster (nv-vac) isolated from *Halorubrum* (*Natronobacterium*) *vacuolatum* harbours 12 *gvp* genes in a slightly different arrangement that are all transcribed as a single unit (Fig. 8) (Mayr and Pfeifer 1997; Pfeifer 2004).

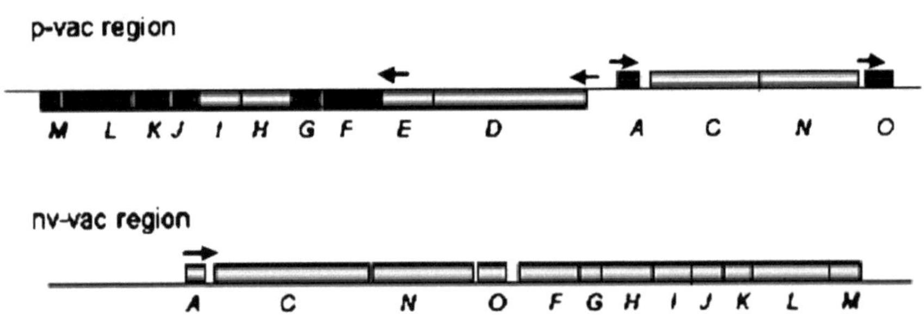

Fig. 8 Genetic map of the p-vac region of *Hbt. salinarum* and the nv-vac region of *Hrr. vacuolatum*. *Arrows* mark the start sites and direction of transcription. The 8 essential *gvp* genes are *shaded in grey* in the p-vac region and constitute the minvac region. The 14 gvp genes of c-vac, mc-vac exhibit the same arrangement as in p-vac

4.1
Gas Vesicle Gene Clusters and Minimal Vac Region

Hbt. salinarum PHH1 contains the two distinct *gvp* gene clusters p-vac and c-vac (Englert et al. 1992a). Expression of the p-vac region in the wild-type strain leads to spindle-shaped gas vesicles with pGvpA as the exclusive major structural protein (Fig. 9). The c-*gvpACNO* cluster gene of c-vac is not expressed in *Hbt. salinarum* PHH1, but the p-vac deletion strain *Hbt. salinarum* PHH4 contains cylinder-shaped gas vesicles composed of cGvpA (Horne and Pfeifer 1989; Englert et al. 1992a). The various GvpA proteins of haloarchaea are highly conserved and differ only in a few positions (Fig. 5). Despite the fact that gas vesicles of *Hbt. salinarum* PHH1 are formed by the expression of p-vac, a portion of the c-vac region, i.e. the c-*gvpDEFGHIJKLM* operon, is also transcribed in this strain. However, only cGvpD involved in the repression of gas vesicle formation is detectable, whereas the transcriptional activator cGvpE required for the activation of the P_{cA} promoter is absent (Krüger and Pfeifer 1996). The presence of the other cGvp proteins encoded by this operon has not yet been determined. The heterologous pGvpE protein of the p-vac region is unable to activate the P_{cA} promoter of the c-*gvpACNO* unit (Gregor and Pfeifer 2001), and because of this, gas vesicles of *Hbt. salinarum* PHH1 only contain pGvp proteins.

Hbt. salinarum NRC-1 (whose genome sequence has been determined) contains two copies of the p-vac-related *gvp1* gene cluster, one on the 192-kbp pNRC100, and another one on the 365-kb pNRC200 (Ng et al. 2000). This minichromosome carries in addition to *gvp1* also the *gvp2* gene cluster that is homologous to c-vac but lacks c-*gvpM*, an essential gene for gas vesicle formation. *Haloferax mediterranei* harbors the related mc-vac region that leads to cylinder-shaped gas vesicles (Englert et al. 1992a), whereas the *Halorubrum vacuolatum* cluster consists of the single nv-*gvpACNOFGHIJKLM* transcription unit (Fig. 8; Mayr and Pfeifer 1997; Pfeifer 2004). The two regulator genes *gvpDE* have not been found so far. The GvpA proteins are relatively highly conserved (> 97% identity), and sequence similarities of > 79% are found for GvpF, G, J, K, M and N, whereas similarities between 54–74% have been determined for GvpC, H, I, L and O. The difference in gas vesicle shape (spindle- versus cylinder-shaped) is in part due to the few amino acid alterations in pGvpA versus cGvpA (Beard et al. 2002b). However, complementation studies with the heterologous transcription units p-*gvpACNO* + mc-*gvpDEFGHIJKLM* (or mc-*gvpACNO* + p-*gvpDEFGHIJKLM*) that lead to the formation of chimeric gas vesicles in *Haloferax volcanii* transformants suggest that the shape is not exclusively determined by GvpA (Offner et al. 1998). More likely, the relative concentration of the other Gvp proteins also influences the gas vesicle shape. The mcD-M/pA-O transformants contain cylinder-shaped gas vesicles despite the fact that the major structural protein is pGvpA that usually forms spindle-shaped gas vesicles, and pD-M/mcA-O transformants harbour shorter cylinder-shaped gas vesicles, although they are formed by mcGvpA (Offner et al. 1998). *Hfx. volcanii* lacks the genes encoding gas vesicles, and transformation with the entire *gvp* gene cluster results in gas-vacuolate (Vac$^+$) transformants.

The importance of each *gvp* gene for gas vesicle formation has been determined using the p-vac region of PHH1 and the related *gvp1* gene cluster of NRC-1 (Offner et al. 2000; DasSarma et al. 1994). In the case of the p-vac region, single p-*gvp* genes were deleted and the remaining p-*gvp* genes analyzed for their ability to drive gas vesicle formation in *Hfx. volcanii* transformants. Each construct used is able to complement to a Vac$^+$ phenotype when all remaining *gvp* genes are added in transformants (Offner et al. 2000). The deletion of *gvpC*, D, E, H, I or N results in transformants still containing light refractile bodies. However, in some cases altered gas vesicle shapes are observed (Fig. 9; see Sect. 4.2 for further descriptions). The remaining genes p-*gvpFGJKLMAO* are all essential for the formation of light refractile bodies and constitute the minimal vac-region minvac (Offner et al. 2000). Transformants harboring these eight *gvp* genes produce minor amounts of gas vesicles.

In the case of the related *gvp1* gene cluster of *Hbt. salinarum* NRC-1, the *gvp* genes have been mutated by the insertion of a *kappa* (κ) element, and a deletion variant of NRC-1 lacking the *gvp1* gene cluster was used as the recipient (DasSarma et al. 1994). However, this strain still contains the *gvpM*-depleted *gvp2* gene cluster that might be still expressed. Compared to the results of our deletion analyses, differences are found in six out of 14 cases (see Sect. 4.2). Some of these could be due to the presence of *gvp2*, but an even more severe point of critique on these analyses is that neither the lack of the respective Gvp protein derived from the mutated *gvp1* gene, nor the presence of the Gvp protein encoded by the *gvp* gene located downstream have been determined to exclude a polar effect of the κ-insertion (DasSarma et al. 1994). A false interpretation has been already uncovered with *gvpM*, that was de-

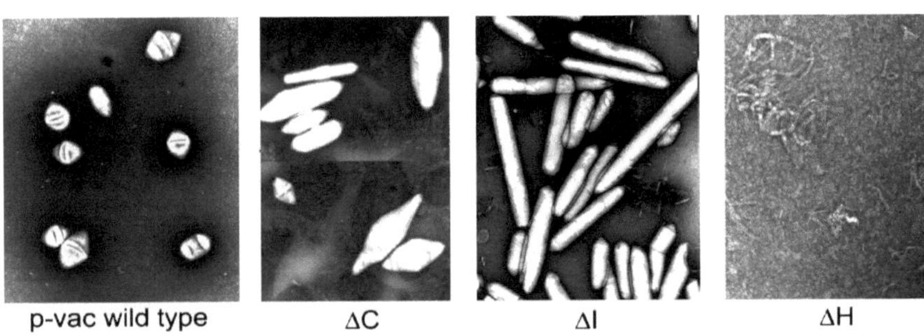

Fig. 9 Electron micrographs of gas vesicles isolated from *Hfx. volcanii* transformants containing the p-vac region, or a p-vac region lacking the p-*gvpC* (ΔC), p-*gvpI* (ΔI), or p-*gvpH* gene (ΔH)

scribed as not essential in the first publication, but later it became clear that the *gvpM* :: κ mutant still contains the GvpM protein (Shukla and DasSarma 2004).

4.2 Functions of the Various Gvp Proteins

The *gvpACNO* gene cluster encodes the two gas vesicle structural proteins GvpA and GvpC. The 8-kDa **GvpA** forms the ribbed structure exhibiting the hydrophobic inner surface. As expected, ΔA transformants are Vac negative. The related hydrophobic proteins GvpJ and GvpM are thus unable to substitute for GvpA. The adjacent *gvpC* gene encodes **GvpC**, the protein that attaches to the outer surface of the gas vesicles and functions as a stabilisator (Englert and Pfeifer 1993; Halladay et al. 1993). GvpC of haloarchaea contains 6- to 7-aa repeats of 32 to 38 aa length and indicates a zinc-finger domain (CCHH) near the C-terminus (Fig. 10). The lack of *gvpC* in ΔC transformants leads to a similar amount of gas vesicles that are longer than the spindle-shaped ones of the wild type and in addition of irregular shape, with varying diameters throughout a single gas vesicle (Fig. 9) (Offner et al. 1996). Obviously, the absence of GvpC during gas vesicle formation does not affect the incorporation of GvpA into the structure. In contrast, *gvpC* :: κ1 insertion mutants contain only tiny bicone structures suggestive of a defect in GvpA incorporation (DasSarma et al. 1994). Since the κ element is inserted near the 3′ terminus of *gvpC* this morphology might be rather due to a lack of GvpN encoded by the adjacent *gvpN* gene. Both ΔN and *gvpN* :: κ transformants contain tiny gas vesicles in minor amounts (Offner et al. 1996; DasSarma et al. 1994). The **GvpN** protein contains a p-loop motif and appears to be an ATPase of the AAA+ chaperone-like superfamily. GvpN could function as a chaperone in the assembly of the gas vesicles since the lack of GvpN impairs the incorporation of GvpA. The 12- to 15-kDa **GvpO** is clearly essential for gas vesicle formation, since ΔO transformants of p-vac and mc-vac are Vac negative (Offner et al. 1996; Englert et al. 1992b). The function of GvpO is not yet known, but a gene homologous to *gvpO* is also present in the *gvp* gene clusters of soil bacteria.

The p-*gvpFGHIJKLM* genes of p-vac are only transcribed during exponential growth, and this "early" expression is similar to the time point of expression in mc-vac, where the long mc-*gvpDEFGHIJKLM* transcript is only detectable during the exponential growth phase (Offner and Pfeifer 1995; Röder and Pfeifer 1996). In later growth phases, shorter transcripts (mc-*gvpDEF*, mc-*gvpDE* and mc-*gvpD*) are present in large amounts that are used to produce the two regulatory proteins GvpD and GvpE in *Hfx. mediterranei*. From the time point of transcription it is justified to assume that the products encoded by *gvpFGHIJKLM* are involved in early steps of gas vesicle formation. Six of these, GvpF, G, J, K, L and M are essential, whereas GvpH and

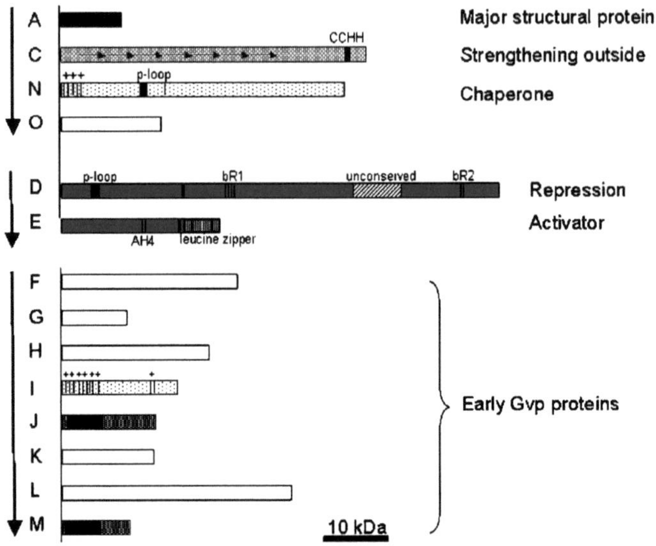

Fig. 10 Schematic representation of the various Gvp proteins derived from the p-vac region of *Hbt. salinarum*. Special features of these proteins are marked. See text for further explanations

GvpI are not required (Offner et al. 2000). Except for GvpK, minor amounts of all these proteins have been detected in gas vesicle preparations (Shukla and DasSarma 2004), but these analyses still need confirmation. **GvpJ** and **GvpM** are hydrophobic proteins with sequence similarities to GvpA, suggesting that both could be structural components of the gas vesicles (Fig. 10). **GvpF** and **GvpL** indicate some sequence similarities, and contain predicted coiled-coil domains enabling an oligomerization (Shukla and DasSarma 2004). GvpF, **GvpG** and **GvpK** are synthesized during exponential growth and prior to the appearance of light refractile bodies in *Hbt. salinarum* or *Hfx. mediterranei* (Zimmermann and Pfeifer 2003; Offner and Pfeifer, unpublished results). All these proteins presumably take part in early nucleation processes required for the formation of the bicone structure.

GvpI and **GvpH** are not essential for the synthesis of light refractile bodies (Offner et al. 2000). ΔI transformants of PHH1 contain very long (up to 2.7 μm) cylinder-shaped gas vesicles, suggesting that the highly basic GvpI protein (pI = 10.8) is required for the termination of GvpA incorporation (Fig. 9). Contrasting these results, *gvpI*::κ mutants are gas vesicle negative (DasSarma et al. 1994). This result could be again explained by a polar effect of the κ-element on the expression of the adjacent *gvpJ*. ΔH transformants contain gas vesicles of similar size as found in the wild type, but they disaggregate into ribs as soon as they are prepared for electron microscopy (Fig. 9) (Offner et al. 2000). These results imply a stabilizing role of GvpH.

In summary, the "early" proteins encoded by *gvpFGHIJKLM* could organize a self-assembling complex or function as scaffolding proteins in initial stages of bicone formation. The particular location of each of these Gvp proteins in cells during gas vesicle formation or on intact gas vesicles has not yet been determined, but antisera against all of these are now available and it should be possible to gain further insights into the locations of these proteins.

4.3
The Regulatory Proteins GvpD and GvpE

GvpD and GvpE are required for the regulation of *gvp* expression (Fig. 10). **GvpE** is a transcriptional activator, whereas GvpD has a role in the repression of gas vesicle formation (Offner and Pfeifer 1995). Most of the *gvp* promoters are induced by GvpE (Krüger and Pfeifer 1996; Zimmermann and Pfeifer 2003; Hofacker et al. 2004). Archaeal promoters consist of a TATA-box centred around position-28 upstream of the transcriptional start site, and the transcription initiation factor TFB, the TATA-box binding protein TBP and a multicomponent RNA polymerase (RNAP) related to the eukaryotic RNAP II are required for transcription initiation (Bell and Jackson 1998). The sequence TGAAACGG-n4-TGAACCAA located upstream of the TFB-responsive element BRE has been determined as a GvpE-responsive element in the P_{mcA} promoter that is more than 400-fold activated in the presence of GvpE (Gregor and Pfeifer 2005). The C-terminal portion of GvpE consists of an amphiphilic helix including a motif indicative of a basic leucine-zipper protein typically involved in the regulation of gene expression in eukaryotes (Krüger et al. 1998; Plösser and Pfeifer 2002).

GvpD is involved in the repression of gas vesicle formation. ΔD transformants are gas vesicle overproducers with a ball-shaped morphology due to numerous gas vesicles that alter the triangular flat-shaped *Haloferax* cells into little balls (Fig. 11) (Englert et al. 1992b). The amount of gas vesicles is reduced to wild-type levels in $\Delta D + D$ transformants (Pfeifer et al. 1994). The GvpD protein contains a p-loop motif near the *N*-terminus that is important for the repression function (Fig. 10) (Pfeifer et al. 2001). GvpE and GvpD are able to interact, and this interaction appears to induce the degradation of GvpE (Zimmermann and Pfeifer 2003; Scheuch and Pfeifer, unpublished data). Homologues of GvpD and GvpE are not present in bacteria, because the transcription apparatus and also the promoter structures are very different in these organisms.

How environmental factors such as high light intensity, UV light, low oxygen, anaerobic conditions or the salt concentration of the medium affect the regulation of the *gvp* gene expression is not known so far. Many of these conditions lead to a reduction in the amount of *gvpA* transcription in *Hbt. salinarum* or *Hfx. mediterranei* (Englert and Pfeifer 1990; Hechler and Pfeifer, unpublished data). Signal transduction pathways aiming at GvpE or its coun-

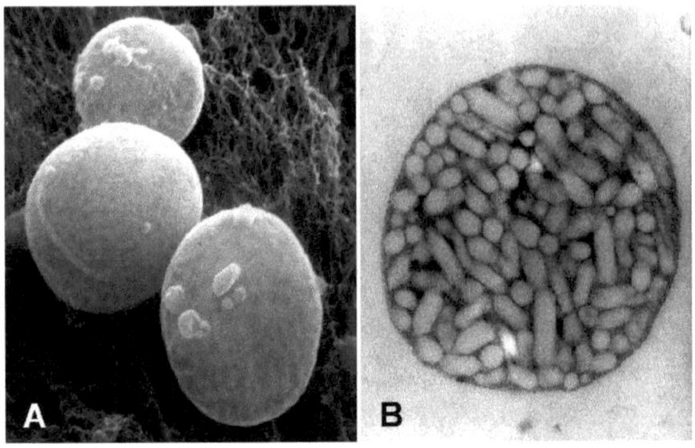

Fig. 11 Electron micrograph of *Hfx. volcanii* transformants harbouring a mc-vac region that incurred a deletion in mc-*gvpD* (ΔD). **A** High resolution scanning electron micrograph of ΔD transformant cells. **B** Transmission electron microscopy of ultrathin sections of ΔD transformants. The diameter of the cells is approximately 1.7 μm in both cases. The micrographs were taken in cooperation with G. Wanner (LMU München)

teractor GvpD could be involved here, but also differences in the constituents of the basal transcription apparatus could play a role. *Hbt. salinarum* contains several genes encoding TBP and TFB that could have a function in the regulation of gene expression. In addition, the activity of RNases appears to depend on the salt concentration leading to a faster degradation of *gvp* mRNAs in cells grown in media containing less than 17% salt (Jäger et al. 2002). Thus, differences in mRNA stability contribute to the salt- and growth phase-dependent presence of *gvp* transcripts.

5 The *gvp* Gene Clusters of Bacteria

The first gene encoding GvpA has been identified in the cyanobacterium *Calothrix* (Damerval et al. 1987). Two related *gvpA* gene copies followed by the *gvpC* gene encoding the stabilizing protein have been identified. Additional *gvp* genes related to the haloarchaeal accessory *gvp* genes were discovered only recently in cyanobacteria. In addition, *gvp* gene clusters have been identified in the soil bacteria *Bacillus megaterium* and several *Actinomyces* species, raising the question of the production and possible function of gas vesicles in these bacteria.

5.1 Gas Vesicle Genes in Cyanobacteria

Two *gvpA* genes and *gvpC* constitute the *gvp* gene cluster identified in the cyanobacterium *Calothrix* (Fig. 12) (Damerval et al. 1987), and similar *gvp*

Archaea				Cyanobacteria			soil bacteria			
Hbt.	Hfx.	Hrr.	Ms.	Ana	Nos	Mic	Bac	Strep	Fran	Rhodo
A	A	A	A	A	A	A	A	O	O	Y
C	C	C	A	A	A	A	P	A		Z
N	N	N	A	C	C	A	Q			A
O	O	O		N	N	C	B			
						N	R			
D	D						N			
E	E									
F	F	F	F	J	J	J	F	F	J	F
G	G	G	G			X	G	G	L	G
H	H	H		K	K	K	L	Y	S	O
I	I	I		F/L	F/L	F/L	S	Z	K	
J	J	J	J	G	G	G	K	J	A	J
K	K	K	K	V	V	V	J	L	F	L
L	L	L	L	W	W	W	T	S	Y	S
M	M	M	M				U	K	Z	K

Fig. 12 Schematic comparison of the *gvp* gene clusters found in Archaea and Bacteria. The vertical columns indicate the arrangement of *gvp* genes in each cluster

gene clusters have been characterized in *Anabaena*, *Pseudoanabaena* and various *Planktothrix* species (Damerval et al. 1987, 1991; Hayes and Powell 1995; Beard et al. 2000a). Two to seven copies of *gvpA* are found in *Planktothrix* (Beard et al. 2000a, 2002a), and three gene copies are present in the *Microcystis aeruginosa* (Mlouka et al. 2004). Multiple copies of *gvpA* encoding the major gas vesicle structural protein appear to be typical for cyanobacteria producing numerous gas vesicles per cell, whereas the smaller haloarchaeal cells only contain a single *gvpA* gene per cluster.

The *gvp* gene cluster of *Microcystis* consists of the two transcription units *gvpA*$_{\mathrm{I}}$,*A*$_{\mathrm{II}}$,*A*$_{\mathrm{III}}$ *CNJX* and *gvpKFG*, followed by *gvpV* in inverse orientation, and *gvpW* is again in the same orientation as *gvpKFG* (Fig. 12). Similar genes and gene organizations are found in *Anabaena/Nostoc* PCC7120, including *gvpW* and *gvpV* but lacking *gvpX* (Kinsman and Hayes 1997; Mlouka et al. 2004). However, the gene encoding the essential GvpM of haloarchaea is missing. An analysis by transformation experiments demonstrating the minimal requirement of these genes for gas vesicle formation is still lacking for cyanobacteria.

5.2 The *gvp* Gene Cluster of Soil Bacteria

Gas vesicle genes have been characterized in the Gram-positive *Bacillus megaterium* and various *Streptomyces, Rhodococcus* and *Frankia* species, demonstrating that homologous *gvp* genes are also found in soil bacteria. *Bacillus megaterium* contains the *gvpAPQBRNFGLSKJTU* gene cluster

(Fig. 12) (Li and Cannon 1998), and transformation of *Escherichia coli* with this cluster (or *gvpB-U*) results in the formation of tiny gas vesicles with an average length of 40 nm. The *gvpA* and *gvpB* genes both encode hydrophobic, GvpA-type proteins. GvpR exhibits 44% sequence similarity to the GvpO of *Streptomyces coelicolor*, and 39% to the GvpO of haloarchaea. GvpS could be a more distant relative of the GvpA-J-M family of hydrophobic gas vesicle proteins and might be a substitute for GvpM in the soil bacteria. The basic GvpQ is somewhat related to GvpI of the haloarchaea. The *gvpTU* genes located at the end of the cluster have no archaeal homologues, but are essential for gas vesicle formation in *E. coli* (Li and Cannon 1998).

Even more astounding is the presence of gas vesicle gene clusters in the mycelium forming *Streptomyces coelicolor* (Offner et al. 2000; Bentley et al. 2002). The *gvpOAFGYZJLSK* gene cluster contains homologues to all essential haloarchaeal *gvp* genes, similar to *Rhodococcus* (van Keulen et al. 2005). The gene cluster of *Frankia* (*gvpOJLSKAFYZ*) lacks *gvpG* (Fig. 12). None of these soil bacteria contain genes encoding the stabilizer protein GvpC or the chaperone GvpN. All contain the "early" *gvp* genes in the arrangement *gvpJLSK*, and only in *Bacillus gvpJ* is found at a different position. In contrast, haloarchaea and the methanogene *Methanosarcina barkeri* harbour these genes as a *gvpJKLM* cluster (Fig. 12).

A major difference to the gas-vesicle forming aquatic bacteria is that the *gvpA* gene of *Streptomyces* encodes a 124 to 170 aa product that is almost twice as large as the 71- to 79-aa GvpA of haloarchaea and cyanobacteria (Fig. 5) (von Keulen et al. 2005). The additional C-terminal portion does not resemble other known sequences. Gas vesicles have not yet been observed in *Streptomyces*, and given the functional constraints of GvpA, this protein might not be used to form gas-filled structures. Further studies are required to determine the time points of expression and also the location of the various Gvp proteins in these mycelial bacteria.

6 Biotechnological Applications

Gas vesicles isolated from the cyanobacterium *Anabaena flos-aquae* have been used to increase the oxygen supply of mammalian cells in culture (Kashyap et al. 1998; Sundararajan and Ju 2000). The gas vesicles are harvested from *Anabaena* cells lysed in a hypertonic sucrose solution, and strengthened and sterilized by glutaraldehyde treatment. These gas vesicles are then used as oxygen carriers to substitute the addition of perfluorocarbons and hydrocarbons usually used to enhance the oxygen supply in bioreactors. These substances have to be dispersed into very fine droplets to achieve high efficiency, and this requires vigorous mechanical agitation which cannot be used in shear-sensitive systems. The addition of 1.8% gas vesicles to a cell culture leads to a 35% higher cell concentration, indicative of a further support of the cell culture with oxygen.

Isolated gas vesicles of haloarchaea have been used for the production of antibodies (Stuart et al. 2001, 2004). The insertion of foreign peptides of 6- to 235-aa length in the C-terminal portion of GvpC still leads to functional gas vesicles that can be isolated by flotation and are used in the absence of adjuvant to immunize mice. The recombinant gas vesicles present the desired peptidyl epitopes at the surface and lead to a long-lived antibody response in rabbits. This method has been used to raise antibodies against various amino acid sequences of SIV.

Acknowledgements This work was supported by the Deutsche Forschungsgemeinschaft (DFG). Arnulf Kletzin is thanked for support in bioinformatic analyses, and Torsten Hechler for several microscopic pictures of haloarchaea. Gerhard Wanner (LMU München, Germany), Sonja Offner and Christoph Englert contributed as former coworkers to the electron microscopy studies.

References

Archer D, King N (1984) Isolation of gas vesicles from *Methanosarcina barkeri*. J Gen Microbiol 130:167–172

Beard S, Handley B, Hayes PK, Walsby AE (1999) The diversity of gas vesicle genes in *Planktothrix rubescens* from Lake Zurich. Microbiology 145:2757–2768

Beard S, Davis P, Iglesias-Rodriguez D, Skulberg O, Walsby AE (2000) Gas vesicle genes in *Planktothrix* spp. from Nordic lakes: strains with weak gas vesicles possess a longer variant of GvpC. Microbiol 146:2009–2018

Beard S, Handley B, Walsby AE (2002a) Spontaneous mutations in gas vesicle genes of *Planktothrix* spp. affect gas vesicle production and critical pressure. FEMS Microbiol Lett 215:189–195

Beard S, Hayes PK, Pfeifer F, Walsby AE (2002b) The sequence of the major gas vesicle protein, GvpA, influences the width and strength of halobacterial gas vesicles. FEMS Microbiol Lett 213:149–157

Belenky M, Meyers R, Herzfeld J (2004) Subunit structure of gas vesicles: A MALDI-TOF mass spectrometry study. Biophys J 86:499–505

Bell SD, Jackson S (1998) Transcription and translation in Archaea: a mosaic of eukaryal and bacterial features. *Trends in Microbiol* 6:222–228

Bentley S, Chater K, Cerdeno-Tarraga A, Challis G, Thompson N, James K, Harris D, Quail M, Kieser H et al. (2002) Complete genome sequence of the model actinomycete *Streptomyces coelicolor* A3(2). Nature 417:141–147

Blaurock A, Walsby AE (1976) Crystalline structure of gas vesicle wall from *Anabaena flos-aquae*. J Mol Biol 105:183–199

Blaurock A, Wober W (1976) Structure of wall of *Halobacterium halobium* gas vesicles. J Mol Biol 106:871–888

Bolhuis H, te Poele, Rodriguez-Valera F (2004) Isolation and cultivation of Walsby's square archaeon. Environ Microbiol 6:1287–1291

Bowen C, Jensen T (1965) Blue-green algae: fine structure of the gas vacuoles. Science 147:1460–1462

Bright D, Walsby AE (1999) The relationship between critical pressure and width of gas vesicles in isolates of *Planktothrix rubescens* from Lake Zurich. Microbiol 145:2769–2775

Buchholz B, Hayes PK, Walsby AE (1993) The distribution of the outer gas vesicle protein, GvpC, on the *Anabaena* gas vesicle, and its ratio to GvpA. J Gen Microbiol 139:2353–2363

Burns D, Camakaris H, Janssen P, Dyall-Smith M (2004) Cultivation of Walsby's square haloarchaeon. FEMS Microbiol Lett 238:469–473

Clarke A, Walsby AE (1978) The occurrence of gas vacuolate bacteria in lakes. Arch Microbiol 118:223–228

Damerval T, Houmard J, Guglielmi G, Csiszar K, Tandeau de Marsac N (1987) A developmentally regulated *gvpABC* operon is involved in the formation of gas vesicles in the cyanobacterium *Calothrix*-7601. Gene 54:83–92

Damerval T, Castets A, Houmard J, Tandeau de Marsac N (1991) Gas vesicle synthesis in the cyanobacterium *Pseudanabaena* sp—Occurrence of a single photoregulated gene. Mol Microbiol 5:657–664

DasSarma S, Damerval T, Jones J, Tandeau de Marsac N (1987) A plasmid-encoded gas vesicle protein gene in a halophilic archaebacterium. Mol Microbiol 1:365–370

DasSarma S, Halladay J, Jones J, Donovan J, Giannasca P, Tandeau de Marsac N (1988) High-frequency mutations in a plasmid-encoded gas vesicle gene in *Halobacterium halobium*. Proc Natl Acad Sci USA 85:6861–6865

DasSarma S, Arora P, Lin F, Molinari E, Yin L (1994) Wild-type gas vesicle formation requires at least 10 genes in the *gvp* gene-cluster of *Halobacterium halobium* plasmid pNRC100. J Bacteriol 176:7646–7652

Dunton P, Walsby AE (2005) The diameter and critical collapse pressure of gas vesicles in *Microcystis* are correlated with GvpCs of different length. FEMS Microbiol Let 247:37–43

Englert C, Pfeifer F (1993) Analysis of gas vesicle gene expression in *Haloferax mediterranei* reveals that GvpA and GvpC are both gas vesicle structural proteins. J Biol Chem 268:9329–9336

Englert C, Horne M, Pfeifer F (1990) Expression of the major gas vesicle protein gene in the halophilic archaebacterium *Haloferax mediterranei* is modulated by salt. Mol Gen Genet 222:225–232

Englert C, Krüger K, Offner S, Pfeifer F (1992a) Three different but related gene clusters encoding gas vesicles in halophilic Archaea. J Mol Biol 227:586–592

Englert C, Wanner G, Pfeifer F (1992b) Functional analysis of the gas vesicle gene cluster of the halophilic Archaeon *Haloferax mediterranei* defines the vac-region boundary and suggests a regulatory role for the GvpD gene or its product. Mol Microbiol 6:3543–3550

Gosink J, Herwig R, Staley J (1997) *Octadecabacter arcticus* gen nov, sp nov, and *O. antarcticus*, sp nov, nonpigmented, psychrophilic gas vacuolate bacteria from polar sea ice and water. Syst Appl Microbiol 20:356–365

Gregor D, Pfeifer F (2001) Use of a halobacterial *bgaH* reporter gene to analyse the regulation of gene expression in halophilic Archaea. Microbiol 147:1745–1754

Gregor D, Pfeifer F (2005) In vivo analyses of constitutive and regulated promoters in halophilic archaea. Microbiol 151:25–33

Halladay J, Jones J, Lin F, Macdonald A, DasSarma S (1993) The rightward gas vesicle operon in *Halobacterium* plasmid pNRC100—identification of the *gvpA* and *gvpC* gene products by use of antibody probes and genetic analysis of the region downstream of *gvpC*. J Bacteriol 175:684–692

Hayes PK, Buchholz B, Walsby AE (1992) Gas vesicles are strengthened by the outer-surface protein, GvpC. Arch Microbiol 157:229–234

Hayes PK, Lazarus C, Bees A, Walker J, Walsby AE (1988) The protein encoded by GvpC is a minor component of gas vesicles isolated from the cyanobacteria *Anabaena flos-aquae* and *Microcystis* sp. Mol Microbiol 2:545–552

Hayes PK, Powell R (1995) The *gvpA/C* cluster of *Anabaena flos-aquae* has multiple copies of a gene encoding GvpA. Arch Microbiol 164:50–57

Hayes PK, Walsby AE (1984) An investigation into the recycling of gas vesicle protein derived from collapsed gas vesicles. J Gen Microbiol 130:1591–1596

Hayes PK, Walsby AE, Walker J (1986) Complete amino acid sequence of cyanobacterial gas vesicle protein indicates a 70-residue molecule that corresponds in size to the crystallographic unit cell. Biochem J 236:31–36

Hofacker A, Schmitz K, Cichonczyk A, Sartorius-Neef S, Pfeifer F (2004) GvpE- and GvpD-mediated transcription regulation of the p-*gvp* genes encoding gas vesicles in *Halobacterium salinarum*. Microbiol 150:1829–1838

Horne M, Pfeifer F (1989) Expression of two gas vacuole protein genes in *Halobacterium halobium* and other related species. Mol Gen Genet 218:437–444

Horne M, Englert C, Pfeifer F (1988) Two genes encoding gas vacuole proteins in *Halobacterium halobium*. Mol Gen Genet 213:459–464

Horne M, Englert C, Wimmer C, Pfeifer F (1991) A DNA region of 9 kbp contains all genes necessary for gas vesicle synthesis in halophilic archaebacteria. Mol Microbiol 5:1159–1174

Houwink A (1956) Flagella, gas vacuoles and cell-wall structure in *Halobacterium halobium*; an electron microscopy study. J Gen Microbiol 15:146–150

Irgens R, Suzuki I, Staley J (1989) Gas vacuolate bacteria obtained from marine waters of Antarctica. Curr Microbiol 18:261–265

Isaksen F, Teske A (1996) *Desulforhodopalus vacuolatus* gen. nov, sp nov, a new moderately psychrophilic sulfate-reducing bacterium with gas vacuoles isolated from a temperate estuary. Arch Microbiol 166:160–168

Jäger A, Samorski R, Pfeifer F, Klug G (2002) Individual *gvp* transcrip segments in *Haloferax mediterranei* exhibit varying half-lives, which are differentially affected by salt concentration and growth phase. Nucleic Acids Res 30:5436–5443

Jones J, Young D, DasSarma S (1991) Structure and organization of the gas vesicle gene cluster on the *Halobacterium halobium* plasmid pNRC100. Gene 102:117–122

Jung D, Achenbach L, Karr E, Takaichi S, Madigan M (2004) A gas vesiculate planktonic strain of the purple non-sulfur bacterium *Rhodoferax antarcticus* isolated from Lake Fryxell, Dry Valleys, Antarctica. Arch Microbiol 182:236–243

Kashyap S, Sundararajan A, Ju L (1998) Flotation characteristics of cyanobacterium *Anabaena flos-aquae* for gas vesicle production. Biotechnol Bioengineer 60:636–641

Karr E, Sattley W, Jung D, Madigan M, Achenbach L (2003) Remarkable diversity of phototrophic purple bacteria in a permanently frozen Antarctic lake. Appl Environ Microbiol 69:4910–4914

Kinsman R, Hayes PK (1997) Genes encoding proteins homologous to halobacterial Gvps N, J, K, F, & L are located downstream of *gvpC* in the cyanobacterium *Anabaena flos-aquae*. DNA Seq 7:97–106

Kinsman R, Walsby AE, Hayes PK (1995) GvpCs with reduced numbers of repeating sequence elements bind to and strengthen cyanobacterial gas vesicles. Mol Microbiol 17:147–154

Klebahn H (1895) Gasvakuolen, ein Bestandteil der Zellen der wasserblütenbildenden Phycochromaceen. Flora (Jena) 80:241

Konopka A, Staley J, Lara J (1975) Gas vesicle assembly in *Microcyclus aquaticus*. J Bacteriol 122:1301–1309

Konopka A, Lara J, Staley J (1977) Isolation and characterization of gas vesicles from *Microcyclus aquaticus*. Arch Microbiol 112:133–140

Krantz M, Ballou C (1973) Analysis of *Halobacterium halobium* gas vesicles. J Bacteriol 114:1058–1067

Krasil'nikov N, Duda V, Pivovarov G (1971) Characteristics of the cell structure of soil anaerobic bacteria forming vesicular caps on their spores. Mikrobiologiya 40:681–685

Krüger K, Hermann T, Armbruster V, Pfeifer F (1998) The transcriptional activator GvpE for the halobacterial gas vesicle genes resembles a basic region leucine-zipper regulatory protein. J Mol Biol 279:761–771

Krüger K, Pfeifer F (1996) Transcript analysis of the c-vac region and differential synthesis of the two regulatory gas vesicle proteins GvpD and GvpE in *Halobacterium salinarium* PHH4. J Bacteriol 178:4012–4019

Larsen H, Omang S, Steensland H (1967) On the gas vacuoles of the halobacteria. Arch Microbiol 59:197–203

Lehmann H, Jost M (1971) Kinetics of the assembly of gas vacuoles in the blue-green alga *Microcystis aeruginosa* Kuetz emend Elkin. Arch Microbiol 79:59–68

Li N, Cannon M (1998) Gas vesicle genes identified in *Bacillus megaterium* and functional expression in *Escherichia coli*. J Bacteriol 180:2450–2458

Mayr A, Pfeifer F (1997) The characterization of the nv-*gvpACNOFGH* gene cluster involved in gas vesicle formation in *Natronobacterium vacuolatum*. Arch Microbiol 168:24–32

McMaster T, Miles M, Walsby AE (1996) Direct observation of protein secondary structure in gas vesicles by atomic force microscopy. Biophys J 70:2432–2436

Mlouka A, Comte K, Castets A, Bouchier C, Tandeau de Marsac N (2004) The gas vesicle gene cluster from *Microcystis aeruginosa* and DNA rearrangements that lead to loss of cell buoyancy. J Bacteriol 186:2355–2365

Mwatha W, Grant W (1993) *Natronobacterium vacuolatum* sp nov, a haloalkaliphilic archaeon isolated from Lake Magadi, Kenya. Int J Syst Bacteriol 43:401–404

Neuwald A, Aravind L, Spouge J, Koonin E (1999) AAA(+): A class of chaperone-like ATPases associated with the assembly, operation, and disassembly of protein complexes. Genome Res 9:27–43

Ng W, Arora P, DasSarma S (1994) Large deletions in class III gas vesicle-deficient mutants of *Halobacterium halobium*. Syst Appl Microbiol 16:560–568

Ng W, Kennedy S, Mahairas G, Berquist B, Pan M, Shukla H, Lasky S, Baliga N, Thorsson V, Sbrogna J et al. (2000) Genome sequence of *Halobacterium* species NRC-1. Proc Natl Acad Sci USA 97:12176–12181

Oesterhelt D (1998) The structure and mechanism of the family of retinal proteins from halophilic archaea. Curr Opin Struct Biol 8:489–500

Offner S, Pfeifer F (1995) Complementation studies with the gas vesicle-encoding p-vac region of *Halobacterium salinarium* PHH1 reveal a regulatory role for the p-*gvpDE* genes. Mol Microbiol 16:9–19

Offner S, Wanner G, Pfeifer F (1996) Functional studies of the *gvpACNO* operon of *Halobacterium salinarium* reveal that the GvpC protein shapes gas vesicles. J Bacteriol 178:2071–2078

Offner S, Ziese U, Wanner G, Typke D, Pfeifer F (1998) Structural characteristics of halobacterial gas vesicles. Microbiol 144:1331–1342

Offner S, Hofacker A, Wanner G, Pfeifer F (2000) Eight of fourteen *gvp* genes are sufficient for formation of gas vesicles in halophilic archaea. J Bacteriol 182:4328–4336

Oliver R, Walsby AE (1984) Direct evidence for the role of light-mediated gas vesicle collapse in the buoyancy regulation of *Anabaena flos-aquae* (Cyanobacteria). Limnol Oceanogr 29:879–886

Overmann J, Lehmann S, Pfennig N (1991) Gas vesicle formation and buoyancy regulation in *Pelodictyon phaeoclathratiforme* (green sulfur bacteria). Arch Microbiol 157:29–37

Parkes K, Walsby AE (1981) Ultrastructure of a gas-vacuolate square bacterium. J Gen Microbiol 126:503–506

Petter H (1931) On bacteria of salted fish. Proc Natl Acad Sci Amsterdam 34:1417–1423

Pfeifer F (2004) Gas vesicle genes in halophilic archaea and bacteria. In: Ventosa A (ed) Halophilic Microorganisms. Springer, Berlin Heidelberg New York, p 229–239

Pfeifer F, Betlach M (1985) Genome organization in *Halobacterium halobium*: A 70-kb island of more (AT)-rich DNA in the chromosome. Mol Gen Genet 198:449–455

Pfeifer F, Blaseio U (1989) Insertion elements and deletion formation in a halophilic archaebacterium. J Bacteriol 171:5135–5140

Pfeifer F, Englert C (1992) Function and biosynthesis of gas vesicles in halophilic archaea. J Bioenerg Biomembr 24:577–585

Pfeifer F, Weidinger G, Goebel W (1981) Genetic variability in *Halobacterium halobium*. J Bacteriol 145:375–381

Pfeifer F, Zotzel J, Kurenbach B, Röder R, Zimmermann P (2001) A p-loop motif and two basic regions in the regulatory protein GvpD are important for the repression of gas vesicle formation in the archaeon *Haloferax mediterranei*. Microbiol 147:63–73

Plösser P, Pfeifer F (2002) A bZIP protein from halophilic archaea: structural features and dimer formation of cGvpE from *Halobacterium salinarum*. Mol Microbiol 45:511–520

Powell RS, Walsby AE, Hayes PK, Porter R (1991) Antibodies to the N-terminal sequence of GvpA bind to the ends of gas vesicles. J Gen Microbiol 137:2395–2400

Röder R, Pfeifer F (1996) Influence of salt on the transcription of the gas-vesicle genes of *Haloferax mediterranei* and identification of the endogenous transcriptional activator gene. Microbiol 142:1715–1723

Shukla HD, DasSarma S (2004) Complexity of gas vesicle biogenesis in *Halobacterium* sp strain NRC-1: Identification of five new proteins. J Bacteriol 186:3182–3186

Staley J (1968) *Prosthecomicrobium* and *Ancalomicrobium*: new prosthecate freshwater bacteria. J Bacteriol 95:1921–1942

Staley J, Irgens R, Brenner D (1987) *Enhydrobacter aerosarcus* gen nov, sp nov, a gas-vacuolated, facultatively anaerobic, heterotrophic rod. Int J Syst Bacteriol 37:289–291

Staley J, Irgens R, Herwig R (1989) Gas vacuolate bacteria from the sea ice of Antarctica. Appl Environ Microbiol 55:1033–1036

Stoeckenius W, Kunau W (1968) Further characterization of particulate fractions from lysed cell envelopes of *Halobacterium halobium* and isolation of gas vacuole membranes. J Cell Biol 38:337–357

Stuart E, Morshed F, Sremac M, DasSarma S (2001) Antigen presentation using novel particulate organelles from halophilic archaea. J Biotechnol 88:119–128

Stuart E, Morshed F, Sremac M, DasSarma S (2004) Cassette-based presentation of SIV epitopes with recombinant gas vesicles from halophilic archaea. J Biotechnol 114:225–237

Sundararajan A, Ju L (2000) Glutaraldehyde treatment of proteinaceous gas vesicles form cyanobacterium *Anabaena flos-aquae*. Biotechnol Prog 16:1124–1128

Thomas R, Walsby AE (1985) Buoyancy regulation on a strain of *Microcystis*. J Gen Microbiol 131:799–809

Van Ert M, Staley J (1971) A new gas vacuolated heterotrophic rod from freshwaters. Arch Microbiol 80:70–77

Van Keulen G, Hopwood D, Dijkhuizen L, Sawers G (2005) Gas vesicles in actinomycetes: old buoys in novel habitats? Trends Microbiol 13:350–354

Waaland JR, Branton D (1969) Gas vacuole development in a blue-green alga. Science 163:1339–1341

Walker J, Hayes PK, Walsby AE (1984) Homology of gas vesicle proteins in cyanobacteria and halobacteria. J Gen Microbiol 130:2709–2715

Walsby AE (1972) Structure and function of gas vacuoles. Bacteriol Rev 36:1–32

Walsby AE (1982) Permeability of gas vesicles to perfluorocyclobutane. J Gen Microbiol 128:1679–1684

Walsby AE (1994) Gas vesicles. Microbiol Rev 58:94–144

Walsby AE (2005) Archaea with square cells. Trends Microbiol 13:193–195

Walsby AE, Bleything A (1988) The dimensions of cyanobacterial gas vesicles in relation to their efficiency in providing buoyancy and withstanding pressure. J Gen Microbiol 134:2635–2645

Walsby AE, Hayes PK (1988) The minor cyanobacterial gas vesicle protein, GvpC, is attached to the outer surface of the gas vesicle. J Gen Microbiol 134:2647–2657

Walsby AE, McAllister G (1987) Buoyancy regulation of microcystis in lake Okaro, NZ. J Mar Freshwater Res 21:521–524

Walsby AE, Revsbech NP, Griffel DH (1992) The gas-permeability coefficient of the cyanobacterial gas vesicle wall. J Gen Microbiol 138:837–845

Yao V, Spudich J (1992) Primary structure of an archaebacterial transducer, a methyl-accepting protein associated with sensory rhodopsin I. Proc Natl Acad Sci USA 89:11915–11919

Zhang W, Brooun A, Mueller M, Alam M (1996) The primary structure of the archaeon *Halobacterium salinarum* blue light receptor sensory rhodopsin II and its transducer, a methyl-accepting protein. Proc Natl Acad Sci USA 93:8230–8235

Zimmermann P, Pfeifer F (2003) Regulation of the expression of gas vesicle genes in *Haloferax mediterranei*: interaction of the two regulatory proteins GvpD and GvpE. Mol Microbiol 49:783–794

Microbiol Monogr (2)
J.M. Shively: Complex Intracellular Structures in Prokaryotes
DOI 10.1007/7171_023/Published online: 20 May 2006

Carboxysomes and Carboxysome-like Inclusions

Sabine Heinhorst[1] (✉) · Gordon C. Cannon[1] · Jessup M. Shively[1,2]

[1]Department of Chemistry and Biochemistry, The University of Southern Mississippi, 118 College Dr. #5043, Hattiesburg, MS 38406-0001, USA
sabine.heinhorst@usm.edu

[2]Department of Genetics and Biochemistry, Clemson University, Clemson, SC 29634-0318, USA

Abstract Carboxysomes and related polyhedral bacterial inclusions are complex structures that are composed of a limited set of related proteins. The importance of these prokaryotic organelles as metabolic organizers in autotrophs as well as heterotrophic bacteria is becoming much more apparent. The carboxysome, which is by far the best characterized representative of these inclusions, is found in a variety of phylogenetically distant autotrophic bacteria and contains the central CO_2 fixing enzyme, ribulose-1,5-bisphosphate carboxylase/oxygenase (RuBisCO). The particle participates in the essential CO_2 concentrating mechanism and is likely protecting RuBisCO from oxygen. By contrast, the functions of polyhedral inclusions in heterotrophic prokaryotes that have been experimentally observed or inferred from comparative genomic analyses are less well understood. This review summarizes the current state of knowledge regarding structure, function and genetics of carboxysomes and related polyhedral microcompartments.

1
Introduction

Since their discovery in 1973 (Shively et al. 1973a,b) and the elucidation of their role in bacterial CO_2 fixation, carboxysomes have been largely regarded as a peculiarity limited to autotrophic microbes and, as such, have not received their due attention. However, these prototypes of bacterial microcompartments have recently been cast into the scientific limelight by the discovery that many heterotrophic bacteria contain genes related to carboxysome shell proteins and are able to form polyhedral particles under special metabolic conditions. Likewise, in a time of concerns about global climate changes caused by increased CO_2 emissions due to human activities, the realization that many carboxysome-forming cyanobacteria and chemolithotrophs are important CO_2 consumers whose metabolic activity contributes significantly to the global carbon cycle (Cannon et al. 2001) has kindled renewed interest in carboxysome structure and its relationship to the function of these bacterial inclusions. Finally, carboxysomes and related bacterial microcompartments have the potential to be exploited in the emerging field of nanotechnology. An understanding of the mechanisms by which simple protein components self-assemble into complex three-dimensional chemical reactors that are capable of improving the catalytic efficiency of enzymes could lead to the development of nanoscale devices for a variety of applications in biomedicine and other fields.

2
Carboxysome Structure, Composition and Function

Electron microscopic examination of bacterial thin sections and of negatively stained carboxysome preparations at various stages of purification revealed particles of more or less regular hexagonal shape and with diameters ranging from 100 to 400 nm. In general, the carboxysomes of chemolithotrophs, which have now been classified as cso- or α-type, are smaller and of a more uniform size (100–120 nm diameter) (Shively and English 1991) than the polyhedral bodies of many cyanobacteria, particularly those of filamentous species, which can be quite large (Price et al. 1998). Although carboxysomes are reminiscent of icosahedral virions, conventional and cryo-electron microscopic studies of carboxysomes from *Halothiobacillus neapolitanus* (Fig. 1) showed that, unlike any known virions, geometry and size of individual facets as well as symmetry relationships in entire particles vary, depending on the viewing angle (Paredes et al. 2001). Similar analyses have yet to be performed with the ccm- or β-type carboxysomes found in some cyanobacteria. Carboxysome geometry and three-dimensional structure remain unresolved questions that require further ultrastructural and crystallographic investigations.

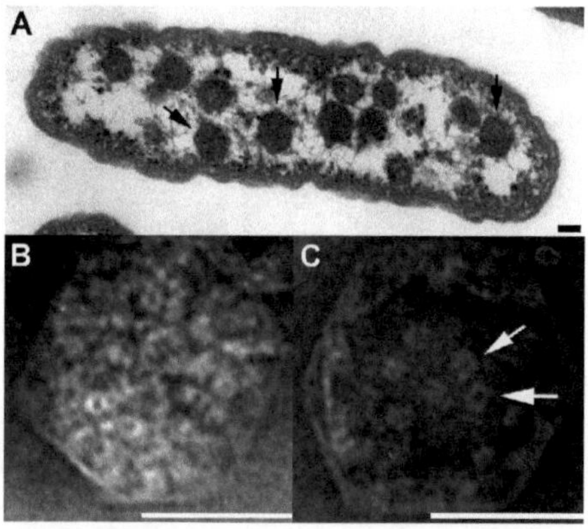

Fig. 1 Transmission electron micrographs of carboxysomes from *Halothiobacillus neapolitanus*. *Panel A:* Thin section of a *H. neapolitanus* cell containing several polyhedral carboxysomes (some marked by *arrows*); *Panel B*: Negatively stained purified carboxysome; *Panel C:* The shell of a broken carboxysome. The small donut-shaped structures (see *arrows*) are individual form I RuBisCO holoenzyme molecules. *Scale bar* = 100 nm

A distinct ultrastructural feature of carboxysomes is their 3–4 nm thick shell that surrounds numerous smaller, regularly shaped entities of almost doughnut shape. These components, which seem to completely fill the carboxysome interior, represent individual molecules of form I ribulose 1,5-bisphosphate carboxylase/oxygenase (RuBisCO) holoenzyme (Fig. 1) (Shively et al. 1973a,b; Lanares and Codd 1981; Holthuijzen et al. 1986c). The large and small subunits of the carboxysomal RubisCO are present in equimolar amounts in *H. neapolitanus* carboxysomes. Together, these two polypeptides represent approximately 60–70 wt % of the particle's protein (Cannon 1982; Cannon and Shively 1983; Cannon et al. 1991) (Table 1). The presence of the remaining Calvin–Benson–Bassham (CBB) cycle enzymes in *H. neapolitanus* carboxysomes, as suggested earlier (Beudeker and Kuenen 1981), could not be corroborated for the particles in this species, and has since been attributed to contamination with cytosolic proteins (Holthuijzen et al. 1986b,c). Assuming that carboxysomes are filled completely with RuBisCO, and considering size and geometry of the microcompartments as well as the enzyme's quaternary structure, approximately 270 copies of RuBisCO holoenzyme are contained in each particle (Shively and English 1991), corresponding to approximately 2160 copies of each subunit type (Table 1). This estimate is in contrast to the suggestion by Holthuijzen et al. (1986b) that carboxysomes are essentially hollow compartments, with the thin layer of RuBisCO holoenzyme molecules that remain attached to the shell of broken carboxysomes (Fig. 1) representing their entire RuBisCO complement. Although intriguing, this idea is not supported by the observed stoichiometries of carboxysome proteins and the amount of soluble RuBisCO that can be recovered from broken carboxysomes (Table 1) (Cannon and Shively 1983).

Table 1 *Halothiobacillus neapolitanus* carboxysome composition

Protein [a]	M_r theor.[b] [kDa]	M_r app.[b] [kDa]	Wt % [c]	Copies per carboxysome	Theor. pI [d]	Comments
CbbL	52.3	55	70 (62)	2160	5.82	270 form 1 RuBisCO copies per carboxysome
CbbS	12.8	12.6		2160	6.06	
CsoS2A	92	130	6	143	9.06	329 CsoS2 copies per carboxysome
CsoS2B		85	6	186		
CsoS3	57	64	2.3	81	6.08	40 CsoS3 dimers per carboxysome?
OrfA	8.9	—	—	—	5.72	
OrfB	8.8	—	—	—	5.15	
CsoS1C	9.9	6.4	13 (13)	2970	5.60	3510 CsoS1 copies per carboxysome
CsoS1A	9.9	6.4			5.58	
CsoS1B	11.3	15	3.7 (4)	540	4.86	

[a] Proteins are listed in the order in which their genes are located in the putative cso operon

[b] M_r theor. is the molecular mass calculated from the protein's deduced amino acid sequence; M_r app. is its apparent molecular mass based on its SDS-PAGE migration behavior

[c] Wt % is the percent of the carboxysome molecular mass (2.35×10^8 kDa) contributed by individual polypeptides. The values in parentheses were calculated by Cannon and Shively (1983)

[d] The pI was calculated from the protein's deduced amino acid sequence

To date, only the chemolithotrophs *H. neapolitanus* and two strains of the genus *Nitrobacter* have yielded preparations of highly purified carboxysomes that are stable in vitro and suitable for rigorous biochemical characterization (Shively et al. 1977; Ebert 1982; Cannon and Shively 1983; Shively and English 1991). In particular the homogeneous carboxysome preparations from *H. neapolitanus* have been useful in shedding light on the particles' protein composition through *N*-terminal amino acid sequencing of individual polypeptide components, which provided identities for the major carboxysomal shell proteins and facilitated identification of carboxysome genes in this and other bacteria (Cannon and Shively 1983; English et al. 1994; Lorbach 1995; Baker 1998). By contrast, obtaining pure carboxysomes from cyanobacteria has proven extremely difficult. Cyanobacterial carboxysomes do not stand up well to purification, and the abundant interior photosynthetic membranes in these bacteria complicate the distinction between bona fide carboxysome components and co-purifying unrelated protein contaminants (Lanares and Codd 1981; Price et al. 1992). Proteomics approaches applied to carboxysome-enriched fractions from *Synechococcus* WH8102 (Gonzales et al. 2005) and *Synechococcus* PCC7942 (Long et al. 2005) confirmed the presence of some candidate proteins in carboxysome-enriched preparations but failed to establish what is likely to be their full shell protein complement. Instead, the field has had to rely mainly on genetic approaches to identify putative carboxysome proteins in cyanobacteria. What is clear, however, is that carboxysomes do not have a lipid component (Cannon and Shively 1983; Holthuijzen et al. 1986c; Codd 1988; Shively and English 1991). Likewise, DNA does not seem to be associated with carboxysomes (Holthuijzen et al. 1986a), contrary to an earlier finding that had prompted the intriguing suggestion of an evolutionary relationship between the polyhedra of autotrophic bacteria and bacteriophages (Westphal et al. 1979). All current biochemical evidence is consistent with the assumption that carboxysomes are composed entirely of proteins. The molecular mass of the pure particles from *H. neapolitanus* was estimated to be approximately 2.4×10^8 Da (Holthuijzen et al. 1986b; Shively and English 1991).

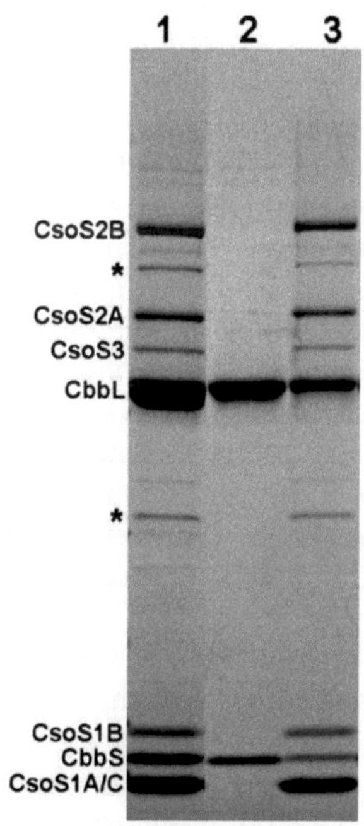

Fig. 2 Stained carboxysome polypeptides separated by SDS-polyacrylamide gel electrophoresis. *Lane 1*: Purified carboxysome proteins; *Lane 2*: Proteins released from broken carboxysomes (supernatant after centrifugation); *Lane 3*: Shell-enriched fraction of broken carboxysomes (pellet after centrifugation). The carboxysome polypeptides are identified on the *left*. As determined by mass spectrometry (unpublished), the bands marked with *asterisks* are SDS-resistant aggregates of CsoS1A and CsoS1C (*lower band*) and a possible third structural variant of CsoS2 (*upper band*), respectively

Following limited mechanical disintegration, fractured carboxysomes from *H. neapolitanus* can be separated by density gradient centrifugation into a RuBisCO-enriched supernatant and a pellet of near-empty "ghost" shells that more or less maintain their polyhedral shape (Fig. 1). The shell polypeptides, which together account for approximately 30 wt % of the total carboxysome protein in this bacterium, separate during denaturing polyacrylamide gel electrophoresis into at least five discernible bands of apparent molecular weights that range from 6400 to 130 000 (Fig. 2, Table 1) (Cannon and Shively 1983; Holthuijzen et al. 1986c; Cannon et al. 1991; Baker et al. 2000).

2.1
The CsoS1, CcmK and CcmO Proteins

By far the most abundant shell components in carboxysomes from *H. neapolitanus* are three closely related small proteins of almost identical primary structure, termed CsoS1A, CsoS1B and CsoS1C (Fig. 2) (English et al. 1994). CsoS1A and CsoS1C are different in only one amino acid and 90% homologous to CsoS1B, which is 12 residues longer than Cso1A and CsoS1C (Shively et al. 1996; Cannon et al. 2003). Together, they represent close to 17 wt % of the total carboxysome protein, which translates into approximately 3500 copies per particle (Table 1). This value is close to the predicted number (3730) of shell polypeptides necessary to form a dodecahedron or icosahedron with the dimensions of a carboxysome (Shively and English 1991). The CsoS1 proteins are the most highly conserved carboxysome shell proteins. Among the thiobacilli *H. neapolitanus*, *Thiobacillus denitrificans*, *Acidithiobacillus ferrooxidans* and *Thiomonas intermedia*, their amino acid sequences are over 70% identical (Shively et al. 1998; Cannon et al. 2002). Furthermore, genes encoding proteins related to the CsoS1 family are found in all prokaryotes that produce carboxysomes or other polyhedral microcompartments (Price et al. 1998; Shively et al. 1998a) (see Sect. 5), and are therefore believed to be the main structural determinants of the shell in all these particles.

The obvious functional homologs of the CsoS1 proteins in β-cyanobacteria are the CcmK proteins and CcmO, which, although closely related to the Csos1 proteins of α-carboxysomes, clearly segregate into separate phylogenetic branches (Price et al. 1998; Shively et al. 1998a; Cannon et al. 2002). Like the highly conserved CsoS1 proteins of α-carboxysomes, the CcmK proteins of *Synechococcus* PCC7942, *Synechocystis* PCC6803 and the marine *Synechococcus* PCC7002 also have very similar primary structures (87–92% identities). Exceptions are the CcmK3 and CcmK4 proteins from *Synechocystis* PCC6803, which are only 47 and 54% identical to their CcmK counterparts in the other two cyanobacteria (Ludwig et al. 2000). It is possible that this low degree of relatedness and the location of the *ccmK3* and *ccmK4* genes at a distance from their other two *ccmK* homologs and from *ccmLMN* are indicative of a role for CcmK3 and CcmK4 in an alternative cyanobacterial microcompartment that is unrelated to CO_2 fixation. A recent transposon mutagenesis study of *Synechocystis* PCC6803 revealed a requirement for *ccmK4* for optimal photoautotrophic growth; however, the exact role of its protein product was not further investigated (Zhang et al. 2004).

It is ironic that despite the difficulties encountered with the purification of cyanobacterial carboxysomes and the limited information currently available about their composition, CcmK2 and CcmK4 from *Synechococcus* PCC 6803 are the first carboxysomal proteins whose three-dimensional structures were elucidated. Kerfeld et al. (2005) described the structure of the folding domain that characterizes CcmK proteins, CsoS1 proteins, and their homologs in other bacterial polyhedral microcompartments. The wedge-shaped bacterial microcompartment (BMC) domains (Fig. 3) interact with other BMC monomers to form hexamers with positively charged central pores of 4 (CcmK4) to 7 Å (CcmK2) diameter. In the crystals, the hexamers pack tightly, forming sheets (CcmK2) or strips (CcmK4) of 2–3 nm thickness. This value agrees well with the 3–4 nm that was reported for the thickness of the carboxysome shell based on electron microscopic observations (Cannon and Shively 1983; Shively and English 1991). The gaps between the hexamers in

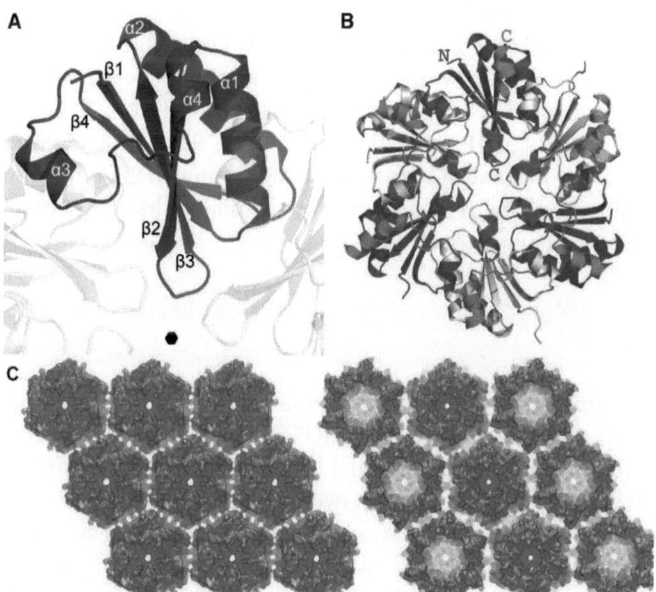

Fig. 3 Three-dimensional structure of CcmK proteins, the major carboxysome shell components. *Panel A:* CcmK monomer showing the BMC domain fold; *Panel B:* CcmK hexamers; *Panel C:* CcmK2 sheets and CcmK4 ribbons. Reprinted with permission from Kerfeld et al., Science 09:936–938 (2005). Copyright (2005) AAAS

the crystals are lined with charged residues and, together with the positively charged central pores, may mediate or control in- and efflux of RuBisCO substrates and products that must cross the carboxysome shell (Kerfeld et al. 2005). It has been suggested that the observed pentamers of CcmK2 form the corners of the polyhedral carboxysome shell and, through their interaction with two-dimensional hexamer arrays, determine the three-dimensional shape of the particle (Kerfeld et al. 2005).

2.2 The OrfA, OrfB and CcmL Proteins

The predicted products of the two-gene repeat *orfA* and *orfB* in the putative α-carboxysome operons are 72–80% and 53–65% conserved, respectively, among the thiobacilli. The homolog of *orfA* and *orfB* in β-cyanobacteria encodes the predicted carboxysome protein CcmL. The CcmL proteins of *Synechococcus* PCC7942 and *Synechocystis* PCC6803 are 67 and 69% identical to that of *Synechococcus* PCC7002 (Ludwig et al. 2000). So far, neither OrfA nor OrfB protein has been detected in greater than trace amounts in purified carboxysomes from *H. neapolitanus*. If they are true shell components, these two proteins do not represent abundant structural components. However, since their homologs are found in all gene clusters associated with carboxysomes and other bacterial polyhedral particles, it must be assumed that these two small proteins of less than 9 kDa (Table 1) have important functions in some aspect of microcompartment biogenesis. A clue for their potential role is provided by mutants of *Synechococcus* PCC 7942 in which the *ccmL* gene is disrupted (Price and Badger 1991; Price et al. 1993). The carboxysomes in the mutant are elongated, suggesting that CcmL might aid in delimiting the size of the arrays formed by the major shell proteins (Price and Badger 1989b, 1991; Price et al. 1993).

2.3 The CsoS2 Protein

The large carboxysome shell protein CsoS2 from *H. neapolitanus* has a calculated M_r of 92.3 kDa but migrates in SDS-polyacrylamide gels as two bands (CsoS2A and CsoS2B) of 85 000 and 130 000 apparent molecular weight, respectively (Fig. 2). Together, these two protein species have an estimated abundance of 330 copies per particle, which corresponds to approximately 12 wt % (Table 1). Immunogold microscopy verified that CsoS2 is associated with the carboxysomal shell (Baker et al. 1999). Since the N-termini of CsoS2A and CsoS2B are identical, both species are reported to be glycosylated and are recognized by the same antibody, the exact structural differences that might explain their aberrant migration behavior are unclear. It is possible that differential glycosylation (Baker et al. 1999), the amino acid repeat motifs present in the deduced primary sequence of CsoS2 and its homologs (Cannon et al. 2003), and/or the protein's unusually high calculated pI of 9.06 (Table 1) contribute to its unique properties. With less than 30% identical residues, the amino acid sequences of the CsoS2 proteins are the least conserved of the carboxysome proteins among the thiobacilli (Cannon et al. 2003), and a recognizable homolog of CsoS2 does not seem to exist in cyanobacterial ccm-type (β) carboxysomes and in other bacterial microcompartments (Price et al. 1998; Cannon et al. 2002).

2.4
CsoS3, the Carbonic Anhydrase of cso-Type (α) Carboxysomes

The amino acid sequence of the 57 kDa CsoS3 protein is only moderately conserved (38–49% identities) among the thiobacilli (Cannon et al. 2003). The protein was assigned to the carboxysome shell by immunogold microscopy (Baker et al. 2000) but is probably not an important structural determinant, since at an estimated 80 copies per particle (2.3 wt %) its abundance is considerably lower than that of CsoS2 and the CsoS1 proteins (Fig. 2, Table 1) (Cannon and Shively 1983). Using a sensitive mass spectrometric assay, native and recombinant CsoS3 protein from *H. neapolitanus* and recombinant CsoS3 homologs from several cyanobacteria carrying cso-type (α) carboxysome genes were recently shown to possess carbonic anhydrase activity (So et al. 2004). Although such an enzyme is postulated to be a crucial component of the microcompartment's proposed carbon concentrating function in cyanobacteria (Price et al. 1998; Badger 2003) and was predicted to be present in the interior of β-carboxysomes (Fridlyand et al. 1996; Kaplan and Reinhold 1999), a carbonic anhydrase activity eluded detection for decades in homogeneous α-carboxysome preparations from *H. neapolitanus* and *Nitrobacter* (Cannon 1982; Ebert 1982; Cannon and Shively 1983). Like other carbonic anhydrases, CsoS3 is inhibited by ethoxyzolamide, albeit to a lower degree (I_{50} approximately 200 μM) than most other carbonic anhydrases (So et al. 2004; So and Espie 2005). The carboxysomal carbonic anhydrase is also sensitive to dithiothreitol, a property that undoubtedly contributed to its elusiveness, since RuBisCO CO_2 fixation assays have traditionally been performed in the presence of a reducing agent. The almost completely divergent amino acid sequences of this protein and its homologs in those cyanobacteria that carry cso-type (α) carboxysomes, combined with an unusually high monomer M_r of 55 to 63 kDa, prompted the initial assignment of the carboxysomal carbonic anhydrase to a novel (ε) class (So et al. 2004).

The crystal structure of CsoS3 from *H. neapolitanus* was elucidated recently (Sawaya et al. 2006). The protein features an *N*-terminal, predominantly α-helical domain that may mediate heterologous interactions with other carboxysome proteins. The structure of the central catalytic domain of CsoS3 is very similar to that of other β-carbonic anhydrases, despite the unique primary structure of the carboxysomal protein. This domain contains a zinc ion coordinated by a histidine and two cysteine residues in an arrangement that is conserved in all β-carbonic anhydrases. However, instead of the characteristic pair of active sites, CsoS3 features a pair of fused (catalytic and *C*-terminal) domains that, although structurally similar, are highly divergent in amino acid sequence. CsoS3 protein forms homodimers that, because of the large number of intermolecular interactions between the complementary monomer surfaces in the crystal and their ability to form extended filaments, are thought to be significant for the protein's enzymatic role and/or its interactions with other carboxysome proteins (Sawaya et al. 2006).

2.5
Potential Carbonic Anhydrases of ccm-Type (β) Carboxysomes

A carbonic anhydrase of low activity has also been reported to associate with β-carboxysomes of some, but possibly not all cyanobacterial species (Lanares et al. 1985; Price and Badger 1989b; Price et al. 1992; So and Espie 1998). The CcaA (IcfA) protein from *Synechococcus* PCC7942 is an ethoxyzolamide- ($I_{50} = 2{-}4\,\mu M$) and DTT-sensitive β-type carbonic anhydrase of 31 kDa that co-purifies with carboxysome-enriched fractions (Price and Badger 1989b; Fukuzawa et al. 1992; Price et al. 1992, 1998; Yu et al. 1992; Badger 2003; So and Espie 2005). Insertion mutants of *Synechococcus* PCC7942 carrying a kanamycin resistance cassette in the *ccaA* (*icfA*) gene require an elevated CO_2 level for growth, accumulate high intracellular $[HCO_3^-]$ and form carboxysomes (Price and Badger 1989b; Fukuzawa et al. 1992; Price et al. 1992; Yu et al. 1992). The mutant phenotype, combined with the observed loss of the cytosolic bicarbonate pool in *Synechococcus* PCC7942 overexpressing an active heterologous carbonic anhydrase (Price and Badger 1989a) led to the proposal that CcaA resides in the carboxysome interior. However, recognizable homologs of the *ccaA* gene have so far been identified only in the genomes of some, but not all β-cyanobacteria and are apparently missing altogether from the gene complement of α-cyanobacteria (Badger 2003; So and Espie 2005). Furthermore, the protein does not interact with RuBisCO in yeast two-hybrid screens (So et al. 2002), and since homogeneous preparations of cyanobacterial β-carboxysomes are not available, the presence and location of this enzyme in these particles has not been established unequivocally.

It is also unclear whether the CcmM protein of β-cyanobacteria, whose mass (55–65 kDa) is similar to that of CsoS3, serves as the functional homolog of CsoS3 in β-carboxysomes. The relatedness of CcmM proteins among *Synechococcus* PCC7942, *Synechocystis* PCC6803 and *Synechococcus* PCC7002 (41–47% identities) is comparable to the degree of similarity between the CsoS3 polypeptides thiobacilli (Ludwig et al. 2000; Cannon et al. 2003). The CcmM and CsoS3 proteins do not have similar primary structures, but the *N*-terminal domain of CcmM shares homology with the catalytic region of γ-carbonic anhydrases, particularly that of the enzyme from the archaeon *Methanosarcina thermophila* (Alber and Ferry 1994). Another interesting feature of the protein is its multiple short *C*-terminal motifs that are homologous to the small subunit of RuBisCO (Price et al. 1993; Ogawa et al. 1994; Ludwig et al. 2000) and have been proposed to possibly mediate interaction of this putative carbonic anhydrase with RuBisCO (Price et al. 1993; Ludwig et al. 2000). Like *ccaA* mutants of Synechococcus PCC7942, insertion mutants of Synechococcus PCC7002 with a defective *ccmM* gene have a high CO_2-requiring (HCR) phenotype and accumulate a high intracellular bicarbonate pool that is utilized inefficiently (Sültemeyer et al. 1997; Ludwig et al. 2000). Unlike their *ccaA* counterparts, however, they do not form carboxysomes. Although it is tempting to propose that CcmM serves a similar function as the α-carboxysomal CsoS3, until actual enzymatic activity of native or recombinant CcmM can be demonstrated, a catalytic role for this protein in the carboxysomes of β-cyanobacteria is uncertain (Price et al. 1998; Badger and Price 2003; So and Espie 2005).

2.6 Carboxysome Function

According to the quantitative model for carbon assimilation by cyanobacteria (Reinhold et al. 1989, 1991; Kaplan and Reinhold 1999), carboxysomes represent the final step of a carbon concentrating mechanism (CCM) that serves to accumulate a large cytosolic pool of inorganic carbon (C_i) in the form of bicarbonate. In some cyanobacteria, intracellular concentrations of dissolved C_i can exceed extracellular levels by up to 1000-fold (Price et al. 1998), and similar CCM steps are also likely to operate in chemolithotrophic bacteria. Dobrinski et al. (2005) recently reported intracellular C_i concentrations for the deep vent sulfur oxidizer *Thiomicrospira crunogena* that are two orders of magnitude higher than extracellular levels. According to the CCM model, the carboxysome, which is thought to be permeable to HCO_3^-, converts the bicarbonate to CO_2 through the action of the carbonic anhydrase in the particle's interior and thereby supplies RuBisCO with near-saturating concentrations of its substrate. A crucial element of this model is the assumption that CO_2 leakage from carboxysomes to the cytosol, which would equilibrate the intracellular C_i pool and eliminate the catalytic advantage RuBisCO is thought to derive from sequestration into microcompartments, is prevented by the particles' CO_2-impermeable protein shell (Reinhold et al. 1989). Alternatively, the arrays of RuBisCO molecules that are thought to surround carbonic anhydrase and utilize the CO_2 in the interior of the particles may create a concentration gradient of the gas from high levels in the carboxysome center to very low levels at its perimeter (Reinhold et al. 1991). At present, there is no direct experimental support for either of the proposed molecular mechanisms of carboxysome function. Furthermore, the finding that the carbonic anhydrase CsoS3 is located in the shell of α-carboxysomes and not, as predicted by the CCM model, in the particles' interior, calls for an expansion of the current model and an inclusion of alternative or additional potential molecular mechanisms of carboxysome function (Cannon et al. 2001). CsoS3, because of its unique orientation in the carboxysome shell, is likely to mediate bicarbonate-to-CO_2 conversion that is directed towards the particle interior (So et al. 2004). The presumed ability of the tightly packed major shell constituents to control and mediate passage of charged molecular species across the carboxysome shell and prevent leakage of the uncharged gas CO_2 into the cytosol (Price et al. 1998; Badger 2003; Kerfeld et al. 2005) must also include oxygen, whose entry into the particles' interior is likely to be limited by the shell as well. Until direct measurements of gas fluxes across the carboxysome shell can be accurately assessed, the possibility has to be considered that both α- and β-carboxysomes provide a catalytic advantage to the sequestered RuBisCO by protecting the enzyme from oxygen in addition to or instead of their currently accepted role in the CCM.

3 Carboxysome Gene Arrangement

Carboxysome genes are clustered to varying degrees in the genomes of the autotrophic bacteria that form these microcompartments (Cannon et al. 2001, 2002, 2003; Badger and Price 2003). On the basis of differences in gene arrangement and on homology relationships between individual genes, two prototypes of carboxysome gene clusters are recognized: In the genomes of chemolithotrophs and α-cyanobacteria, all carboxysome genes cluster in putative operons, while the presumed carboxysome genes usually are less tightly linked and tend to be arranged in more than one cluster in β-cyanobacteria (Cannon et al. 2001, 2002, 2003; Badger and Price 2003). The hallmark of the chemolithotroph/α-cyanobacterial carboxysome genes is the arrangement of shell protein genes in the order *csoS2, csoS3, orfA, orfB*, $csoS1_n$ ($n = 1$–3) (Cannon et al. 2001, 2002, 2003). These genes are usually preceded by *cbbL* and *cbbS*, which encode the large and small subunit, respectively, of form IA RuBisCO, and by a *cbbR* gene upstream from *cbbL* in some bacteria. Of the thiobacilli, *T. denitrificans* is the only species in which *cbbL* and *cbbS* are missing from the putative carboxysome operon (Cannon et al. 2003). Whether the apparent inability of this species to form carboxysomes is related to lack of a physical connection between the genes for RuBisCO and those encoding carboxysome shell proteins is not known. It is also unclear whether the carboxysome shell proteins are expressed in this species.

The *ccmK, ccmL, ccmM, ccmN* and *ccmO* genes, which are thought to encode carboxysome shell components in β-cyanobacteria, were identified in mutants that are unable to form polyhedral inclusions and require elevated

CO_2 levels for growth (Price et al. 1993; Badger et al. 2002). Although direct biochemical evidence for an association of any of the Ccm proteins with the carboxysome is lacking, the CsoS1 homolog CcmK almost certainly represents an important structural element of the shell (Kerfeld et al. 2005). CcmO, another protein that is related to CsoS1 proteins, is believed to be important for microcompartment assembly in an as yet unknown function (Marco et al. 1994). CcmL may delimit carboxysome size (Price and Badger 1991; Price et al. 1993), and CcmM may be the predicted carbonic anhydrase located in the interior of β-carboxysomes (Price et al. 1998; Kaplan and Reinhold 1999; Badger and Price 2003; So and Espie 2005). A function has yet to be assigned to the *ccmN* gene product, which has a predicted M_r of 26 kDa and is moderately conserved (38% identical residues) among the cyanobacteria *Synechococcus* PCC7942, *Synechocystis* PCC6803 and *Synechococcus* PCC7002) (Ludwig et al. 2000). The CcmN protein is thought to play a role in an aspect of carboxysome assembly (Price et al. 1998) that, at this point, does not exclude a structural function. Like α-carboxysomes, β-carboxysomes are filled with RuBisCO molecules (McKay et al. 1993). However, the *rbcL* and *rbcS* genes that encode the form IB RuBisCO of β-carboxysomes, are usually not physically linked to the presumed shell protein gene clusters in β-cyanobacterial genomes (Cannon et al. 2001, 2002; Badger et al. 2002).

Comparative genomic analyses of carboxysome-forming autotrophic prokaryotes strongly suggest that the entire carboxysome operon was transferred laterally from proteobacteria to cyanobacteria (Hess et al. 2001) or vice versa (Badger et al. 2002; Badger and Price 2003). Although the divergence of α- and β-cyanobacteria may have preceded the acquisition of α-carboxysomes, parallel evolution of both cyanobacterial carboxysome types cannot be excluded. Furthermore, it remains to be determined whether the two carboxysome types enhance the catalytic ability of RuBisCO by the same molecular mechanisms in all autotrophic prokaryotes that contain these microcompartments.

4 Regulation of Carboxysome Gene Expression

Carboxysome biogenesis is a complex process that likely requires regulatory mechanisms at various levels to ensure coordinated expression of its protein components in response to environmental and intracellular cues. Although sensing and signaling pathways, and the molecular entities that regulate carboxysome gene expression at the transcriptional and/or translational level have not yet been fully elucidated, a variety of biochemical, molecular biological and bacterial genome data mining approaches are being used to establish the molecular details of these mechanisms.

4.1 Carbon Availability

Since carboxysome structure and RubisCO activity are intimately linked to effective inorganic carbon utilization and, therefore, growth of autotrophic bacteria, it is not surprising that the number of microcompartments and the activity of the sequestered RuBisCO vary with C_i availability in most bacteria that form the polyhedral particles (Purohit et al. 1976; Beudeker et al. 1980; Cannon 1982; McKay et al. 1993; Orus et al. 2001; Harano et al. 2003; Yoshizawa et al. 2004). In *Synechococcus* sp. PCC7942, the abundance of RuBisCO and of the putative carboxysome proteins CcmK, CcmL, CcmM, CcmN, CcmO, CcaA appears to be regulated predominantly at the transcriptional level, as determined by a combination of approaches that include traditional Northern and Western blotting and real-time PCR measurements (Omata et al. 2001; Harano et al. 2003; Woodger et al. 2003). Following downshifts of cultures from a high (ambient to 2%) to a low ($\leq 0.03\%$) CO_2 environment, *rbcL*, *ccmK*, *ccmL*, *ccmM*, *ccmN*, *ccmO* and *ccaA* transcript abundance increases rapidly, reaching two- (*rbcL*) to five-fold (*ccmK*) elevated levels within 30–60 min following the shift (Omata et al. 2001; Woodger et al. 2003). Using similar experimental approaches as well as whole-genome microarray hybridizations, less than two-fold or no changes in the transcript levels of these genes were detectable in *Synechocystis* PCC6803 within 30 to 60 min following shifts from CO_2-enriched to ambient or CO_2-free air (Omata et al. 2001; McGinn et al. 2003) and at later time points (Wang et al. 2004). At present, there is no clear consensus about the direction of the observed changes in this bacterium. The small change in transcript levels of some and lack of response of other putative carboxysome genes suggest that these genes are regulated independently, probably at the level of translation in *Synechocystis* PCC 6803. Alternatively, and probably more likely, the observed lack of gene induction at low C_i may be related to the more or less constant number of carboxysomes that exist over a wide range of CO_2 levels in this bacterium (So et al. 2002).

In the obligate lithotrophic hydrogen oxidizer *Hydrogenovibrio marinus*, expression of its three sets of RubisCO genes differentially responds to specific ranges of inorganic carbon levels, and results from reverse transcription-PCR analysis suggest that this regulation takes place predominantly at the level of transcription in this bacterium. Induction of the RubisCO gene set that is part of the carboxysome gene cluster closely matches the appearance of the polyhedral bodies at a moderately low (0.15%) CO_2 level, suggesting that the enzyme and the carboxysome shell proteins may be coordinately regulated and possibly constitute an operon (Yoshizawa et al. 2004).

Although not directly related to carbon availability, it is noteworthy that the abundance of the large and small subunit of RuBisCO is downregulated when *A. ferrooxidans* is grown on sulfur, and upregulated when

Fe^{2+} serves as the energy source (Ramirez et al. 2004). Of the two forms of RuBisCO expressed in this bacterium (Heinhorst et al. 2002), the RuBisCO polypeptides that were identified in the proteomics screen are the products of the *cbbLS1* genes, which are part of the putative carboxysome operon.

4.2 Light

Whether light plays a role in the regulation of carboxysome gene expression in cyanobacteria is an as yet unresolved issue. The response of the *ccmKLMN* and *rbcLS* genes may depend on the cyanobacterial species, magnitude of the change in light intensity, length of exposure to high light intensity, and additional growth parameters such as C_i availability. Using a combination of real-time PCR and microarray hybridizations, induction of RubisCO and/or putative carboxysome shell protein genes following exposure of *Synechocystis* PCC6803 to light of high intensity was reported by some (Watson and Tabita 1996; Hihara et al. 2001; Gill et al. 2002; Huang et al. 2002), while others saw no significant change in the mRNA levels of these genes (McGinn et al. 2003) or even a slight decline in transcript abundance (Tu et al. 2004).

4.3 The Operon Question

The expression of individual carboxysome genes, most of which are present in one copy in the gene clusters, must be differentially regulated to give rise to the vastly different levels of individual protein constituents needed to form the polyhedral particle (Table 1, Fig. 2) (Cannon and Shively 1983; Holthuijzen et al. 1986c; Shively and English 1991). The number of RuBisCO polypeptides far exceeds that of most carboxysomal shell proteins (Table 1), and while some shell components (CsoS1 proteins, CsoS2AB) are quite abundant, others (CsoS3) represent only minor polypeptide species (Cannon and Shively 1983; Shively and English 1991) (Table 1, Fig. 2). Although it is not known how the relative stoichiometries of carboxysome constituents are coordinated, the carboxysomal RuBisCO genes (*rbcL/cbbL* and *rbcS/cbbS*) appear to be co-transcribed (Watson and Tabita 1996; Baker et al. 1998; Vogel et al. 2003), and in some α-cyanobacteria form a tricistronic message that includes an upstream *csoS1* (*ccmK*) gene (Watson and Tabita 1996; Vogel et al. 2003). Experimental mapping of transcription start sites in the *csoS1-rbcL-rbcS* operon of the marine α-cyanobacteria *Prochlorococcus* MED4 and *Synechococcus* WH 7803, combined with computational analyses revealed the existence of the conserved sequence motif GAACAT on the 5′ ends of these transcripts that coincides with their experimentally determined transcription start sites (Watson and Tabita 1996; Vogel et al. 2003) and the inferred start sites in *Prochlorococcus* MIT9313 and *Synechococcus* WH8102, two other marine α-cyanobacteria. Likewise, the – 35 and – 10 regions upstream from *ccmK*, which are strong candidates for promoter elements, are also practically identical in these four species (Vogel et al. 2003). No information is available about possible additional promoters downstream from the *rbcS* genes, which might govern expression of the remaining downstream carboxysome genes in these bacteria.

In *H. neapolitanus*, the transcription start site for the *cbbL* gene was determined by primer extension analysis (Baker 1998), and putative – 10 and – 35 promoter regions were identified upstream from *cbbL* with the aid of a promoter prediction program (Cannon et al. 2003). The same program predicted the existence of additional promoters within the putative carboxysome operon. Experimental evidence for their existence, however, is scarce at this point and is based solely on the sizes of transcripts identified by Northern blotting. A *csoS1A* probe detected, in addition to a major transcript of 400 bases, larger RNA species in *H. neapolitanus* that could potentially encode two or all three CsoS1 proteins (English et al. 1994). Likewise, a *H. neapolitanus* mutant that carries a kanamycin resistance cassette in its *cbbL* gene and in which the downstream *cbbS* gene is not transcribed, expresses mRNA species, albeit at highly reduced levels, that hybridize to *csoS1A*. These transcripts, of which the largest ($>$ 1 kb) could potentially encode all three CsoS1 proteins, must be derived from an internal promoter within the carboxysome gene cluster (English et al. 1994; Baker et al. 1998). The significance of such a promoter for the control of carboxysome gene expression in wild-type bacteria is not known.

In *Synechococcus* PCC 7942, the pattern of *ccm* transcripts changes drastically upon induction by low C_i levels (Omata et al. 2001). The appearance of much larger RNA species following a shift of the culture to CO_2-limiting conditions prompted the suggestion that *ccmKLMN* may function as an operon only if carbon availability is limited and that expression of these genes may not be coordinately regulated when the carbon source is abundant.

The currently available evidence derived from transcript analyses and carboxysome protein expression studies suggests that promoters may exist downstream from the RubisCO genes in the putative carboxysome operon of *H. neapolitanus* and within the *ccmKLMN* cluster/operon of *Synechococcus* PCC 7942. However, sizes, abundance and coding capacity of mRNA species have not been analyzed thoroughly, and potential alternative regulatory mechanisms must be considered, such as the presence of transcription terminators at the 3′-ends of some genes in the putative operon (Baker et al. 1998; Cannon et al. 2003), processing of the primary transcript to smaller RNA entities with varying half lives, and differential translation efficiencies related to ribosomal binding sites located upstream from individual carboxysome genes (Baker et al. 1998, 2000; Cannon et al. 2003).

4.4 Potential LysR-type Regulators

The CbbR protein, a member of the LysR-family of transcriptional regulators, controls the expression of genes (*cbb*) encoding the Calvin–Benson–Bassham cycle enzymes in many autotrophic prokaryotes (Gibson and Tabita 1996; Bowien and Kusian 2002). Its gene, *cbbR*, is usually located upstream from the *cbb* operon and transcribed in the opposite direction. A *cbbR* homolog is also located in close proximity to the carboxysome genes in the sulfur bacteria *A. ferrooxidans* and *T. denitrificans* (Cannon et al. 2003), whereas in other thiobacilli and in some cyanobacteria it is located at a greater distance (Baker et al. 1998). Despite attempts to elucidate the roles of CbbR homologs in cyanobacterial mutants (Omata et al. 2001), the significance of CbbR proteins for the regulation of carboxysome gene expression is not understood. In *H. marinus* single and double mutants that carry disrupted *cbbR1* and/or *cbbRm* genes, levels of carboxysomal RubisCO (CbbLS-2) are unaffected at low CO_2 concentrations, suggesting the existence of different, as yet unknown regulators that govern the expression of the enzyme and, possibly, of the other carboxysome genes. Unlike in the wild-type strain, however, the carboxysomal RubisCO protein is also present at elevated CO_2 levels in the *cbbR1 cbbRm* double mutant, in which the expression of the other two RubisCO gene sets is severely reduced (Toyoda et al. 2005). These results suggest a regulatory crosstalk that coordinates the expression of the three RuBisCO gene sets and may include the carboxysome genes in this bacterium.

5 Polyhedral Bodies of Other Bacteria

Polyhedral microcompartments do not only exist in autotrophic prokaryotes but are also found in heterotrophic bacteria. *Salmonella enterica* serovar Typhimurium LT2 forms polyhedral bodies when grown on 1,2-propanediol or ethanolamine as a carbon source (Shively et al. 1998a; Bobik et al. 1999; Havemann and Bobik 2003). Interestingly, of the more than 20 polypeptides encoded by the *S. enterica* propanediol utilization (*pdu*) operon, several (PduA, PduJ, PduK, PduT, PduB, PduB′, PduU) are related to varying degrees to the major carboxysome shell constituents, the CsoS1 and CcmK proteins (Chen et al. 1994; Shively et al. 1998a; Bobik et al. 1999; Havemann et al. 2002; Havemann and Bobik 2003; Kerfeld et al. 2005). Likewise, the *S. enterica* and *Escherichia coli eut* operons, which encode the proteins needed for ethanolamine degradation, contain homologs of carboxysome shell proteins. The EutN protein is related to CcmL and the OrfA and OrfB polypeptides; EutS, EutM, EutL and EutK have sequence similarity to CcmK and CsoS1 proteins (Stojiljkovic et al. 1995; Shively et al. 1998a; Kofoid et al.

1999). Data mining of microbial genome sequences, combined with traditional biochemical and electron microscopy analysis has led to the realization that polyhedral microcompartments and the potential to form such structures are much more common among obligately heterotrophic bacteria than was originally assumed (Havemann and Bobik 2003; Kerfeld et al. 2005, our unpublished observation). Genes encoding proteins that are related to the carboxysomal CcmK/CsoS1 group and to CcmL/OrfAB almost always occur together in the genomes of at least 19 bacterial genera that are represented by over 36 species (our unpublished observation). In the Pfam database (Bateman et al. 2004), their protein products are grouped into bacterial microcompartment (accession number PF00936) and EutN_CcmL (accession number PF03319) families, respectively. In fact, these genes are so widespread among prokaryotes that the term "enterosomes", which was coined originally for the polyhedral microcompartments of enteric bacteria (Cannon et al. 2001), is no longer appropriate for the inclusions these protein products form in a variety of prokaryotes.

Since heterotrophs do not possess the genes for the RuBisCO subunits and are unable to fix CO_2, their polyhedral bodies must serve other, more diverse metabolic purposes. Recently, the term "metabolosome" was suggested to better reflect the roles these bacterial compartments play in a variety of heterotrophic metabolic pathways (Brinsmade et al. 2005). The protein complement of the microcompartments formed in *S. enterica* during growth on propanediol includes the four enzymes necessary for efficient utilization of that carbon source (Havemann and Bobik 2003); the polyhedral bodies connected to ethanolamine degradation are thought to contain ethanolamine ammonia lyase (EutBC), and possibly acetaldehyde dehydrogenase (EutE) and the putative alcohol dehydrogenase EutG (Brinsmade et al. 2005). These protein compartments have been postulated to have a protective function and/or provide a catalytic advantage for the sequestered enzymes. In addition, or alternatively, microcompartmentalization may serve to prevent release of metabolic intermediates that are toxic for the cell (Shively et al. 1998a; Bobik et al. 1999; Kofoid et al. 1999; Johnson et al. 2001; Price-Carter et al. 2001). A *S. enterica* deletion mutant lacking functional *eutLKMN* genes is unable to grow on ethanolamine as a carbon and energy source and does not form the characteristic electron dense structures visible in transmission electron micrographs of the wild type that are believed to represent metabolosomes (Brinsmade et al. 2005). Growth on ethanolamine of a mutant in which the entire *eut* operon was deleted is restored by transformation with overexpression constructs for ethanolamine ammonia lyase (EutBC) and its reactivase (EutA), suggesting that one function of the metabolosome in this bacterium is concentration of the ethanolamine degrading enzyme by compartmentalization, presumably leading to a more efficient metabolic conversion of this carbon source.

6 Conclusions and Outlook

In the last few years, great strides have been made towards increasing our understanding of carboxysomes and related bacterial polyhedral inclusions. The wide distribution of genes for shell protein homologues among heterotrophic bacteria has led to the realization that polyhedral microcompartments should be viewed as organizers or nano-size reactors for key reactions in special metabolic pathways. As such, they represent novel regulatory strategies of microbial metabolism (Cannon et al. 2001). For most of these metabolic organizers, the molecular mechanisms by which they function are either not known or are inferred mainly from indirect evidence obtained through genetic approaches. The carboxysome of autotrophic bacteria, on the other hand, has proven amenable to biochemical and protein structural analyses, which are prerequisites for direct studies of the particle's function in autotrophic metabolism.

The identification of a novel carbonic anhydrase as a shell component of α-carboxysomes has raised questions about the molecular mechanism by which these polyhedral inclusions are currently thought to contribute to the CCM of autotrophs, and about the identity and location of the carbonic anhydrase that may be the enzyme's functional homolog in cyanobacterial β-carboxysomes.

Elucidation of the three-dimensional structures of the major carboxysomal shell proteins (Kerfeld et al. 2005) has led to suggestions about the role of the protein coat in mediating and regulating passage of RuBisCO substrates and products. If the shell proves to be a general diffusional barrier for gases, including O_2, this property would represent a bacterial solution to the problem of RuBisCO's oxygen sensitivity. The self-assembly of CcmK proteins into hexamers that form large two-dimensional arrays suggests that the multimers may represent protomers in the sequence of events that lead to the formation of carboxysomes. Whether shell assembly precedes insertion of RuBisCO molecules (Price and Badger 1991) or follows pre-assembly of polyhedral RuBisCO arrays (Orus et al. 1995, 2001) is not clear at present. Likewise, the suggested role of the *C*-terminal portion of CcmM in recruiting RuBisCO to β-carboxysomes or in organizing the protein in the microcompartments (Ludwig et al. 2000) needs to be further investigated. The ability to overexpress bona fide and inferred constituents of carboxysomes in *E. coli* to yield large amounts of recombinant proteins for structural and functional analyses (So et al. 2004; Kerfeld et al. 2005; Sawaya et al. 2006), combined with selective modification of carboxysome proteins by targeted mutagenesis and assessment of the effects the mutant proteins have on assembly and properties of the particles in vivo and in vitro, will yield answers regarding structure and function of these very interesting protein assemblies and will help to elucidate their role in metabolism throughout the microbial world.

Acknowledgements SH and GCC acknowledge financial support from the National Science Foundation (MCB 0444568 and DMR 0213883).

References

Alber BE, Ferry JG (1994) A carbonic anhydrase from the archaeon *Methanosarcina thermophila*. Proc Natl Acad Sci USA 91:6909–6913

Badger MR (2003) The roles of carbonic anhydrases in photosynthetic CO_2 concentrating mechanisms. Photosynthesis Res 77:83–94

Badger MR, Price GD (2003) CO_2 concentrating mechanisms in cyanobacteria: molecular components, their diversity and evolution. J Exp Bot 54:609–622

Badger MR, Hanson D, Price GD (2002) Evolution and diversity of CO_2 concentrating mechanisms in cyanobacteria. Funct Plant Biol 29:161–173

Baker SH (1998) Molecular investigations of carbon dioxide fixation in thiobacilli. PhD Dissertation, Clemson University, Clemson, SC

Baker SH, Jin S, Aldrich HC, Howard GT, Shively JM (1998) Insertion mutation of the form I *cbbL* gene encoding ribulose bisphosphate carboxylase/oxygenase (RuBisCO) in *Thiobacillus neapolitanus* results in expression of form II RuBisCO, loss of carboxysomes, and an increased CO_2 requirement for growth. J Bacteriol 180:4133–4139

Baker SH, Lorbach SC, Rodriguez-Buey M, Williams DS, Aldrich HC, Shively JM (1999) The correlation of the gene *csoS2* of the carboxysome operon with two polypeptides of the carboxysome in *Thiobacillus neapolitanus*. Arch Microbiol 172:233–239

Baker SH, Williams DS, Aldrich HC, Gambrell AC, Shively JM (2000) Identification and localization of the carboxysome peptide CsoS3 and its corresponding gene in *Thiobacillus neapolitanus*. Arch Microbiol 173:278–283

Bateman A, Coin L, Durbin R, Finn RD, Hollich V, Griffith-Jones S, Khanna A, Marshall M, Moxon S, Sonnhammer ELL, Studholme DJ, Yeats C, Eddy SR (2004) The Pfam protein families database. Nucl Acids Res 32:D138–D141

Beudeker RF, Kuenen JG (1981) Carboxysomes: Calvinosomes? FEBS Lett 131:269–274

Beudeker RF, Cannon GC, Kuenen JG, Shively JM (1980) Relations between D-ribulose-1,5-bisphosphate carboxylase, carboxysomes, and CO_2 fixing capacity in the obligate chemolithotroph *Thiobacillus neapolitanus* grown under different limitations in the chemostat. Arch Microbiol 124:185–189

Bobik TA, Havemann GD, Busch RJ, Williams DS, Aldrich HC (1999) The propanediolutilization (*pdu*) operon of *Salmonella enterica* serovar Typhimurium LT2 includes genes necessary for formation of polyhedral organelles involved in coenzyme B(12)-dependent 1,2-propanediol degradation. J Bacteriol 181:5967–5975

Bowien B, Kusian B (2002) Genetics and control of CO_2 assimilation in the chemoautotroph *Ralstonia eutropha*. Arch Microbiol 178:85–93

Brinsmade SR, Paldon T, Escalante-Semerena JC (2005) Minimal functions and physiological conditions required for growth of *Salmonella enterica* on ethanolamine in the absence of the metabolosome. J Bacteriol 187:8039–8046

Cannon GC (1982) Carboxysomes and CO_2 fixation in *Thiobacillus neapolitanus*. PhD Dissertation, Clemson University, Clemson, SC

Cannon GC, Shively JM (1983) Characterization of a homogeneous preparation of carboxysomes from *Thiobacillus neapolitanus*. Arch Microbiol 134:52–59

Cannon GC, English RS, Shively JM (1991) In situ assay of ribulose-1,5-bisphosphate carboxylase/oxygenase in *Thiobacillus neapolitanus*. J Bacteriol 173:1565–1568

Cannon GC, Bradburne CE, Aldrich HC, Baker SH, Heinhorst S, Shively JM (2001) Microcompartments in prokaryotes: carboxysomes and related polyhedra. Appl Environ Microbiol 67:5351–5361

Cannon GC, Heinhorst S, Bradburne CE, Shively JM (2002) Carboxysome genomics: a status report. Funct Plant Biol 29:175–182

Cannon GC, Baker SH, Soyer F, Johnson DR, Bradburne CE, Mehlman JL, Davies PS, Jiang QL, Heinhorst S, Shively JM (2003) Organization of carboxysome genes in the thiobacilli. Curr Microbiol 46:115–119

Chen P, Andersson DI, Roth JR (1994) The control region of the *pdu/cob* regulon in *Salmonella typhimurium*. J Bacteriol 176:5474–5482

Codd GA (1988) Carboxysomes and ribulose bisphosphate carboxylase/oxygenase. Adv Microb Physiol 29:115–164

Dobrinski KP, Longo DL, Scott KM (2005) The carbon-concentrating mechanism of the hydrothermal vent chemolithoautotroph *Thiomicrospira crunogena*. J Bacteriol 187:5761–5766

Ebert A (1982) Ribulose-1,5-bisphosphate carboxylase in *Nitrobacter*. PhD Dissertation, University of Hamburg, Hamburg, Germany

English RS, Lorbach SC, Qin X, Shively JM (1994) Isolation and characterization of a carboxysome shell gene from *Thiobacillus neapolitanus*. Mol Microbiol 12:647–654

Fridlyand L, Kaplan A, Reinhold L (1996) Quantitative evaluation of the role of a putative CO_2-scavenging entity in the cyanobacterial CO_2-concentrating mechanism. Biosystems 37:229–238

Fukuzawa H, Suzuki E, Komukai Y, Miyachi S (1992) A gene homologous to chloroplast carbonic anhydrase (*icfA*) is essential to photosynthetic carbon dioxide fixation by *Synechococcus* PCC7942. Proc Natl Acad Sci USA 89:4437–4441

Gibson JL, Tabita FR (1996) The molecular regulation of the reductive pentose phosphate pathway in proteobacteria and cyanobacteria. Arch Microbiol 166:141–150

Gill R, Katsulakis E, Schmitt W, Taroncher-Oldenburg G, Misra J, Stephanopoulos G (2002) Genome-wide dynamic transcriptional profiling of the light-to-dark transition in *Synechocystis* sp. strain PCC 6803. J Bacteriol 184:3671–3681

Gonzales AD, Light YK, Zhang Z, Iqbal T, Lane TW, Martino A (2005) Proteomic analysis of the CO_2-concentrating mechanism in the open-ocean cyanobacterium *Synechococcus* WH8102. Can J Bot 83:735–745

Harano K, Ishida H, Kittaka R, Kojima K, Inoue N, Tsukamoto M, Satoh R, Iwaki T, Wadano A (2003) Regulation of the expression of ribulose-1,5-bisphosphate carboxylase/oxygenase (EC 4.1.1.39) in a cyanobacterium, *Synechococcus* PCC7942. Photosynth Res 78:59–65

Havemann GD, Bobik TA (2003) Protein content of polyhedral organelles involved in coenzyme B12-dependent degradation of 1,2-propanediol in *Salmonella enterica* serovar Typhimurium LT2. J Bacteriol 185:5086–5095

Havemann GD, Sampson EM, Bobik TA (2002) PduA is a shell protein of polyhedral organelles involved in coenzyme B(12)-dependent degradation of 1,2-propanediol in *Salmonella enterica* serovar typhimurium LT2. J Bacteriol 184:1253–1261

Heinhorst S, Baker SH, Johnson DR, Davies PS, Cannon GC, Shively JM (2002) Two copies of form 1 RuBisCO genes in *Acidithiobacillus ferrooxidans* ATCC 23270. Curr Microbiol 45:115–117

Hess WR, Rocap G, Ting C, Larimer FW, Lamerdin J, Stilwagon S, Chisholm SW (2001) The photosynthetic apparatus of *Prochlorococcus*: insights through comparative genomics. Photosynth Res 70:53–72

Hihara Y, Kamei A, Kanehisa M, Kaplan A, Ikeuchi M (2001) DNA microarray analysis of cyanobacterial gene expression during acclimation to high light. Plant Cell 13:793–806
Holthuijzen YA, Maathuis JFL, Kuenen JG, Konings RNH, Konings WN (1986a) Carboxysomes of *Thiobacillus neapolitanus* do not contain extrachromosomal DNA. FEMS Microbiol Lett 42:193–198
Holthuijzen YA, VanBreeman JFL, Konings WN, VanBruggen EFJ (1986b) Electron microscopic studies of carboxysomes of *Thiobacillus neapolitanus*. Arch Microbiol 144:258–262
Holthuijzen YA, VanBreeman JFL, Kuenen JG, Konings WN (1986c) Protein composition of the carboxysomes of *Thiobacillus neapolitanus*. Arch Microbiol 144:398–404
Huang L, McCluskey MP, Ni H, LaRossa RA (2002) Global gene expression profiles of the cyanobacterium *Synechocystis* sp. strain PCC 6803 in response to irradiation with UV-B and white light. J Bacteriol 184:6845–6858
Johnson CL, Pechonik E, Park SD, Havemann GD, Leal NA, Bobik TA (2001) Functional genomic, biochemical, and genetic characterization of the *Salmonella pduO* gene, an ATP:cobalamine adenosyltransferase gene. J Bacteriol 183:1577–1584
Kaplan A, Reinhold L (1999) CO_2 concentrating mechanisms in photosynthetic microorganisms. Annu Rev Plant Physiol Plant Mol Biol 50:539–570
Kerfeld CA, Sawaya MR, Tanaka S, Nguyen CV, Phillips M, Beeby M, Yeates TO (2005) Protein structures forming the shell of primitive bacterial organelles. Science 309:936–938
Kofoid E, Rappleye C, Stojiljkovic I, Roth J (1999) The 17-gene ethanolamine (*eut*) operon of *Salmonella typhimurium* encodes five homologues of carboxysome shell proteins. J Bacteriol 181:5317–5329
Lanares T, Codd GA (1981) Ribulose 1,5-bisphosphate carboxylase and polyhedral bodies of *Chlorogloeopsis*. Planta 153:279–285
Lanares T, Hawthornthwaite AM, Codd GA (1985) Localization of carbonic anhydrase in the cyanobacterium *Chlorogloeopsis fritschii*. FEMS Microbiol Lett 26:285–288
Long BM, Price GD, Badger MR (2005) Proteomic assessment of an established technique for carboxysome enrichment from *Synechococcus* PCC7942. Can J Bot 83:746–757
Lorbach SC (1995) Molecular physiology of the thiobacilli. PhD Dissertation, Clemson University, Clemson, SC
Ludwig M, Sültemeyer D, Price GD (2000) Isolation of *ccmKLMN* genes from the marine cyanobacterium *Synechococcus* sp. PCC7002 and evidence that CcmM is essential for carboxysome assembly. J Phycol 36:1109–1118
Marco E, Martinez I, Ronen-Tarazi M, Orus MI, Kaplan A (1994) Inactivation of *ccmO* in *Synechococcus* sp. PCC 7942 results in a mutant requiring high levels of CO_2. Appl Environ Microbiol 60:1018–1020
McGinn PJ, Price GD, Maleszka R, Badger MR (2003) Inorganic carbon limitation and light control the expression of transcripts related to the CO_2-concentrating mechanism in the cyanobacterium *Synechocystis* sp. strain PCC6803. Plant Physiol 132:218–229
McKay RML, Gibbs SP, Espie GS (1993) Effect of dissolved inorganic carbon on the expression of carboxysomes, localization of Rubisco and the mode of carbon transport in cells of the cyanobacterium *Synechococcus* UTEX 625. Arch Microbiol 159:21–29
Ogawa T, Armichay D, Gurevitz M (1994) Isolation and characterization of the *ccmM* gene required by the cyanobacterium *Synechocystis* PCC6803 for inorganic carbon utilization. Photosynth Res 39:183–190
Omata T, Gohta S, Takahashi Y, Harano Y, Maeda S-I (2001) Involvement of a CbbR homolog in low CO_2-induced activation of the bicarbonate transporter operon in cyanobacteria. J Bacteriol 183:1891–1898

Orus MI, Rodriguez ML, Martinez F, Marco E (1995) Biogenesis and ultrastructure of carboxysomes from wild type and mutants of *Synechococcus* sp. Strain PCC 7942. Plant Physiol 107:1159–1166

Orus MI, Rodriguez-Buey ML, Marco E, Fernandez-Valiente E (2001) Changes in carboxysome structure and grouping and in photosynthetic affinity for inorganic carbon in Anabaena strain PCC 7119 (cyanophyta) in response to modification of CO_2 and Na^+ supply. Plant Cell Physiol 42:46–53

Paredes AM, Soyer F, Aldrich HC, Ludtke S, Tsuruta H, Chiu W, Shively JM (2001) Am Soc Microbiol. 101th Annual Meeting May 20–24 Orlando, FL

Price GD, Badger MR (1989a) Expression of human carbonic anhydrase in the cyanobacterium *Synechococcus* PCC7942 creates a high CO_2-requiring phenotype. Evidence for a central role for carboxysomes in the CO_2 concentrating mechanism. Plant Physiol 91:505–513

Price GD, Badger MR (1989b) Isolation and characterization of high CO_2-requiring-mutants of the cyanobacterium *Synechococcus* PCC 7942: Two phenotypes that accumulate inorganic carbon but are apparently unable to generate CO_2 within the carboxysome. Plant Physiol 91:514–525

Price GD, Badger MR (1991) Evidence for the role of carboxysomes in the cyanobacterial CO_2-concentrating mechanism. Can J Bot 69:963–973

Price GD, Coleman JR, Badger MR (1992) Association of carbonic anhydrase activity with carboxysomes isolated from the cyanobacterium *Synechococcus* PCC7942. Plant Physiol 100:784–793

Price GD, Howitt SM, Harrison K, Badger MR (1993) Analysis of a genomic DNA region from the cyanobacterium *Synechococcus* sp. strain PCC7942 involved in carboxysome assembly and function. J Bacteriol 175:2871–2879

Price GD, Sültemeyer D, Klughammer B, Ludwig M, Badger MR (1998) The functioning of the CO_2 concentrating mechanism in several cyanobacterial strains: a review of general physiological characteristics, genes, proteins and recent advances. Can J Bot 76:973–1002

Price-Carter M, Tingey J, Bobik TA, Roth JR (2001) The alternative electron acceptor tetrathionate supports B12-dependent anaerobic growth of *Salmonella enterica* serovar typhimurium on ethanolamine or 1,2-propanediol. J Bacteriol 183:2463–2475

Purohit K, McFadden BA, Cohen AL (1976) Purification, quaternary structure composition, and properties of D-ribulose-1,5-bisphosphate carboxylase from *Thiobacillus neapolitanus.* J Bacteriol 127:505–515

Ramirez P, Guiliani N, Valenzuela L, Beard SJ, Jerez CA (2004) Differential protein expression during growth of *Acidithiobacillus ferrooxidans* on ferrous iron, sulfur compounds, or metal sulfides. Appl Environ Microbiol 70:4491–4498

Reinhold L, Zviman M, Kaplan A (1989) A quantitative model for carbon fluxes and photosynthesis in cyanobacteria. Plant Physiol Biochem 27:945–954

Reinhold L, Kosloff R, Kaplan A (1991) A model for inorganic carbon fluxes and photosynthesis in cyanobacterial carboxysomes. Can J Bot 69:984–988

Sawaya MR, Cannon GC, Heinhorst S, Tanaka S, Williams EB, Yeates TO, Kerfeld CA (2006) The structure of β-carbonic anhydrase from the carboxysomal shell reveals a distinct subclass with one active site for the price of two. J Biol Chem 281(11):7546–7555

Shively JM, English RS (1991) The carboxysome, a prokaryotic organelle: a mini review. Can J Bot 69:957–962

Shively JM, Ball F, Brown DH, Saunders RE (1973a) Functional organelles in prokaryotes: Polyhedral inclusions (carboxysomes) in *Thiobacillus neapolitanus.* Science 182:584–586

Shively JM, Ball FL, Kline BW (1973b) Electron microscopy of the carboxysomes (polyhedral bodies) of *Thiobacillus neapolitanus*. J Bacteriol 116:1405–1411

Shively JM, Bock E, Westphal K, Cannon GC (1977) Icosahedral inclusions (carboxysomes) of *Nitrobacter agilis*. J Bacteriol 132:673–675

Shively JM, Lorbach SC, Jin S, Baker SH (1996) Carboxysomes: the genes of *Thiobacillus neapolitanus*. In: Lidstrom ME, Tabita FR (eds) Microbial Growth on C1 Compounds. Kluwer, Dordrecht, The Netherlands, p 56–63

Shively JM, Bradburne CE, Aldrich HC, Bobik TA, Mehlman JL, Jin S, Baker SH (1998a) Sequence homologs of the carboxysomal polypeptide CsoS1 of the thiobacilli are present in cyanobacteria and enteric bacteria that form carboxysomes/polyhedral bodies. Can J Bot 76:906–916

Shively JM, van Keulen G, Meijer WG (1998b) Something from almost nothing: carbon dioxide fixation in chemoautotrophs. Annu Rev Microbiol 52:191–230

So AKC, Espie GS (1998) Cloning, characterization and expression of carbonic anhydrase from the cyanobacterium *Synechocystis* PCC6803. Plant Mol Biol 37:205–215

So AKC, Espie GS (2005) Cyanobacterial carbonic anhydrases. Can J Bot 83:721–734

So AKC, John-McKay ME, Espie GS (2002) Characterization of a mutant lacking carboxysomal carbonic anhydrase from the cyanobacterium Synechocystis PCC6803. Planta 214:456–467

So AKC, Espie GS, Williams EB, Shively JM, Heinhorst S, Cannon GC (2004) A novel evolutionary lineage of carbonic anhydrase (epsilon class) is a component of the carboxysome shell. J Bacteriol 186:623–630

Stojiljkovic I, Baumler AJ, Heffron F (1995) Ethanolamine utilization in *Salmonella typhimurium*: nucleotide sequence, protein expression, and mutational analysis of the *cchA cchB eutE eutJ eutG eutH* gene cluster. J Bacteriol 177:1357–1366

Sültemeyer D, Klughammer B, Ludwig M, Badger MR, Price GD (1997) Random insertional mutagenesis used in the generation of mutants of the marine cyanobacterium *Synechococcus* PCC7002 with an impaired CO_2 concentrating mechanism. Aust J Plant Physiol 24:317–327

Toyoda K, Yoshizawa Y, Arai H, Ishii M, Igarashi Y (2005) The role of two CbbRs in the transcriptional regulation of three ribulose-1,5-bisphosphate carboxylase/oxygenase genes in *Hydrogenovibrio marinus* strain MH-110. Microbiol 151:3615–3625

Tu C, Shrager J, Burnap RL, Postier BL, Grossman AR (2004) Consequences of a deletion in *dspA* on transcript accumulation in *Synechocystis* sp. strain PCC6803. J Bacteriol 186:3889–3902

Vogel J, Axmann IM, Herzel H, Hess WR (2003) Experimental and computational analysis of transcriptional start sites in the cyanobacterium *Prochlorococcus* MED4. Nucl Acids Res 31:2890–2899

Wang H-L, Postier BL, Burnap RL (2004) Alterations in global patterns of gene expression in *Synechocystis* sp. PCC 6803 in response to inorganic carbon limitation and the inactivation of *ndhR*, a LysR family regulator. J Biol Chem 279:5739–5751

Watson GM, Tabita FR (1996) Regulation, unique gene organization, and unusual primary structure of carbon fixation genes from a marine phycoerythrin-containing cyanobacterium. Plant Mol Biol 32:1103–1115

Westphal K, Bock E, Cannon GC, Shively JM (1979) Deoxyribonucleic acid in *Nitrobacter* carboxysomes. J Bacteriol 140:285–288

Woodger FJ, Badger MR, Price GD (2003) Inorganic carbon limitation induces transcripts encoding components of the CO_2-concentrating mechanism in *Synechococcus* sp. PCC7942 through a redox-independent pathway. Plant Physiol 133:2069–2080

Yoshizawa Y, Toyoda K, Arai H, Ishii M, Igarashi Y (2004) CO_2-responsive expression and gene organization of three ribulose-1,5-bisphosphate carboxylase/oxygenase enzymes and carboxysomes in *Hydrogenovibrio marinus* strain MH-110. J Bacteriol 186:5685–5691

Yu J-W, Price GD, Song L, Badger MR (1992) Isolation of a putative carboxysomal carbonic anhydrase gene from the cyanobacterium *Synechococcus* PCC7942. Plant Physiol 100:794–800

Zhang S, Laborde SM, Frankel LK, Bricker TM (2004) Four novel genes required for optimal photoautotrophic growth of the cyanobacterium *Synechocystis* sp strain PCC6803 identified by in vitro transposon mutagenesis. J Bacteriol 186:875–879

Microbiol Monogr (2)
J.M. Shively: Complex Intracellular Structures in Prokaryotes
DOI 10.1007/7171_024/Published online: 13 June 2006

Magnetosomes in Magnetotactic Bacteria

André Scheffel · Dirk Schüler (✉)

Max Planck Institute for Marine Microbiology, Celsiusstrasse 1, 28359 Bremen, Germany
dschuele@mpi-bremen.de

Abstract The ability of magnetotactic bacteria (MTB) to orient and migrate along magnetic field lines is based on magnetosomes, which are membrane-enclosed intracellular crystals of a magnetic iron mineral. The biomineralization of magnetosomes is a process with genetic control over the accumulation of iron, the deposition of the magnetic crystal within a specific compartment, as well as their intracellular assembly and alignment into chain-like structures. Magnetite crystals produced by MTB have uniform species-specific morphologies and sizes, which are mostly unknown from inorganic systems. In addition, magnetosome chain formation is an example of the highest structural level achievable in a prokaryotic cell. In this work, we give an overview of the biology of MTB and the structure and functions of bacterial magnetosomes. In addition we summarize the current knowledge of the physico-chemical and molecular genetic basis of magnetosome biomineralization and chain formation.

1 Introduction

The intriguing capability of magnetotactic bacteria (MTB) to navigate along geomagnetic field lines is due to the presence of unique inclusions, the magnetosomes, which are chain-like intracellular structures that consist of nanometer-scale, magnetic iron-mineral crystals enveloped by a membrane vesicle. Of the many types of organelles and subcellular structures that have evolved in prokaryotic organisms, the magnetosome provides one of the rare examples in which an inorganic crystalline constituent is enclosed within a membrane vesicle that provides a nanoreactor for the controlled growth of an iron oxide or sulfide crystal. The bacterial magnetosome chain affords an example of one of the most complex subcellular structures found in a prokaryotic cell. The formation of this multi-particle magnetic receptor is achieved by a remarkable degree of control over the biomineralization of a perfectly shaped mineral crystal, which is assembled into a highly ordered chain-like structure. Bacterial magnetosomes have attracted a broad interdisciplinary interest which goes beyond microbiology and biochemistry. However, understanding of magnetosome organization and formation has remained poor for many years after the discovery of MTB by Richard Blakemore (Blakemore 1975), mostly due to the unavailability of techniques to isolate, cultivate and manipulate MTB in the laboratory. The first pure culture of MTB, a magnetotactic spirillum, was isolated in R. Wolfe's lab by R. Blakemore and colleagues (Blakemore et al. 1979). Studies on this organism later led to the first description of "magnetosomes", as proposed by Balkwill et al. (1980). Another milestone achieved by the same group in the understanding of magnetosome formation was the identification of a biological membrane termed the "magnetosome membrane" (MM) enclosing the crystals, and it was established that empty MM vesicles are present before magnetite formation (Gorby et al. 1988). These and several more recent findings later led to the conclusion that bacterial magnetosomes represent true prokaryotic membrane-bounded organelles, displaying a comparable degree of complexity as eukaryotic counterparts (Komeili et al. 2004). In recent years, tremendous progress has been made by several groups in the elucidation of biochemistry, genetics, and cell biology of bacterial magnetosome synthesis, which is being highlighted in this work.

2 Diversity and Physiology of Magnetotactic Bacteria

Magnetosome formation occurs in several distinct phylogenetic lineages of MTB, which represent a heterogeneous group of aquatic prokaryotes with a variety of morphological types. All known MTB are motile by means of flagella and have a cell wall resembling that of typical Gram-negative bacteria. While most MTB from natural populations appear to be affiliated with the Alphaproteobacteria, examples of magnetosome formation have also been described for some representatives of the Deltaproteobacteria (DeLong et al. 1993; Kawaguchi et al. 1995), Gammaproteobacteria (Simmons et al. 2004), and the *Nitrospira* phylum (Spring et al. 1994; Flies et al. 2005). On the basis of this diversity, it was suggested that MTB are polyphyletic with respect to their 16S RNA and may descend from multiple evolutionary origins (DeLong et al. 1993; Amann et al. 2004). Commonly found morphotypes include coccoid cells as well as rods, vibrios and spirilla of various dimensions and sizes (Schüler 1999). Notably, some unusual morphologies were observed among MTB, including the giant magnetotactic rod "*Magnetobacterium bavaricum*" containing up to nearly 1000 magnetosome particles (Vali et al. 1987; Spring et al. 1993), and a multicellular magnetotactic prokaryote originally described by Farina et al. (1990) and Mann et al. (1990). Despite their high abundance and ubiquitous occurrence, most MTB have proven recalcitrant to axenic cultivation and only very few species have been isolated in pure culture (Bazylinski and Frankel 2004). The few isolated strains are metabolically diverse and versatile. Most magnetotactic isolates from the Alphaproteobacteria are sensitive towards high oxygen concentration and grow either microaerobically using oxygen or anaerobically using nitrate or nitrous oxide as electron acceptors, whereas the magnetite-forming Deltaproteobacterium *Desulfovibrio magneticus* grows anaerobically by reduction of fumarate or sulfate (for a review see Bazylinski and Frankel 2004). Many MTB metabolize short-chained organic acids, but several MTB have been recently shown to be capable of chemo-lithoautotrophic growth (Williams et al. 2006; Bazylinski et al. 2004). Difficulties in isolating and cultivating MTB arise from their lifestyle, which is adapted to complex chemical gradients that are typically encountered in stratified sediments. Accordingly, only a few species of MTB are available in pure culture, and therefore, our knowledge on magnetosome synthesis mostly comes from studies on *Magnetospirillum* species and closely related MTB within the Alphaproteobacteria.

3
Function of Magnetosomes

There is compelling evidence that the primary function of bacterial magnetosomes is to serve as navigational devices in magnetotaxis by interaction with the Earth's magnetic field. Magnetotaxis refers to motility directed by a magnetic field, although this was suggested to be a misnomer, as in contrast to a true tactic response, magnetotactic cells do not actually sense magnetic field gradients (Bazylinski and Frankel 2004). Instead, it has become apparent that passive magnetic orientation is effective in combination with other navigational mechanisms such as aerotaxis, so that alternative definitions such as "magneto-aerotaxis" in fact seem to more precisely describe this phenomenon (Frankel et al. 1997).

The preferred motility direction found in natural populations of MTB is northward in the northern hemisphere, whereas it is southward in the southern hemisphere. Because of the inclination of the geomagnetic field, migration in these preferred directions would cause cells in both hemispheres to swim downward. In the environment, MTB occur in highest numbers at or closely below the oxic-anoxic transition zone (OATZ) of marine and freshwater environments (Simmons et al. 2004; Flies et al. 2005; Bazylinski 1998), which in most cases is located a few mm below the sediment surface, or alternatively may occur in stratified water columns. Thus, by facilitating vertical up and down orientation, magnetotaxis seems to aid bacteria in the navigation along the multiple and steep chemical gradients commonly found in that zone.

Although magnetosomes seem to be clearly optimized for navigation with respect to their chemical composition, size, shape, number, and organization, magnetosome formation is evidently not essential for growth and survival of a cell under laboratory conditions, as demonstrated by analysis of nonmagnetic mutants (Blakemore et al. 1979; Ullrich et al. 2005; Dubbels et al. 2004; Schübbe et al. 2003; Matsunaga et al. 2005). Besides magnetotaxis, other possible explanations for the intracellular iron deposition have been considered, including for instance, a function in iron homeostasis (Mann et al. 1990; Blakemore and Blakemore 1991). In fact, growth of magnetosome-free mutants was more sensitive towards increased extracellular iron concentrations than growth of the magnetic wild type, which seems to hint towards a potential role of magnetite precipitation in intracellular detoxification of potentially harmful levels of soluble iron (Schübbe et al. 2003). It might be envisioned that a contribution to iron homeostasis by the precipitation of an inert iron mineral could have been one of the initial steps driving the evolution of more elaborate chains of magnetosome crystals which in addition than provided an additional selective advantage by functioning as a navigational device. Other suggested functions for magnetite formation were in energy conservation or redox cycling (Spring et al. 1993; Guerin and Blakemore 1992), although there is so far no convincing evidence that magnetite formation might yield a significant contribution to cellular energy metabolism.

4
Structure and Magnetic Characteristics of Magnetosome Crystals

All magnetotactic bacteria synthesize inorganic ferrimagnetic crystals of either magnetite (Fe_3O_4) or the iron sulfide greigite (Fe_3S_4) (Farina et al. 1990; Frankel et al. 1983). In addition, the iron sulfides mackinawite (tetragonal

FeS) and cubic FeS have been found in iron sulfide-type magnetosomes as putative precursors for greigite (Posfai et al. 1998). The chemical signature of magnetosome crystals within a species is genetically determined since bacteria continue to synthesize magnetite under conditions that favor greigite mineralization. Only one organism has been reported to simultaneously produce iron oxide and iron sulfide magnetosomes (Bazylinski et al. 1993). In this uncultured bacterium magnetosome chains are assembled from magnetite and greigite particles with each mineral occurring in its mineral-specific crystal morphology as found in single mineral containing organisms. The formation of both mineral types within the same cell is remarkable since the physico-chemical conditions favoring formation of either mineral are vastly different with respect to the redox potential. While iron-sulfide producing bacteria seem generally confined to anaerobic, H_2S-containing habitats, magnetite producers mainly populate the oxic-anoxic transition zone with traces of oxygen (Flies et al. 2005; Bazylinski et al. 1995).

Crystal size distribution (CSD), morphology and chemical purity of magnetosome crystals are subject to a species-specific genetic control (Posfai et al. 1998; Devouard et al. 1998). Unlike synthetically produced magnetite grains, which usually show log-normal CSDs, magnetosome particles are characterized by a negatively skewed CSD having a sharp limit towards larger crystal sizes (Devouard et al. 1998).

Mature magnetite and greigite crystals produced by MTB typically fall within the stable single magnetic domain range between 30–120 nm (Bazylinski et al. 1994; Moskowitz 1995), and thus are of the optimum size for magnetotaxis. While crystals of this size range are permanently magnetic, smaller sizes would not efficiently contribute to the cellular magnetic moment, as those crystals are superparamagnetic at ambient temperatures, which means they do not show a persistent magnetization. Contrary, in crystals larger than 120 nm domain walls can build up, thereby forming multiple magnetic domains of opposite magnetic orientation, which reduces the total magnetic remanence of the crystal. Nevertheless, unusually large magnetosomes of up to 200 nm in length were reported from an uncultivated magnetic diplococcus (Lins and Farina 1998). While sizes and also habits of magnetic crystals are species-specific and uniform within a single cell, there is a considerable diversity of magnetosome morphologies found in different MTB (Bazylinski et al. 1994; Petersen et al. 1989; Vali and Kirschvink 1991) (Fig. 1). High-resolution transmission electron microscopy and selected-area electron diffraction analysis of crystal habits synthesized by bacteria revealed that magnetite crystal morphology is based on combinations of the isometric crystallographic forms $\{1\ 1\ 1\}$, $\{1\ 1\ 0\}$ and $\{1\ 0\ 0\}$ (Devouard et al. 1998). Reported non-isometric morphologies such as prismatic and bullet-, or hook-shaped crystals might result from anisotropic growth due to chemical gradients or specific inhibition of growth sites by organic molecules. Morphologies of greigite crystals include mainly $\{1\ 1\ 1\}$ and less frequently $\{1\ 0\ 0\}$ forms and crystals

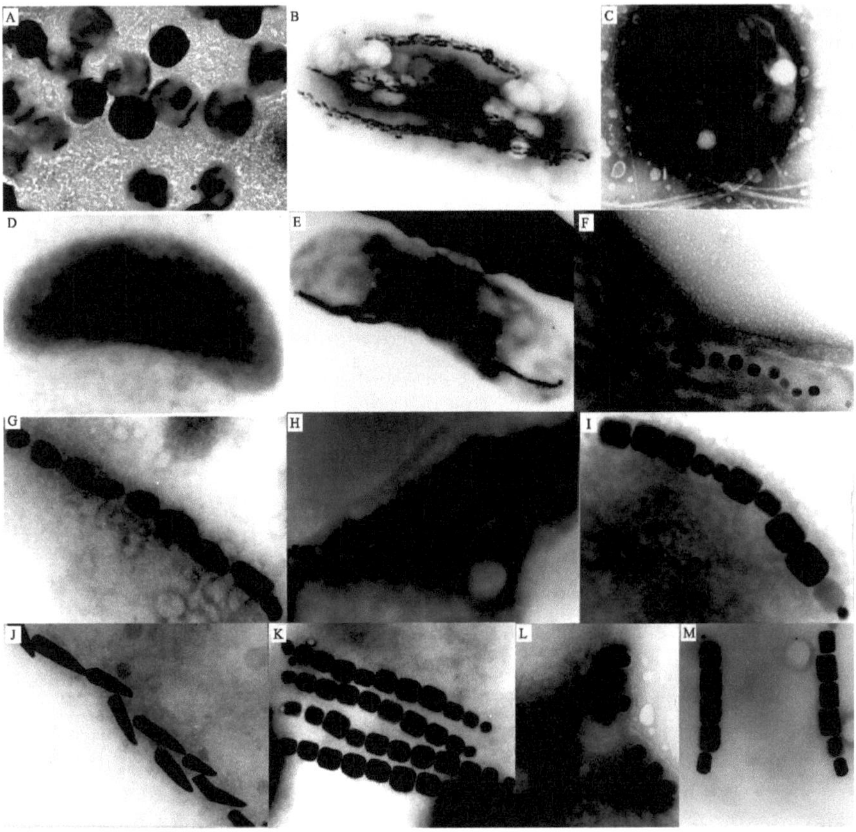

Fig. 1 Diversity of magnetotactic bacteria and magnetosomes. **A–E** Transmission electron micrographs of a variety of morphological forms found in MTB, including coccoid cell forms (**A, C**), large rod-shaped bacteria with one or several magnetosome chains (**B, E**), a magnetic vibrio (**D**) and spirilla (**E**). **F–M** Electron micrographs of crystal morphologies and organization of magnetosomes found in various MTB. With the exception of (**F**), all examples are from uncultivated MTB, which were magnetically collected from environmental samples

were often found elongated along [1 0 0] axes (Posfai et al. 1998; Bazylinski et al. 1994).

5 The Magnetosome Membrane

Magnetic crystals synthesized in *Magnetospirillum* species are enveloped by a lipid bilayer described as the magnetosome membrane (MM) (Gorby et al. 1988). There is growing evidence that the presence of a MM seems to be common to many, if not all magnetite-forming MTB, although it has been suspected that formation of magnetosomes in "*M. bavaricum*" and in some other magnetotactic bacteria is associated with a proteinaceous templating structure rather than membrane-bounded vesicles (Hanzlik et al. 2002; Taylor and Barry 2004). However, compartmentalization of the mineral synthesis by the MM provides strict control over the process of mineral formation by the precise regulation of the chemo-physical conditions within the vesicle. Magnetosome biomineralization is preceded by the formation of the magnetosome vesicles, which have been found to exist prior to magnetite formation and which already tend to assemble into imperfect chains in iron-deprived cells (Gorby et al. 1988; Komeili et al. 2004; Scheffel et al. 2005). Empty vesicles are approximately 40 nm in diameter, which is roughly equivalent to the size of mature magnetite crystals (Gorby et al. 1988).

5.1 Structure and Composition of the Magnetosome Membrane

Intact MM-enclosed magnetite crystals, which can be readily isolated and purified from disrupted cells by means of magnetic separation and ultracentrifugation, tend to form chains in vitro (Fig. 2A). Isolated magnetosomes can also spontaneously reorganize in the form of flux-closed rings (Philipse and Maas 2002), which besides magnetic interactions is apparently facilitated by direct contact mediated by the MM (Fig. 2B). The membrane can be solubilized by hot SDS treatment or organic solvents, which results in the immediate agglomeration from suspensions of the particles (Fig. 2C). Biochemical analysis of isolated magnetosomes from *M. gryphiswaldense* revealed that phosphatidylethanolamine and phosphatidylglycerol were the most abundant lipids found in the MM and other cell membranes, while amide-linked fatty acids, which are common in the outer membranes of gram-negative

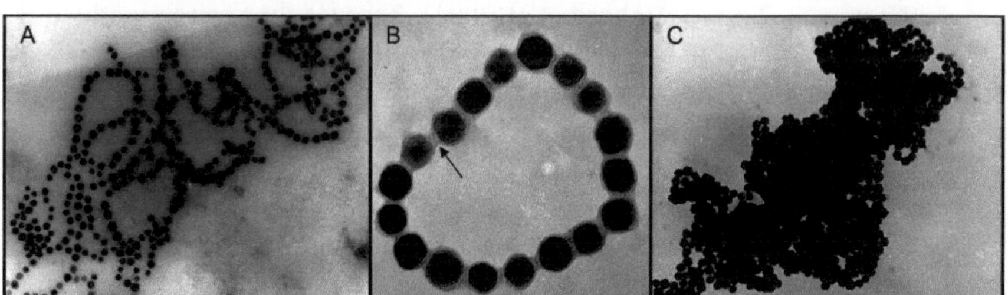

Fig. 2 **A** Isolated magnetosome particles with intact MM spontaneously tend to form chains. **B** Membrane-enclosed particles frequently reorganize in the form of flux-closed rings, which seem to be stabilized and interconnected by organic material of the MM that form junctions between crystals (*arrow*). **C** Removal of the MM by SDS treatment results in the instant agglomeration of membrane-free magnetite crystals

bacteria, were not detected in the MM (Grünberg et al. 2004; Baeuerlein 2003). This similarity in lipid composition between the MM and the cytoplasmic membrane (CM) and the observed localization adherent to the CM seems to point towards a common cellular origin. Recently, Komeilli and co-workers presented electron-microscopic evidence that magnetosome vesicles in fact derive from invaginations of the inner membrane (Komeili et al. 2005), which suggest that MM and CM at least transiently form a continuum with each other (Gorby et al. 1988; Bazylinski and Frankel 2004; Scheffel et al. 2005).

5.2 Magnetosome Membrane Proteins

Most of our knowledge about the protein composition of the MM comes from the analysis of the *M. gryphiswaldense* magnetosome subproteom. The application of various proteomic techniques revealed a complex and highly specific subset of proteins present at various quantities. There is growing evidence that homologous MM constituents are present in other related MTB, although the protein patterns of total MM preparations from different MTB appear to be dissimilar (Matsunaga et al. 2005; Grünberg et al. 2001, 2004; Okuda et al. 1996). Analysis of extracted MM from *M. gryphiswaldense* led to the identification of 18 specific magnetosome membrane proteins (MMPs) (Ullrich et al. 2005; Grünberg et al. 2001, 2004) (Table 1). Several of the identified polypeptides in MM preparations represent oligomers or post-translational modifications of the same protein, as for instance proteolytic processing, or covalently bound c-type heme as revealed by peroxidase staining. Glycosylation, which is a modification commonly found in proteins associated with other biomineralizing systems, was not detected in any MMP so far (Grünberg et al. 2004).

Beside a set of major and apparently specific MMPs, a number of minor constituents were occasionally found bound to isolated magnetosomes (Matsunaga et al. 2005; Grünberg et al. 2004). Because they were mostly identified in small amounts and represent highly abundant cellular proteins, they most likely represent contaminations from other subcellular compartments. For example, Mms16 which was detected in MM preparations from different *Magnetospirillum* strains, was originally suggested to be involved in the formation of magnetosome membrane vesicles in *Magnetospirillum* strain AMB-1 (Okamura et al. 2001). However, recent experimental evidence in *M. gryphiswaldense* demonstrates that Mms16 is not associated with magnetosomes in vivo. Moreover, its function seems entirely unrelated to magnetosome formation, and instead, it is associated with formation of PHB granules. Thus, it most likely represents a contamination, which becomes attached to magnetosomes during cell disruption (Schultheiss et al. 2005; Handrick et al. 2004).

Table 1 Characteristics of identified MMPs from *M. gryphiswaldense* (Compiled from Ullrich et al. 2005; Grünberg et al. 2001, 2004)

Protein	Length (aa)	Deduced Mr (kDa)/pI	Closest homologue in non-magnetic organisms	Characteristics	Determined or putative function
MamA	217	24.01/5.64	MM2348/*M. mazei*	TPR-motifs rel. abundance > 10%	"Activation" of magnetosomes (Komeili et al. 2004)
MamB	297	31.96/5.25	YdfM *B. subtilis*	CDF transporter	Iron transport
MamC	125	12.40/4.88	—	rel. abundance > 16%	Unknown
MamD	314	30.20/9.68	—	Leu/Gly-rich motif	Unknown
MamE	772	78.00/8.69	MLL5022 *R. loti*	PDZ domains protease domain	Serine protease
MamF	111	12.30/9.57	—	rel. abundance > 15%	Unknown
MamG	84	7.70/9.28	—	Leu/Gly-rich motif	Unknown
MamJ	466	48.51/3.80	—	Asp/Glu-rich repeats	Magnetosome chain formation (Scheffel et al. 2005)
MamM	318	34.50/5.82	BH 1238 *B. halodurans*	CDF transporter	Inorganic ion transport
MamN	437	46.14/6.70	TM0934 *T. maritima*	Membrane protein	Unknown
MamO	632	65.40/6.51	CC1282 *C. crescentus*	PDZ domains protease domain	Serine protease
MamQ	272	30.00/8.48	LemA *T. maritima*	Membrane protein	Unknown
MamR	72	8.10/8.48	—	Hydrophilic protein	Unknown
MamS	180	18.71/7.02	—	Membrane protein	Unknown
MamT	174	18.88/10.05	—	Heme binding	Unknown
MamW	138	15.01/12.8	—	Membrane protein	Unknown
Mms6	136	14.26/9.79	ApdA *R. rubrum*	Leu/Gly-rich motif	Iron binding (Arakaki et al. 2003)
MM22	196	20.00/7.14	*E. faecalis* V583	Membrane protein	Unknown

For *M. gryphiswaldense* several major MMPs have been assigned to a number of known protein families based on sequence similarity, which will be described in the following text. The highly abundant MamA, which was also identified in other *Magnetospirillum* species, contains 4–5 copies of the tetratricopeptide repeat (TPR) motif (Komeili et al. 2004; Grünberg et al. 2001; Okuda et al. 1996). TPR motifs, which have been identified in a growing number of heterogeneous proteins with diverse functions, might contribute to protein-protein interactions (Blatch and Lassle 1999). It therefore has been speculated that MamA may have an essential role on magnetosome formation by functioning as a MM receptor that interacts with cytoplasmic proteins or is involved in the assembly of multiprotein complexes within the MM (Okuda et al. 1996; Okuda and Fukumori 2001). However, MamA-deficient mutants of *Magnetospirillum* sp. AMB-1 were not affected in the formation of magnetosome vesicles, but produced magnetite crystals identical in shape and alignment to those in the wild type, albeit at reduced numbers. These results indicate that MamA is not essential for magnetosome biomineralization, but has a so-far uncharacterized function, which might be in the "activation" of magnetosome vesicles as suggested by Komeili and co-workers (Komeili et al. 2004).

A second conserved group of proteins present in the MM of various MTB are members of the cation diffusion facilitators (CDF) family (Paulsen and Saier 1997). CDF proteins function as efflux pumps of toxic divalent cations, such as zinc, cadmium, cobalt, and other heavy-metal ions. The MMPs MamB and MamM display high sequence similarity to the CDF3 subfamily, which was postulated to comprise putative iron-transporting members of the family (Nies 2003; Grass et al. 2005). Consequently, magnetosome-directed transport of iron was attributed to MamB and MamM and preliminary data from mutant analysis seems to support this assumption (Grünberg et al. 2001; Junge et al., unpublished data).

The proteins MamE and MamO display sequence similarity to HtrA-like serine proteases. While MamE was found in the MM of *M. gryphiswaldense* and *Magnetospirillum* sp. AMB-1, a MamO homolog has not yet been identified in AMB-1 (Matsunaga et al. 2005). Instead, the HtrA-like Mms67 protein was identified in the magnetosome membrane of AMB-1 (Matsunaga et al. 2000; Matsunaga et al. 2005), indicating that multiple paralogs of this protein family seem to be present in the MM. Colocated with *mamE* and *mamO* on the same operon is the *mamP* gene in *M. gryphiswaldense* that encodes a further protein of this class. Members of the HtrA family contain PDZ domains combined with a conserved trypsin-like protease domain and are reported to act as molecular chaperones and heat-shock induced proteases, which degrade misfolded proteins in the periplasm, or have other regulatory functions (Clausen et al. 2002). In magnetosome formation MamE and MamO were speculated to act in the processing, maturation and targeting of MMPs during MM assembly (Grünberg et al. 2004), but experimental evidence is missing so far.

Several MMPs seem to be universally present in other MTB, but lack any sequence homology with proteins of known function in non-magnetic organisms, which prevents any predictions of their putative functions. Among those MMPs, MamC and MamF are the most abundant MM-associated proteins in *M. gryphiswaldense*. A sequence feature common to MamD, Mms6 and MamG is a conspicuous hydrophobic sequence motif that is rich in stretches of repeated leucine and glycine residues. Similar LG-rich repetitive motifs were found in framework proteins known to self aggregate forming therefore supermolecular structures sometimes associated with biomineralization (Schüler 2004). Experimental data shows that the Mms6 protein of the *Magnetospirillum* strain AMB-1 exhibits iron-binding activity and affects the morphology of growing magnetite crystals in an in vitro magnetite crystallization assay (Arakaki et al. 2003). This suggests that Mms6 might be involved in the control of crystallization also in vivo. The *C*-terminal region of Mms6 is rich in acidic amino acid residues with carboxyl and hydroxyl groups (aspartate, glutamate, tyrosine, serine), which are known to have a strong affinity to metal ions, and thus the observed activity of Mms6 was speculated to be determined by its acidic C-terminus. As clusters of polyelectrolytic groups are commonly found in numerous biomineralization proteins, this sequence feature is assumed to be involved in protein-mineral interactions (Baeuerlein 2003). Another MMP that contains several domains of highly biased amino acid composition is MamJ. Its central domain contains two identical, highly acidic Glu-rich stretches, in which acidic amino acid residues account for 32.4% of the sequence (Scheffel et al. 2005). Thus, MamJ was initially assumed to be involved in magnetite crystallization (Schüler 2004). However, recent data obtained from a *mamJ* deletion mutant of *M. gryphiswaldense* have revealed that MamJ rather is involved in the formation of magnetosome chains as will be discussed below (Scheffel et al. 2005).

6
Genetics and Molecular Organization of Magnetosome Genes

To identify genes controlling magnetosome formation different techniques have been applied, which include reverse genetics, analysis of spontaneous non-magnetic mutants and transposon mutagenesis studies.

6.1
The Magnetosome Genes are Clustered within a Genomic Magnetosome Island

Magnetosome genes were identified by reverse genetics in several strains of *Magnetospirillum*. In *M. gryphiswaldense* all identified MMPs are encoded within a hypervariable 130 kb genomic region (Ullrich et al. 2005; Schübbe et al. 2003), which display the following characteristics: (i) it harbors all identified magnetosome (*mam*/*mms*) genes; (ii) it contains 42 transposase genes; (iii) most other ORFs are hypothetical genes. The 130 kb region contains numerous direct and inverted repeats, which mostly correspond to similar copies of transposase genes. Its G + C content is slightly distinct from that of the rest of the genome and displays a more heterogeneous distribution. Three tRNA genes are present within this region, which is also bounded by an integrase gene fragment. In summary, all these features are strongly reminiscent of those described for genome islands in other bacteria (Dobrindt et al. 2004). Genome islands usually comprise large genomic regions, which, for instance, are present as pathogenicity islands (PAIs) in the genomes of pathogenic strains, but absent from the genomes of their non-pathogenic counterparts. They often encode "accessory" gene functions, tend to be genetically unstable and can transfer horizontally. Thus, it seems that the 130 kb region encoding magnetosome genes represents a large genomic "magnetosome island", which may have been distributed by lateral gene transfer. It is currently not known if the presence of a MAI is common to all MTB. However, there is evidence that the genomic organization of magnetosome genes is conserved in *Magnetospirillum* species and to a lesser degree in the magnetic coccus MC-1 (Komeili et al. 2004; Schübbe et al. 2003; Matsunaga et al. 2005; Grünberg et al. 2001).

Within the MAI the majority of identified MMPs are encoded within the *mam* (*magnetosome membrane*) and *mms* (*magnetic particle membrane-specific*) gene clusters, along with several genes of unknown function, which are located within less than 35 kb of the MAI (Schübbe et al. 2003; Grünberg et al. 2001). In *M. gryphiswaldense*, the *mamAB* cluster encompasses 17 collinear open reading frames extending 16.4 kb of DNA (Fig. 3). The 2.1 kb *mamGFDC* cluster is located 15 kb upstream of the *mamAB* cluster and comprises 4 ORFs. The 3.6 kb *mms* cluster is located 368 bp upstream of the *mamGFDC* cluster and contains 5 ORFs. The collinear organization and close spacing of genes within the three clusters suggested that they each might be transcribed as polycistronic operons from a single promoter. Cotranscription of genes within the *mamAB*, *mamDC* and *mms* clusters was recently demonstrated indicating the presence of long polycistronic transcripts extending over more than 16 kb. The transcription start points of the *mamAB*, *mamDC*, and *mms* operons were mapped closely upstream of the first genes of the operons. The presence

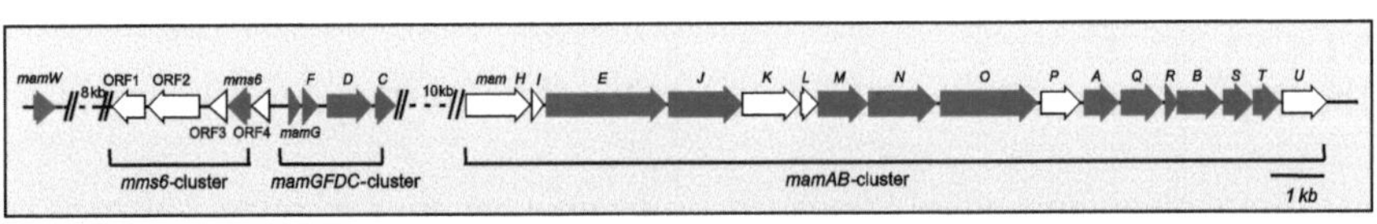

Fig. 3 Molecular organization of a region from the genomic magnetosome island of *M. gryphiswaldense*. Genes encoding MMP (*red arrows*) are clustered within 35 kb and beside the monocistronic *mamW* comprise the *mamGFDC*, *mamAB*, and *mms6* operons

of transcripts was found to be independent from growth phase, however, the expression of the *mam* and *mms* genes was up-regulated under magnetite-forming conditions, i.e. during microaerobiosis and in the presence of iron.

6.2 The Magnetosome Island is a Highly Unstable Genomic Region and Undergoes Spontaneous Rearrangements

Substantial sequence polymorphism between clones from different subcultures of *M. gryphiswaldense* was reported for the MAI (Ullrich et al. 2005), suggesting that this region undergoes frequent rearrangements during serial subcultivation in the laboratory. Spontaneous mutants affected in magnetosome formation arise at a frequency of up to 10^{-2} after prolonged storage of cells at 4 °C or exposure to oxidative stress (Ullrich et al. 2005). All non-magnetic mutants exhibited extended and multiple deletions within the MAI and had lost either parts or the entire *mms*- and *mam*-gene clusters encoding magnetosome proteins. Mutations are polymorphic with respect to the sites and extents of deletions, but were all found associated with the loss of various copies of IS elements. These observations indicate that the genomic MAI undergoes frequent transposition events, which lead to subsequent deletions by homologous recombination under conditions of physiological stress (Ullrich et al. 2005; Schübbe et al. 2003). This could be interpreted in terms of adaptation to physiological stress and might contribute to the genetic plasticity and mobilization of the MAI.

The spontaneous loss of the ability to synthesize magnetosomes seems to be a characteristic trait common to diverse MTB and has been reported for different species (Blakemore et al. 1979; Ullrich et al. 2005; Dubbels et al. 2004; Schübbe et al. 2003; Matsunaga et al. 2005). For instance, spontaneous non-magnetic mutants were regularly observed in marine magnetotactic vibrio MV-1. All mutants failed to express a major copper-containing periplasmic protein (ChpA) presumably involved in iron uptake (Dubbels et al. 2004). Although mutants were genetically heterogeneous, they all exhibited two point mutations (transversions) at identical wobble positions within *chpA*, which apparently did prevent the translation of its transcript by an unknown mechanism. This finding suggests that there is no universal reason causing the observed genetic instability of the magnetic phenotype, but that different mechanisms are responsible for the appearance of spontaneous non-magnetic mutants in different MTB.

6.3 Mutational Analysis of Magnetosome Formation

Other attempts to identify genes controlling magnetosome formation have involved conventional mutagenesis approaches. However, genetic analysis by mutagenesis is complicated by the aforementioned genetic instability of the magnetic phenotype, which requires careful controls, e.g. by complementation analysis, to confidently correlate the observed phenotype to the introduced mutation. In several studies involving transposon mutagenesis, genetic screens for magnetosome genes were performed (Matsunaga et al. 2005; Matsunaga et al. 1992; Wahyudi et al. 2001, 2003; Calugay et al. 2004). For example, of 5762 Tn mutants, 69 were found to be defective for magnetosomes, as recently reported by Matsunaga and co-workers. Disrupted genes were found distributed over the genome and could be categorized to functional groups including signal transduction, energy metabolism, cell envelope biogenesis, cell motility, and unknown functions (Matsunaga et al. 2005). However, the roles of these genes awaits further clarification since none of the mutants could have been complemented so far. Curiously, neither of the previously identified *mam* and *mms* genes, which are essential for magnetosome synthesis, nor other genes located within these operons were among the affected genes in the studies mentioned above. In contrast, all transposon mutants obtained in an independent study on the same organism resulted from insertions in the *mamAB* cluster (Komeili et al. 2004), which further corroborates the functional link of this region to magnetosome biomineralization. Since methods for targeted mutagenesis by allelic replacement have become available for two *Magnetospirillum* species (Komeili et al. 2004; Schultheiss and Schüler 2003; Schultheiss et al. 2004), gene disruption studies for several genes within the *mamAB* cluster revealed their specific functions in activation and assembly of magnetosomes (Komeili et al. 2004; Scheffel et al. 2005).

7
Subcellular Organization and Assembly of Magnetosomes Chains

In most MTB magnetosome crystals are arranged in one or several chains parallel to the axis of motility in order to sum up their individual magnetic moments. By aligning several crystals into a linear chain with the easy axis of magnetization running parallel to the chain, the resulting magnetic torque is sufficiently large to align the cell body in the geomagnetic field (Dunin-Borkowski et al. 1998, 2001; Proksch et al. 1995; Kirschvink 1982). Magnetostatic calculations on magnetosome chains suggest that this arrangement would collapse without some form of rigid support (Kirschvink 1982). Chain stability was proposed to result from a cellular structure that physically connects the chain to the cell envelope, and which also transfers the magnetic torque to the cell body (Gorby et al. 1988; Hanzlik et al. 2002).

Recently, details of subcellular organization supporting this notion were revealed by the application of cryo-electron tomography (cryo-ET), which is a novel electron-microscopic imaging technique that provides three-dimensional information about pleomorphic structures (Baumeister et al. 1999). Electron tomograms of vitrified *M. gryphiswaldense* cells clearly show a network of filamentous structures traversing the cell closely adjacent to the cytoplasmic membrane (Fig. 4). Most magnetosomes were found arranged intimately along this filamentous structure, which is suggested to be part of the cytoskeleton (Scheffel et al. 2005; Komeili et al. 2005). Empty magnetosome vesicles and those with immature magnetite crystals were predominantly located at the ends of chains (Scheffel et al. 2005), whereas chains of magnetosomes are arranged adjacent to the cytoplasmic membrane, apparently following the cell curvature in a helical manner (Gorby et al. 1988; Scheffel et al. 2005). Such a close connection of the magnetosome chain with a cytoskeletal "magnetosome filament" meets both the requirements of chain stabilization and transfer of the magnetic torque to the cell body. It was speculated that the magnetosome filament might be formed by the *mamK* gene product because of its striking sequence similarity to actin-like MreB proteins (Scheffel et al. 2005; Schüler 2004; Errington 2003), and recent observations on a Δ*mamK* mutant, in which the magnetosome filament is absent, seems to support this notion (Komeili et al. 2005). The tight association of the magnetosome vesicles with the magnetosome filament apparently depends on the presence of the MMP MamJ (Scheffel et al. 2005). Mutant cells of *M. gryphiswaldense* lacking MamJ display a substan-

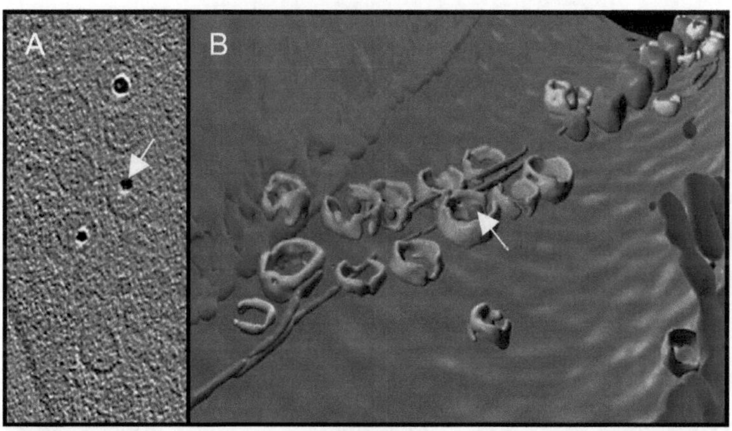

Fig. 4 Tomographic reconstruction of a *M. gryphiswaldense* cell. **A** A section in the *z*-direction from the reconstructed volume showing empty MM vesicles, and those which are partially filled with growing, immature magnetite crystals, are closely attached to the magnetosome filament, which represents a cytoskeletal structure that stabilizes magnetosome chains. **B** Three-dimensional visualization of a part of a *M. gryphiswaldense* cell with the cytoplasmic membrane (*blue*), empty vesicles (*yellow*), growing and mature magnetite crystals (*red*), and the magnetosome filament (*green*) (micrographs and visualization by M. Gruska and J.M. Plitzko, MPI Martinsried)

tially reduced magnetic orientation, although they contain magnetosome crystals identical in number, size, and morphology to those in the wild type. However, in the MamJ-deficient strain magnetosome crystals are no longer aligned in a straight chain as in the wild type, but instead arrange in compact three-dimensional clusters (Fig. 5A,B). Empty magnetosome vesicles were found mostly scattered throughout the cytoplasm, were widely spaced from each other, and dissociated from the magnetosome filament as revealed by cryo-ET. This indicates that magnetosomes are attached to the magnetosome filament by MamJ in wild-type cells, whereas in cells lacking MamJ mature magnetosome crystals are free to agglomerate once they are in close proximity, thereby lowering the magnetostatic energy of their alignment (Scheffel et al. 2005).

Beside linear chain structures, magnetosome arrangements like clumps or individual scattered crystals have also been reported in MTB from natural environments. For instance, in cells of a many-celled magnetotactic prokaryote magnetosomes arrange in short chains or clusters (Farina et al. 1983).

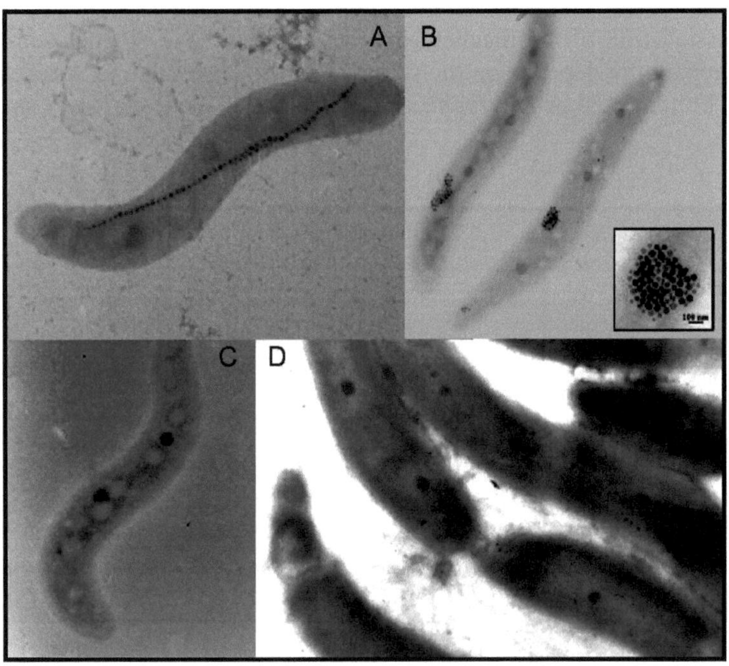

Fig. 5 Wild type (**A**) and various mutant cells (**B, C, D**) of *M. gryphiswaldense*, which are affected in magnetosome organization and crystal formation. **B** *ΔmamJ* deletion mutant. Cells lacking the MamJ protein do not form magnetosome chains, but instead form clusters of magnetosomes. **C** Spontaneous non-magnetic mutant, which is devoid of any magnetite crystals. **D** Spontaneous mutant MSR-1K which forms smaller and irregular crystals (Ullrich et al. 2005)

Similarly, in uncultivated MTB of the "*Bilophococcus*"-type (Moench 1988) magnetosomes are clustered in arrays. Yet, despite of the deviations from the chain-like organization, magnetosomes still generate a total magnetic moment sufficient to facilitate magnetotaxis (Posfai et al. 1998).

8 Pathway of Magnetite Biomineralization and Magnetosome Formation

As magnetosome biomineralization is preceded by magnetosome vesicle formation, mineral synthesis may proceed simultaneously in multiple vesicles (Scheffel et al. 2005). Magnetite formation depends on the presence of iron and the prevalence of micro-oxic conditions (Blakemore et al. 1985; Schüler and Baeuerlein 1998). When cells were grown under oxygen-controlled conditions in a fermenter, magnetite biomineralization occurred only at low dissolved oxygen concentrations below 10 mbar, while higher concentrations were inhibitory to mineral formation (Heyen and Schüler 2003). The amount of magnetite formed did correlate with the dissolved oxygen concentration with most magnetite formed by cells at a partial pressure as low as 0.25 mbar (Heyen and Schüler 2003). The formation of magnetite is coupled to the uptake of vast amounts of iron from the environment which can accumulate to an amount between 3% to 10% of dry weight (Baeuerlein 2003). Details of biogenic magnetite formation are still not fully understood, but a hypothetical model can be postulated based on available physiological, molecular genetic and ultrastructural data (Fig. 6). The high-affinity uptake system of *M. gryphiswaldense* was found to be saturated at extracellular ferric iron concentrations of 15–20 μM Fe without any evidence for involvement of siderophores (Schüler and Baeuerlein 1996). In contrast, *Magnetospirillum* strain AMB-1 has been reported to excrete both hydroxamate- and catechol-type siderophores for the assimilation of large amounts of ferric iron (Calugay et al. 2003; Paoletti and Blakemore 1986). Siderophore-mediated iron uptake of ferric iron was also suggested for the magnetic strain MV-1 (Bazylinski and Frankel 2004). However, in the natural habitat of MTB at or closely below the OATZ, a considerable amount of soluble ferrous is likely to be present. Thus, a periplasmic multicopper-containing Fe(II) oxidase identified in strain MV-1 was alternatively speculated to support the uptake of ferrous iron (Dubbels et al. 2004). In *M. magnetotacticum* an Fe(II)-nitrate oxidoreductase was reported to be involved in the subsequent oxidation of ferrous iron for the formation of the mixed-valence iron oxide magnetite ($Fe^{+3}(Fe^{+2}Fe^{+3})O_4$) (Yamazaki et al. 1995). Cytoplasmic iron is then transported to the magnetosome vesicle, where a part is then thought to form a highly reactive Fe(III) oxide, probably ferrihydrite, which may react with dissolved Fe^{2+} to form magnetite by a via-solution process (Cornell and Schwertmann 2003). Alternatively, coprecipitation of ferrous and ferric iron at alkaline pH was

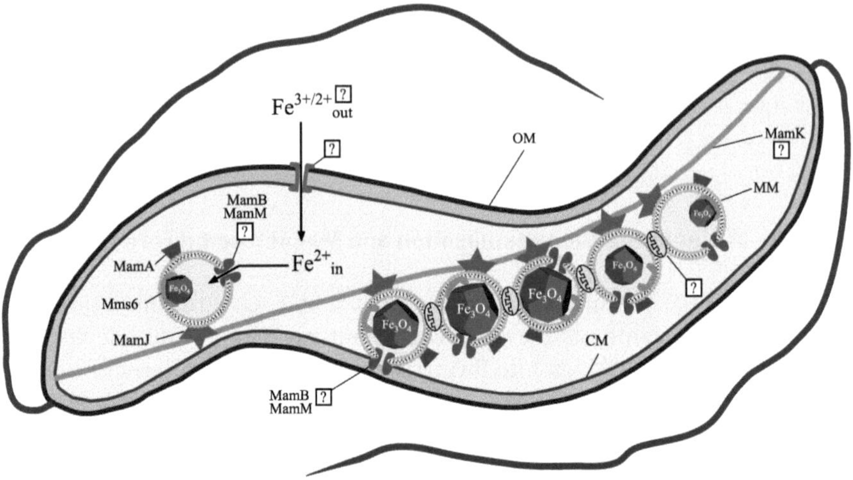

Fig. 6 Hypothetical model for the pathway of magnetite biomineralization and magnetosome chain formation, which is based on identified and putative functions of individual cellular constituents. Ferric or ferrous iron is transported into the cell by an energy-dependent process via an unknown uptake system (*purple*). Supersaturating amounts of ferrous iron are then transported by the MamB/MamM-proteins (*blue*) from the cytoplasm into the magnetosome vesicles, which are activated for magnetite formation by the MamA (*dark green*) protein. Alternatively, iron transport may proceed via MamB/MamM directly from the periplasmic space into the MM vesicles. Intravesicular iron is then thought to be reoxidized and bound to the Mms6 protein (*pink*), which promotes nucleation and crystal growth. MM vesicles, which are attached to the magnetosome filament (*green*) through the MamJ protein (*red*), align themselves at midcell along this cytoskeletal structure. Mature particles are interconnected by an unidentified constituent of the magnetosome membrane (*yellow*)

suggested as a potential mechanism for magnetite particle formation based on results of in vitro crystallization experiments (Faivre et al. 2004).

9
Biotechnological Applications of Magnetosomes

Biogenic magnetic particles produced by MTB have inspired numerous ideas for application due to their unique morphological, magnetic, and chemical characteristics (Schüler and Frankel 1999; Lang and Schüler 2005; Matsunaga et al. 2004). Typical sizes of the monocrystalline particles are in a range, which is not easily accessible by chemical synthesis. Moreover, the highly controlled pathway of biosynthesis provides a morphological diversity and, in some bacteria, shape anisotropy, which are not commonly known from inorganically produced magnetite crystals. In addition, size and morphology of magnetosome crystals are under genetic control, which could be used for the genetic engineering of magnetic nano-crystals with "tailored" characteristics. One of the most interesting features of bacterial magnetosomes is the presence of a biological membrane with defined biochemical composition. The encapsulation of the magnetic crystal within the MM provides a natural "coating", which ensures superior dispersibility of the particles and provides an excellent target for modification and functionalization. Taken together, the characteristics of biogenic magnetic particles offer additional advantages to many applications compared to the use of conventional abiogenic nanomagnets. So far, magnetosome particles have been successfully applied as diagnostic and therapeutic tools in magnetic resonance imaging and hyperthermia treatments (Reszka 2000; Hergt et al. 2005). In other experiments, peptides, enzymes, polymers or oligonucleotides were immobilized to the magnetosome surface (reviewed by Lang and Schüler 2005). For instance, antibody-magnetosome conjugates have been used as biosensors to detect environmental pollutants, hormones and toxic substances while others were employed for the specific magnetic separation of cell targets (Tanaka and Matsunaga 2000; Tanaka et al. 2004; Kuhara et al. 2004). Dendrimer-modified magnetosome particles have been used in automated DNA extraction procedures and mRNA isolation was facilitated by oligo(dT)-modified magnetosomes (Yoza et al. 2003; Sode et al. 1993). In summary, although in several cases technological uses of bacterial magnetosomes have been demonstrated, they still have a largely untapped potential for future nanobiotechnological applications.

10 Conclusions

Since their discovery more than thirty years ago, magnetosomes have emerged as one of the most intriguing and complex intracellular structures found in bacteria and have been suggested as a model to study organelle formation in prokaryotes (Komeili et al. 2004). Although our knowledge about magnetosome structure, function, and synthesis has increased tremendously over the last few years, many open questions remain to be answered. Since the unexpected genetic complexity of magnetosome formation has been recognized, it now has become a major challenge to reveal the individual functions of the numerous genes and proteins that are suspected to be involved in the synthesis of magnetosome chains. Another intriguing problem is the question of the phylogenetic origin of the trait of magnetosome formation in diverse MTB, especially in the light of putative horizontal transfer of the magnetosome island, which might have contributed to shape the genomes of various magnetic bacteria. As "magnetosome" particles similar to those of MTB have been found in higher organisms, the biological mechanism of magnetosome formation might also have relevance for the understanding of magnetoreceptive organelles in eukaryotic organisms (Kirschvink et al. 2001), and even their use as markers for the biological origin of magnetite nanocrystals (Thomas-Keprta et al. 2002).

Acknowledgements We would like to acknowledge our students, colleagues, and numerous collaborators. Research in the authors lab is supported by the Max Planck Society, the Deutsche Forschungsgemeinschaft, and the German BMBF.

References

Arakaki A, Webb J, Matsunaga T (2003) A novel protein tightly bound to bacterial magnetic particles in *Magnetospirillum magneticum* Strain AMB-1. J Biol Chem 278:8745–8750

Baeuerlein E (2003) Biomineralization of Unicellular Organisms: An Unusual Membrane Biochemistry for the Production of Inorganic Nano- and Microstructures. Angew Chem Int Ed 42:614–641

Balkwill D, Maratea D, Blakemore R (1980) Ultrastructure of a magnetotactic spirillum. J Bacteriol 141:1399–1408

Baumeister W, Grimm R, Walz J (1999) Electron tomography of molecules and cells. Trends in Cell Biology 9:81–85

Bazylinski DA, Heywood BR, Mann S, Frankel RB (1993) Fe_3O_4 and FeS_4 in a bacterium. Nature 366:218

Bazylinski DA, Garratt-Reed A, Frankel RB (1994) Electron-microscopic studies of magnetosomes in magnetotactic bacteria. Microsc Res Tech 27:389–401

Bazylinski DA, Frankel RB, Heywood BR, Mann S, King JW, Donaghay PL, Hanson AK (1995) Controlled biomineralization of magnetite (Fe_3O_4) and greigite (Fe_3S_4) in a magnetotactic bacterium. Appl Environ Microbiol 61:3232–3239

Bazylinski D (1998) Ultrastructure and function of the bacterial magnetosome. Abstracts of the General Meeting of the American Society for Microbiology 981998:15

Bazylinski DA, Dean AJ, Williams TJ, Long LK, Middleton SL, Dubbels BL (2004) Chemolithoautotrophy in the marine, magnetotactic bacterial strains MV-1 and MV-2. Arch Microbiol 182:373–387

Bazylinski DA, Frankel RB (2004) Magnetosome formation in prokaryotes. Nat Rev Microbiol 2:217–230

Blakemore R (1975) Magnetotactic bacteria. Science 190:377–379

Blakemore R, Maratea D, Wolfe RS (1979) Isolation and pure culture of a freshwater magnetic spirillum in chemically defined medium. J Bacteriol 140:720–729

Blakemore RP, Short KA, Bazylinski DA, Rosenblatt C, Frankel RB (1985) Microaerobic conditions are required for magnetite formation within *Aquaspirillum magnetotacticum*. Geomicrobiol J 4:53–72

Blatch GL, Lassle M (1999) The tetratricopeptide repeat: a structural motif mediating protein-protein interactions. BioEssays 21:932–939

Calugay RJ, Miyashita H, Okamura Y, Matsunaga T (2003) Siderophore production by the magnetic bacterium *Magnetospirillum magneticum* AMB-1. FEMS Microbiol Lett 218:371–375

Calugay RJ, Okamura Y, Wahyudi AT, Takeyama H, Matsunaga T (2004) Siderophore production of a periplasmic transport binding protein kinase gene defective mutant of *Magnetospirillum magneticum* AMB-1. Biochem Biophys Res Commun 323:852–857

Clausen T, Southan C, Ehrmann M (2002) The HtrA family of proteases: implications for protein composition and cell fate. Mol Cell 10:443–455

DeLong EF, Frankel RB, Bazylinski DA (1993) Multiple evolutionary origins of magnetotaxis in bacteria. Science 259:803–806

Devouard B, Posfai M, Hua X, Bazylinski DA, Frankel RB, Buseck PR (1998) Magnetite from magnetotactic bacteria: size distributions and twinning. Am Mineral 83:1387–1398

Dobrindt U, Hochhut B, Hentschel U, Hacker J (2004) Genomic islands in pathogenic and environmental microorganisms. Nat Rev Microbiol 2:414–424

Dubbels BL, DiSpirito AA, Morton JD, Semrau JD, Neto JN, Bazylinski DA (2004) Evidence for a copper-dependent iron transport system in the marine, magnetotactic bacterium strain MV-1. Microbiology 150:2931–2945

Dunin-Borkowski RE, McCartney MR, Frankel RB, Bazylinski DA, Posfai M, Buseck PR (1998) Magnetic microstructure of magnetotactic bacteria by electron holography. Science 282:1868–1870

Dunin-Borkowski RE, McCartney MR, Posfai M, Frankel RB, Bazylinski DA, Buseck PR (2001) Off-axis electron holography of magnetotactic bacteria: magnetic microstructure of strains MV-1 and MS-1. Eur J Mineral 13:671–684

Errington J (2003) Dynamic proteins and a cytoskeleton in bacteria. Nat Cell Biol 5:175–178

Faivre D, Arginier P, Menguy N, Zuddas P, Pachana K, Gloter A, Laval JY, Guyot F (2004) Mineralogical and isotopic properties of inorganic nanocrystalline magnetites. Geochim Cosmochim Acta 68:4395–4403

Farina M, Lins de Barros HGP, Esquivel DMS, Danon J (1983) Ultrastructure of a magnetotactic microorganism. Biology of the Cell 48:85–88

Farina M, Esquivel DMS, Lins de Barros HGP (1990) Magnetic iron-sulfur crystals from a magnetotactic microorganism. Nature 343:256–258

Flies C, Peplies J, Schüler D (2005a) A combined approach for the characterization of uncultivated magnetotactic bacteria from various aquatic environments. Appl Environ Microbiol 71:2723–2731

Flies C, Jonkers H, deBeer D, Bosselmann K, Böttcher M, Schüler D (2005b) Diversity and vertical distribution of magnetotactic bacteria along chemical gradients in freshwater microcosms. FEMS Microbiol Ecol 52:185–195

Frankel RB, Papaefthymiou GC, Blakemore RP, O'Brien W (1983) Fe_3O_4 precipitation in magnetotactic bacteria. Biochim Biophys Acta 763:147–159

Frankel RB, Bazylinski DA, Johnson MS, Taylor BL (1997) Magneto-aerotaxis in marine coccoid bacteria. Biophys J 73:994–1000

Gorby YA, Beveridge TJ, Blakemore RP (1988) Characterization of the bacterial magnetosome membrane. J Bacteriol 170:834–841

Grass G, Otto M, Fricke B, Haney CJ, Rensing C, Nies DH, Munkelt D (2005) FieF (YiiP) from *Escherichia coli* mediates decreased cellular accumulation of iron and relieves iron stress. Arch Microbiol 183:9–18

Grünberg K, Wawer C, Tebo BM, Schüler D (2001) A large gene cluster encoding several magnetosome proteins is conserved in different species of magnetotactic bacteria. Appl Environ Microbiol 67:4573–4582

Grünberg K, Müller EC, Otto A, Reszka R, Linder D, Kube M, Reinhardt R, Schüler D (2004) Biochemical and proteomic analysis of the magnetosome membrane in *Magnetospirillum gryphiswaldense*. Appl Environ Microbiol 70:1040–1050

Guerin W, Blakemore R (1992) Redox cycling of iron supports growth and magnetite synthesis by *Aquaspirillum magnetotacticum*. Appl Environ Microbiol 58:1102–1109

Handrick R, Reinhardt S, Schultheiss D, Reichart T, Schüler D, Jendrossek V, Jendrossek D (2004) Unraveling the function of the *Rhodospirillum rubrum* activator of polyhydroxybutyrate (PHB) degradation: the activator is a PHB-granule-bound protein (phasin). J Bacteriol 186:2466–2475

Hanzlik M, Winklhofer M, Petersen N (2002) Pulsed-field-remanence measurements on individual magnetotactic bacteria. J Magn Magn Mater 248:258–267

Hergt R, Hiergeist R, Zeisberger M, Schüler D, Heyen U, Hilger I, Kaiser WA (2005) Magnetic properties of bacterial magnetosomes as diagnostic and therapeutic tools. J Magn Magn Mater 293:80–86

Heyen U, Schüler D (2003) Growth and magnetosome formation by microaerophilic *Magnetospirillum* strains in an oxygen-controlled fermentor. Appl Microbiol Biotechnol 61:536–544

Kawaguchi R, Burgess JG, Sakaguchi T, Takeyama H, Thornhill RH, Matsunaga T (1995) Phylogenetic analysis of a novel sulfate-reducing magnetic bacterium, Rs-1, demonstrates its membership of the Delta-proteobacteria. FEMS Microbiol Lett 126:277–282

Kirschvink JL (1982) Paleomagnetic evidence for fossil biogenic magnetite in western Crete. Earth Planet Sci Lett 59:388–392

Kirschvink JL, Walker MM, Diebel CE (2001) Magnetite-based magnetoreception. Curr Opin Neurobiol 11:462–467

Komeili A, Vali H, Beveridge TJ, Newman DK (2004) Magnetosome vesicles are present before magnetite formation, and MamA is required for their activation. Proc Natl Acad Sci USA 101:3839–3844

Komeili A, Li Z, Newman DK, Jensen GJ (2005) Magnetosomes are cell membrane invaginations organized by the actin-like protein MamK. Sciencexpress 10.1126/science.1123231

Kuhara M, Takeyama H, Tanaka T, Matsunaga T (2004) Magnetic cell separation using antibody binding with protein a expressed on bacterial magnetic particles. Anal Chem 76:6207–6213

Lang C, Schüler D (2006) Biomineralization of magnetosomes in bacteria: nanoparticles with potential applications. In: Microbial bionanotechnology – biological self-assembly systems and biopolymer-based nanostructures, BHA Rehm, ed. (Wymondham, Horizon Bioscience), pp 107–124

Lins U, Farina M (1998) Magnetosome size distribution in uncultured rod-shaped bacteria as determined by electron microscopy and electron spectroscopic imaging. Microsc Res Tech 42:459–464

Mann S, Sparks NHC, Frankel RB, Bazylinski DA, Jannasch HW (1990a) Biomineralization of ferrimagnetic greigite (Fe_3O_4) and iron pyrite (FeS_2) in a magnetotactic bacterium. Nature 343:258–260

Mann S, Sparks N, Board R (1990b) Magnetotactic bacteria: microbiology, biomineralization, palaeomagnetism and biotechnology. Adv Microb Physiol 31:125–181

Matsunaga T, Nakamura C, Burgess J, Sode S (1992) Gene transfer in magnetic bacteria: transposon mutagenesis and cloning of genomic DNA fragments required for magnetosome synthesis. J Bacteriol 174:2748–2753

Matsunaga T, Tsujimura N, Okamura H, Takeyama H (2000) Cloning and characterization of a gene, *mpsA*, encoding a protein associated with intracellular magnetic particles from *Magnetospirillum* sp. strain AMB-1. Biochem Biophys Res Commun 268:932–937

Matsunaga T, Okamura Y, Tanaka T (2004) Biotechnological application of nano-scale engineered bacterial magnetic particles. J Mater Chem 14:2099–2105

Matsunaga T, Okamura Y, Fukuda Y, Wahyudi AT, Murase Y, Takeyama H (2005) Complete genome sequence of the facultative anaerobic magnetotactic bacterium *Magnetospirillum* sp. strain AMB-1. DNA Research 12:157–166

Moench T (1988) *Bilophococcus magnetotacticus* gen nov sp nov, a motile, magnetic coccus. Antonie van Leeuwenhoek 54:483–496

Moskowitz BM (1995) Biomineralization of magnetic minerals. Rev Geophys 33:123–128

Nies DH (2003) Efflux-mediated heavy metal resistance in prokaryotes. FEMS Microbiol Rev 27:313–339

Okamura Y, Takeyama H, Matsunaga T (2001) A magnetosome-specific GTPase from the magnetic bacterium *Magnetospirillum magneticum* AMB-1. J Biol Chem 276:48183–48188

Okuda Y, Denda K, Fukumori Y (1996) Cloning and sequencing of a gene encoding a new member of the tetratricopeptide protein family from magnetosomes of *Magnetospirillum magnetotacticum*. Gene 171:99–102

Okuda Y, Fukumori Y (2001) Expression and characterization of a magnetosome-associated protein, TPR-containing MAM22, in *Escherichia coli*. FEBS Lett 491:169–173

Paoletti L, Blakemore R (1986) Hydroxamate produktion by *Aquaspirillum magnetotacticum*. J Bacteriol 167:153–163

Paulsen IT, Saier MH (1997) A novel family of ubiquitous heavy metal ion transport proteins. J Membr Biol 156:99–103

Philipse AP, Maas D (2002) Magnetic colloids from magnetotactic bacteria: Chain formation and colloidal stability. Langmuir 18:9977–9984

Posfai M, Buseck PR, Bazylinski DA, Frankel RB (1998a) Iron sulfides from magnetotactic bacteria: structure, composition, and phase transitions. Am Mineral 83:1469–1481

Posfai M, Buseck PR, Bazylinski DA, Frankel RB (1998b) Reaction sequence of iron sulfide minerals in bacteria and their use as biomarkers. Science 280:880–883

Proksch RB, Schaffer TE, Moskowitz BM, Dahlberg ED, Bazylinski DA, Frankel RB (1995) Magnetic force microscopy of the submicron magnetic assembly in a magnetotactic bacterium. Appl Phys Lett 66:2582–2584

Scheffel A, Gruska M, Faivre D, Linaroudis A, Plitzko JM, Schüler D (2005) An acidic protein aligns magnetosomes along a filamentous structure. Nature advance online publication 10.1038/nature04382

Schultheiss D, Schüler D (2003) Development of a genetic system for *Magnetospirillum gryphiswaldense*. Arch Microbiol 179:89–94

Schultheiss D, Kube M, Schüler D (2004) Inactivation of the flagellin gene *fla*A in *Magnetospirillum gryphiswaldense* results in nonmagnetotactic mutants lacking flagellar filaments. Appl Environ Microbiol 70:3624–3631

Schultheiss D, Handrick R, Jendrossek D, Hanzlik M, Schüler D (2005) The presumptive magnetosome protein Mms16 is a PHB-granule bound protein (phasin) in *Magnetospirillum gryphiswaldense*. J Bacteriol 187:2416–2425

Schübbe S, Kube M, Scheffel A, Wawer C, Heyen U, Meyerdierks A, Madkour M, Mayer F, Reinhardt R, Schüler D (2003) Characterization of a spontaneous nonmagnetic mutant of *Magnetospirillum gryphiswaldense* reveals a large deletion comprising a putative magnetosome island. J Bacteriol 185:5779–5790

Schüler D, Baeuerlein E (1996) Iron-limited growth and kinetics of iron uptake in *Magnetospirillum gryphiswaldense*. Arch Microbiol 166:301–307

Schüler D, Baeuerlein E (1998) Dynamics of iron uptake and Fe_3O_4 biomineralization during aerobic and microaerobic growth of *Magnetospirillum gryphiswaldense*. J Bacteriol 180:159–162

Schüler D (1999) Formation of magnetosomes in magnetotactic bacteria. J Mol Microbiol Biotechnol 1:79–86

Schüler D, Frankel RB (1999) Bacterial magnetosomes: microbiology, biomineralization and biotechnological applications. Appl Microbiol Biotechnol 52:464–473

Schüler D (2004) Molecular analysis of a subcellular compartment: The magnetosome membrane in *Magnetospirillum gryphiswaldense*. Arch Microbiol 181:1–7

Simmons SL, Sievert SM, Frankel RB, Bazylinski DA, Edwards KJ (2004) Spatiotemporal distribution of marine magnetotactic bacteria in a seasonally stratified coastal salt pond. Appl Environ Microbiol 70:6230–6239

Sode K, Kudo S, Sakaguchi T, Nakamura N, Matsunaga T (1993) Application of bacterial magnetic particles for highly selective messenger-RNA recovery system. Biotechnol Tech 7:688–694

Spring S, Amann R, Ludwig W, Schleifer KH, Van Gemerden H, Petersen N (1993) Dominating role of an unusual magnetotactic bacterium in the microaerobic zone of a freshwater sediment. Appl Environ Microbiol 59:2397–2403

Spring S, Amann R, Ludwig W, Schleifer K-H, Schüler D, Poralla K, Petersen N (1994) Phylogenetic analysis of uncultured magnetotactic bacteria from the alpha-subclass of Proteobacteria. Systematic & Applied Microbiology 17:501–508

Tanaka T, Matsunaga T (2000) Fully automated chemiluminescence immunoassay of insulin using antibody-protein A-bacterial magnetic particle complexes. Anal Biochem 72:3518–3522

Tanaka T, Takeda H, Ueki F, Obata K, Tajima H, Takeyama H, Goda Y, Fujimoto S, Matsunaga T (2004) Rapid and sensitive detection of 17 beta-estradiol in environmental water using automated immunoassay system with bacterial magnetic particles. J Biotechnol 108:153–159

Taylor AP, Barry JC (2004) Magnetosomal matrix: ultrafine structure may template biomineralization of magnetosomes. J Microsc (Oxf) 213:180–197

Thomas-Keprta KL, Clemett SJ, Bazylinski DA, Kirschvink JL, McKay DS, Wentworth SJ, Vali H, Gibson Jr. EK, Romanek CS (2002) Magnetofossils from ancient Mars: a robust biosignature in the martian meteorite ALH84001. Appl Environ Microbiol 68:3663–3672

Ullrich S, Kube M, Schübbe S, Reinhardt R, Schüler D (2005) A hypervariable 130 kb genomic region of *Magnetospirillum gryphiswaldense* comprises a magnetosome island, which undergoes frequent rearrangements during stationary growth. J Bacteriol 187:7176–7184

Vali H, Forster O, Amarantidis G, Petersen N (1987) Magnetotactic bacteria and their magnetofossils in sediments. Earth Planetary Sci Lett 86:2–4

Wahyudi AT, Takeyama H, Matsunaga T (2001) Isolation of *Magnetospirillum magneticum* AMB-1 mutants defective in bacterial magnetic particle synthesis by transposon mutagenesis. Appl Biochem Biotechnol 91–93:147–154

Wahyudi AT, Takeyama H, Okamura Y, Fukuda Y, Matsunaga T (2003) Characterization of aldehyde ferredoxin oxidoreductase gene defective mutant in *Magnetospirillum magneticum* AMB-1. Biochem Biophys Res Commun 303:223–229

Williams TJ, Zhang CL, Scott JH, Bazylinski DA (2006) Evidence for autotrophy via the reverse tricarboxylic acid cycle in the marine magnetotactic coccis strain MC-1. Appl Environ Microbiol 72(2):1322–1329

Yamazaki T, Oyanagi H, Fujiwara T, Fukumori Y (1995) Nitrite reductase from the magnetotactic bacterium *Magnetospirillum magnetotacticum* – a novel cytochrome Cd(1) with Fe(II)-nitrite oxidoreductase activity. Eur J Biochem 233:665–671

Yoza B, Arakaki A, Matsunaga T (2003a) DNA extraction using bacterial magnetic particles modified with hyperbranched polyamidoamine dendrimer. J Biotechnol 101:219–228

Yoza B, Arakaki A, Maruyama K, Takeyama H, Matsunaga T (2003b) Fully automated DNA extraction from blood using magnetic particles modified with a hyperbranched polyamidoamine dendrimer. J Biosci Bioeng 95:21–26

Microbiol Monogr (2)
J.M. Shively: Complex Intracellular Structures in Prokaryotes
DOI 10.1007/7171_025/Published online: 13 June 2006

Structure, Function and Formation of Bacterial Intracytoplasmic Membranes

Robert A. Niederman

Rutgers University, Department of Molecular Biology and Biochemistry,
Nelson Biological Laboratories, 604 Allison Road, Piscataway, NJ 08854-8082, USA
rniederm@rci.rutgers.edu

Abstract A variety of prokaryotes increase their membrane surface area for energy transduction processes by developing an intracytoplasmic membrane (ICM), in the form of tubules, interconnected vesicles, and single, paired or stacked lamellae. Recent images of the ICM of the photosynthetic bacterium *Rhodobacter sphaeroides*, obtained by atomic force microscopy, have provided the first submolecular surface views of any complex multi-component membrane. These topographs revealed rows of dimeric core light-harvesting 1 (LH1) (RC) complexes, interconnected by the peripheral light-harvesting 2 (LH2) complex, which also existed in separate clusters. In addition, polarized light spectroscopy demonstrated that this optimal functional arrangement is extended into a long-range pattern of membrane organization. Functional insights provided by the detailed structures of the light-harvesting, RC and cytochrome bc_1 complexes are also discussed, including how LH1 is organized to facilitate ubiquinone exchange. It was shown that LH1-RC core structures are inserted initially into the cytoplasmic membrane, which upon addition of LH2, invaginates to form the ICM, with LH2 packing between rows of dimeric core complexes, and ultimately forming separate bulk LH2 clusters. The ICM of the ecologically important methanotrophs, and chemolithotrophic nitrifying bacteria that convert ammonia to nitrite, is also discussed. The recent determination of the crystal structure of the major ICM protein, methane monooxygenase, and the complete genome sequence of *Methylococcus capsulatus*, are providing further insights into the molecular details of both methane oxidation and utilization of the resulting methanol as a sole source of carbon and energy.

1
Introduction

As introduced elsewhere in this volume (see the work by Shively, 2006, in this volume), many types of bacteria form intracytoplasmic membrane (ICM) structures, including anoxygenic phototrophic bacteria of the α- and γ-proteobacterial subdivisions, cyanobacteria, and aerobes that include the methanotrophs and the chemolithoautotrophic nitrifying bacteria. While, in general, the ICM serves to increase the amount of intracellular membrane available for specific energy transduction processes, the Planctomycetes form prokaryotic ICM types with completely distinct structures and functions, viz., the anammoxosomes, a unique single membrane-bounded envelope in which anaerobic ammonium oxidation occurs, and double membrane bounded nucleoids and pirellulosomes, which provide compartmentation for nucleoid material (see the work by Fuerst and Fuerst et al., 2006, in this volume).

Clearly, the intracytoplasmic photosynthetic membranes of phototropic bacteria are the most thoroughly investigated and best characterized prokaryotic internal membranes, and their ICM surfaces represent the first intact energy transducing membrane to be visualized at submolecular resolution (Scheuring et al. 2005). Accordingly, this review will focus mainly on ICM structure, function and assembly in the much studied anoxygenic photosynthetic prokaryote, *Rhodobacter sphaeroides*, a member of the α-3 subclass of the proteobacteria, and on several other related phototrophs. In addition, ICM formation in this group will be contrasted with the development of the intracytoplasmic photosynthetic lamellar (thylakoid) membranes of cyanobacteria, and lastly, recent studies of the ICM and related topics in the ecologically important methanotrophs and the chemolithoautotrophic nitrifying bacteria will be reviewed.

2 Intracytoplasmic Membranes of Anoxygenic Phototrophic Bacteria

Much of this review is devoted to the structure, function and biogenesis of the intracytoplasmic photosynthetic membranes of purple non-sulfur bacteria, which represent the best studied and most completely understood prokaryotic internal membranes. These structures have served as an important model system for understanding the more complex thylakoids of oxygenic phototrophs, as well as other energy transducing membrane systems.

2.1 Structure of Intracytoplasmic Photosynthetic Membrane

The original discovery of intracytoplasmic photosynthetic membranes in anoxygenic phototrophic bacteria is reviewed elsewhere (Niederman and Gibson 1978; Drews and Niederman 2002; Shively, 2006, in this volume) and it is now known that these structures are continuous with the peripheral cytoplasmic membrane (CM). Depending upon the species of anoxygenic phototroph, transmission electron micrographs of thin-cell section have shown that the ICM can take the form of interconnected vesicles (see Figs. 4 and 5 below), tubules, or lamellae of various types that include thylakoid-like stacks in both regularly organized and irregular arrangements (Drews and Golecki 1995). Moreover, it has been demonstrated in *Rba. sphaeroides* that the normal vesicular ICM is replaced by a largely tubular ICM in the absence of the peripheral light harvesting complex (LH2) (Hunter et al. 1988; Kiley et al. 1988).

In past years, the architecture of the ICM, as well as the distribution of photosynthetic complexes has been extensively investigated by freeze-fracture electron microscopy (reviewed by Drews and Golecki 1995). While a uniform distribution of intramembrane particles over the ICM surfaces of freeze-fracture replicas was observed, along with evidence for the origin of the ICM by invagination of the CM, it is only with the advent of atomic force microscopy (AFM) that photosynthetic complexes can now be visualized directly.

2.1.1 Local Surface Architecture of Intracytoplasmic Photosynthetic Membrane as Revealed by Atomic Force Microscopy

AFM has evolved into a unique and powerful tool for acquiring high-resolution images of membrane proteins directly in native membranes. The recent application of this highly informative structural probe to the native ICM of purple photosynthetic bacteria has provided the first views of the in situ molecular surface of any multi-component biological membrane system (reviewed by Scheuring et al. 2005). The AFM images of membrane patches derived from ICM vesicles (chromatophores) of *Rba. sphaeroides*, shown in Fig. 1, allow the direct visualization of the local native surface ICM architecture of this well-studied photosynthetic bacterium (Bahatyrova et al. 2004a). They reveal a unique organization of ICM domains, as well as crucial new features of the relative positions and functional associations of LH2, together with the core light harvesting complex (LH1) and the photochemical reaction center (RC). Two types of organization were found for the LH2 complexes;

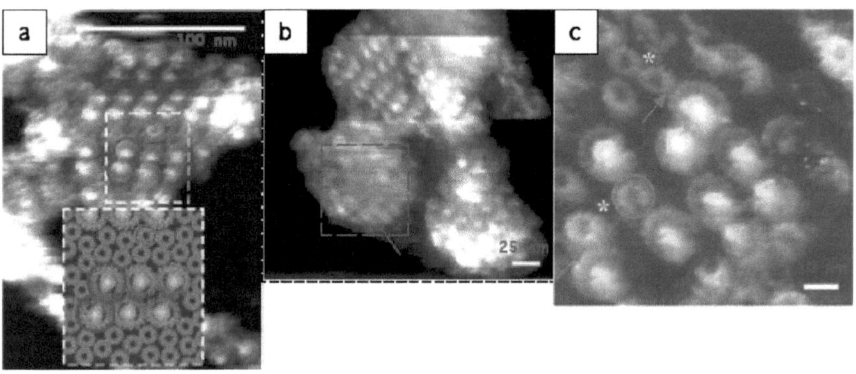

Fig. 1 AFM of native *Rba. sphaeroides* photosynthetic membrane (Bahatyrova et al. 2004). These views show ICM vesicle (chromatophore) patches adhered to mica surfaces. **a** View of LH1-RC core complex arrays surrounded by LH2 complexes. The central protruding feature within the cores corresponds to the RC-H subunit, indicating that the cytoplasmic face of the ICM lies at the surface. *Inset at bottom*: existing structural data (Allen et al. 1987; McDermott et al. 1995; Roszak et al. 2003) was used to model region denoted by the *central dashed box*; a typical LH1-RC dimer is delineated in both images by a *red outline*, and a representative LH2 complex by a *green circle*. **b** Membrane patches showing two types of arrangement of complexes. The *red-boxed region* is composed mainly of LH2 complexes; *arrow points* to an LH2 ring within this large LH2-only domain. No core complexes are seen in this region. **c** Three-dimensional representation of a small region of membrane showing LH1-RC core complex dimers and monomers with associated LH2. *Green arrows* indicate contact points for energy transfer between LH2 and LH1-RC complexes. A nine-unit structure is visible in LH2 rings marked by *asterisks*. The encircled LH2 complex is sandwiched between two LH1-RC complexes

in the first type (Fig. 1a), 10–20 molecules of LH2 form light capture domains that interconnect linear arrays of LH1-RC dimeric core complexes. In the second, LH2 exists in separate domains as larger clusters (Fig. 1b), apparently representing the light-responsive LH2 complement, formed during chromatic adaptation to low intensity illumination (see Sect. 2.3.3 below). The close packing also observed for the linear LH1-RC arrays, often separated by only narrow LH2 energy channels (Fig. 1c), demonstrates an elegant economy in photosynthetic membrane arrangement. The physical continuity between individual LH2 complexes and between LH2 and LH1-RC core dimers, assures that excitations are transferred from LH2 to LH1, and subsequently to the RC, to initiate the primary photochemical events. Moreover, the LH1 complex is ideally positioned to serve as a collection hub for excitations obtained from LH2, which are then rapidly transferred to the RC. Those LH2 complexes situated between rows of LH1-RC dimers, form a relatively invariant complement of pigment–protein complexes that optimally fulfill the basic requirement for efficient trapping and transmission of light energy.

The LH1-RC core complexes are usually found in dimeric associations, arranged in rows of up to six dimers (Fig. 1a,b), such that an LH1 excitation can migrate along a series of dimers until an "open" RC is found. These

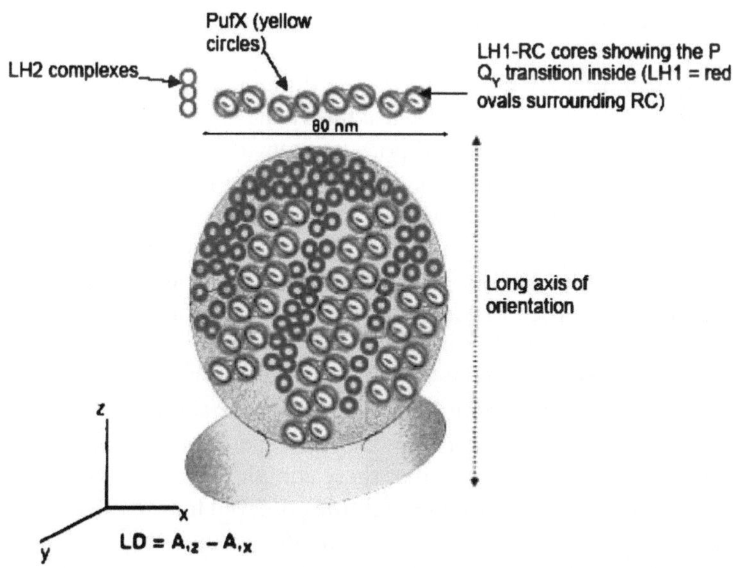

Fig. 2 Model showing the organization of the LH1-RC core and LH2 complexes in a single mature native *Rba. sphaeroides* ICM invagination (chromatophore) (Frese et al. 2004), based upon LD measurements and AFM images. The geometry of the LD measurements is shown relative to the chromatophore long axis. A slightly distorted spherical shape is assumed which is drawn along the axis connecting the starting point and the distal-most point of the ICM invagination. Complexes and membrane are drawn to scale

AFM images also explain why mutant cells lacking LH2 form elongated tubular membranes, which in freeze-fracture replicas show striking patterns of closely packed, highly ordered rows of dimeric LH1-RC particles arrayed helically along their entire length (Westerhuis et al. 2002).

It is notable that the structure-based interpretation of the AFM images shown in the inset of Fig. 1a are reminiscent of a model based on fluorescence quenching in which the photosynthetic units of *Rba. sphaeroides* were depicted as clusters of LH1 complexes surrounding and interconnecting RCs, with LH2 arranged peripherally in large "lakes" (Monger and Parson 1977). This proposal was further refined by singlet–singlet annihilation measurements at 4 K, that were explained by domains of photosynthetic complexes consisting of $\sim$ 4 RCs embedded in core assemblies of $\sim$ 100 LH1 bacteriochlorophyll *a* (BChl) molecules with LH2 BChls separating them (Vos et al. 1986), similar to the arrangement directly visualized by AFM (Fig. 1a). At room temperature, thermal separation of the core cluster no longer exists, and many RCs became connected as a result of energy equilibration among $>$ 1000 BChls, again in accord with the overall antenna arrangement seen in AFM (Fig. 1b).

2.1.2
Long-Range Organization of Intracytoplasmic Photosynthetic Membrane as Revealed by Linear Dichroism

Linear dichroism (LD), which probes the difference in absorption between horizontally and vertically polarized light by oriented complexes, facilitates the determination of transition-dipole moment orientations of LH and RC BChl, relative to the sample orientation axis. These measurements were performed on oriented membranes, since successful application of LD requires that the material under investigation assumes a macroscopically ordered state. The same ICM vesicle (chromatophore) preparations used for AFM were examined in order to assess whether the local order observed in the AFM images was extended into a long-range pattern of photosynthetic unit organization (Frese et al. 2004).

Near-IR LD spectra of both LH1 and LH2 were similar to their absorption spectra in the oriented chromatophore preparation, indicating a similar dipole-moment orientation, and that these complexes possessed the requisite macroscopic order. The reduced magnitude observed for the LD signal of LH2 relative to LH1, however, was shown to be due to an abundance of LH2 in curved regions of the oriented chromatophore vesicles, since LD signals of a similar magnitude were obtained for both LH complexes in flat membrane patches produced by treatment of the chromatophores with the low detergent concentrations used in the preparation of AFM samples.

A negative LD contribution arising from RC accessory BChls, demonstrated previously in $LH2^-$ membranes (Frese et al. 2000), was observed in these wild-type chromatophores. This LD signal had an opposite sign in RC only and LH1-RC strains lacking PufX, an LHI-RC core organizing protein apparently interspersed near an opening in the LH1 ring (Qian et al. 2005), to facilitate ubiquinone–ubiquinol exchange (Barz et al. 1995). The reversal of this RC accessory BChl LD signal is due to a special pufX-induced, non-random orientation of the RC within the LH1 ring and an alignment of all RC-LH1 cores in regular arrays (Frese et al. 2000). The striking observation of this effect in LH2-rich wild-type membranes (Frese et al. 2004), establishes that the PufX-induced order, first observed in LH1-RC-PufX only membranes (Frese et al. 2000), represents the native long-range organization of core arrays that coexists with large domains of LH2 in the native ICM. In addition, LD spectra of the RC-BChl (P) "special pair" transition moments in the native wild-type membrane also assumed a special PufX-induced orientation.

Figure 2 presents a model, based upon AFM topographs and LD measurements, which illustrates the supramolecular organization of the native *Rba. sphaeroides* ICM. A single ICM vesicle is depicted as a semispherical structure with a preferential long axis (see also Fig. 4), densely packed with the pigment–protein complexes. Other salient features of the model are: (*i*) the unique, PufX-dependent RC orientation within the ordered LH1 rings; (*ii*) the arrangement of the core RC-LH1-PufX dimers in rows, interspersed with single or double colinearly aligned groups of LH2 complexes; (*iii*) LH2 complexes not confined to areas between the dimeric core particle, are abundant in highly curved ICM regions. Although the LH1-RC and LH2 arrays were not always aligned in a colinear fashion in the AFM images, this appears to be a consequence of the opening up of the spherical chromatophore membranes by the mild detergent treatment and the flattening of the resulting membrane patches as they adhered to the mica surface. It is also noteworthy that a clustering of LH2 in curved ICM domains is supported by the correlation between the rise in LH2 levels and the increased curvature observed in the ICM during chromatic adaptation to lowered illumination levels (Sturgis and Niederman 1996).

2.2 Structure-Function Relationships in Photosynthetic Complexes of the Intracytoplasmic Membrane

The light reactions that drive photophosphorylation in *Rba. sphaeroides* are typical of purple non-sulfur bacteria, and occur in the ICM (Fig. 3). Radiant energy harvested by LH2 is transferred to LH1, which funnels these excitations to the RC-BChl "special pair", where they are trapped in a transmembrane charge separation. This initiates a cycle of electron-transfer reactions between the primary iron-quinone acceptor (Q_A), the bc_1 complex, cytochrome c_2 and the photooxidized special pair. The resulting electrochemical proton gradient is coupled to the synthesis of ATP. The determination of

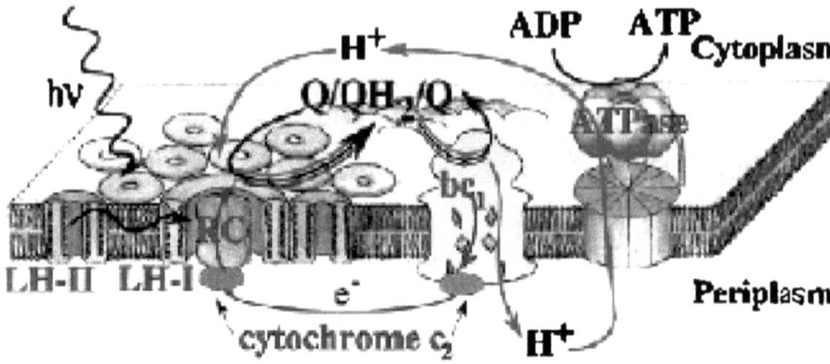

Fig. 3 Schematic depiction of photon absorption and excitation energy transfer (*black undulating arrows*), electron transport (*blue arrows*), and proton movement (*red arrows*) in photosynthetic apparatus of *Rba. sphaeroides* ICM (Hu et al. 1998). See text for details. Reprinted with permission of the authors

the structures of the RC, LH2, LH1-RC core and ubiquinol–cytochrome c_2 oxidoreductase (cytochrome bc_1) complexes has formed the basis for a detailed understanding of the primary photochemical events, as well as the structural organization of ICM components catalyzing them (Sect. 2.1.1).

2.2.1 The Peripheral Light-Harvesting (LH2) Complex – a Nonameric or Octameric Ring of $\alpha\beta$-Heterodimers

The determination of the atomic-resolution structure of the LH2 complex of *Rhodopseudomonas acidophila* by X-ray crystallography (McDermott et al. 1995) was the first detailed structural information to emerge for any integral membrane light-harvesting complex, and represents a major landmark in photosynthetic research. The oligomeric complex forms a ring-like assembly consisting of the transmembrane α helices of nine β and nine α apoprotein subunits (Fig. 1a inset), which make up the respective outer and inner walls, and provide the scaffold for the non-covalently bound BChls and carotenoids. Each $\alpha\beta$-heterodimer binds one B800 BChl and two B850 BChls, with absorption maxima near 800 and 850 nm, respectively, and one–two closely associated molecules of the carotenoid rhodopin glucoside nearly spanning the membrane. A ring of 18 strongly coupled B850 BChls, with overlapping bacteriochlorins oriented perpendicular to the membrane plane, is sandwiched between the two concentric α-helical cylinders in the periplasmic half of the ICM bilayer. This circular pigment arrangement constitutes an annulus for rapid delocalization of excitations, facilitating energy transfer both within the same LH2 ring or to adjacent rings, and ultimately to LH1. The nine B800 BChls are positioned on the outer surface between the β-polypeptide helices and arranged with their tetrapyrrole moieties nearly parallel to the membrane plane. They form a weakly-coupled outer ring of monomeric pigments, in the cytoplasmic half of the bilayer, which transfers excitation energy to the B850 BChls. These structure-function relationships are dramatically illustrated by the intermolecular LH2 BChl energy transfer times derived from time-resolved excitation dynamics studies: B800 → B800, 0.5 ps; B800 → B850, 0.9 ps; B850 → B850, 50–150 fs; LH2 → LH1, 2–4 ps, largely reflecting the distances and mutual orientations between these pigments (see review by Cogdell et al. 2004 for further details).

A similar structure, consisting of eight rather than nine $\alpha\beta$-heterodimers has been determined for the crystalline LH2 complex of *Phaeospirillum molischianum* (Koepke et al. 1996), while in *Rba. sphaeroides*, the LH2 complex consists of an $(\alpha\beta)_9$ ring structure (Walz et al. 1998), as in *Rps. acidophila* and a number of other purple bacteria.

2.2.2
Core Light-Harvesting (LH1) Complex – Variations on an Ellipse of 16 $\alpha\beta$-Heterodimers

The LH1 complex of *Rba. sphaeroides* was the first bacterial core antenna to be isolated in a spectrally intact form (Broglie et al. 1980), and it has subsequently proved to have a number of unusual spectroscopic and structural properties that have undergone extensive investigation. It was soon recognized that the strong near-IR absorption band of the LH1 BChls (maximum at 875 nm) is spectrally heterogeneous, with the long-wavelength component initially designated as the "B896" sub-antenna species (Kramer et al. 1984). Further studies of the fluorescence properties of the *Rba. sphaeroides* LH1 complex, suggested that this spectral inhomogeneity arises from the clustering of strongly coupled BChl dimers with small, random differences in local pigment environments (summarized in Westerhuis et al. 1999). The "B896" component was ultimately attributed to low energy excitonic transitions arising from interactions within inhomogeneous pigment clusters, localized in the curvilinear array of excitonically coupled chromophores.

The pioneering structural work of Karrasch et al. (1995), in which an 8.5 – Å projection map was obtained from 2-D crystals of the reconstituted LH1 complex of *Rhodospirillum rubrum*, revealed a ring of 16 apparent $\alpha\beta$-heterodimers, sufficiently large to enclose the RC. It was unclear, however, how the ubiqiunonol/ubiquinone, produced at their respective RC-Q_B and cytochrome bc_1 sites, could be exchanged across the seemingly impenetrable barrier of LH1 α-helices confronted in this structure.

The most detailed LH1 structure that has appeared thus far, is from the LH1-RC core complex of *Rhodopseudomonas palustris*, determined by X-ray crystallography at 4.8 Å resolution (Roszak et al. 2003), revealing the polypeptide backbone and allowing placement of the curvilinear array of B875 BChls between the inner and outer apoprotein walls (Fig. 1a inset). The RC is at the center and is surrounded by an incomplete oval modeled as 15 transmembrane LH1 $\alpha\beta$-heterodimers coordinating 30 BChls. This monomeric, elliptical structure is interrupted by the single trans-membrane helix W, thought to facilitate ubiquinone movements, as shown for its apparent functional homologue pufX in *Rba. sphaeroides* (Barz et al. 1995). The W helix is located next to the RC groove through which the tail of Q_B projects, and it was proposed that this "16th helix" also plays a role in positioning the RC within the overall structure.

The W helix has not been characterized further, and the most definitive subsequent information on the breach in the LH1 ellipse as a site for communication with the Q_B site, comes from the recent 8.5 Å projection map arising from 2-D crystals of the core LH1-RC dimer of *Rba. sphaeroides* (Qian et al. 2005), in which a plausible location for the well-described PufX protein is suggested. Twenty-eight LH1 $\alpha\beta$-heterodimers were fitted into the ellipsoidal, double-lobed structure, which enclosed the α-helices of two RCs with their Q_B sites adjacent to the putative PufX position near the openings in each of the LH1 helical arrays. This strategic placement of PufX, together with evidence for adjacent, highly mobile LH1 α-helices, implicates this region, as well as the flexibility demonstrated for the LH1 ring (Bahatyrova et al. 2004b) in the optimization of quinone exchange during light-driven cyclic electron flow. The possibility that the LH1 ring of *Rsp. rubrum* also represents a flexible structure that could facilitate transient quinone exchange in the absence of PufX, has been suggested by the various elliptical forms observed for the LH1-RC core complex from this organism in AFM topographs (Fotiadis et al. 2004).

2.2.3
Reaction Center Complex – a Scaffold for Binding the Cofactors that Catalyze the Primary Photochemical Events

The 3-D atomic resolution structure of the photosynthetic RC of the purple non-sulfur bacterium *Blastochloris viridis* was the first structure to be determined for any integral membrane protein and led to the awarding of the 1988 Noble prize in chemistry to Hartmut Michel, Hans Deisenhofer and Robert Huber (Deisenhofer and Michel 1989). The structure of the homologous *Rba. sphaeroides* RC was subsequently obtained; its integral membrane protein components also consist of L, M and H subunits with respective masses of 31, 34 and 28 kDa (Allen et al. 1987). The L and M subunits each have five transmembrane *a* helical spans while the H subunit crosses the membrane once near the *N*-terminus, with the bulk of its mass projecting out into the cytoplasm (Fig. 1a inset).

The transmembrane *a* helices of the L and M protein subunit form a pseudo two-fold symmetry axis perpendicular to the ICM plane, which provides a scaffolding for the arrangement of the cofactors that consist of the Bchl special pair (the primary donor, P), two monomeric accessory BChls, two bacteriopheophytins, the two ubiquinone molecules Q_A and Q_B, an iron atom adjacent to Q_A, and the carotenoid spheroidene. These cofactors are arranged into two branches (A on the M-side, and B on the L-side) in which only the A branch is active in electron transfer.

Once excitations from LH1 reach the RC special pair in 20–40 ps, the energy is used to release an electron which is transferred to bacteriopheophytin$_A$ in $\sim$ 2 ps, possibly via the accessory BChl$_A$. The electron then arrives on Q_A, the primary quinone acceptor in 200 ps; the secondary quinone Q_B is then reduced by the gating of two electrons through Q_A after a second turnover. After picking up two protons from the cytoplasm, the doubly reduced Q_B is released from the RC into the ubiqinone pool (Fig. 3).

These intraprotein RC electron transfer reactions should not be thought of in isolation, since Francia et al. (2004) have elegantly demonstrated that the dimeric LH1-RC core structures of *Rba. sphaeroides* provide a favorable environment for a localized quinone pool. It was further suggested that the LH1 complex plays a role in optimizing the yield of secondary charge separation, by energetically stabilizing the $P^+Q_B^-$ charge separated state.

2.2.4
Cytochrome *bc*$_1$ Complex – Minimal Functional Unit of Catalytic Proteins

The cytochrome *bc*$_1$ complex, which catalyzes electron transfer from ubiquinol to a soluble *c*-type cytochrome, with concomitant transmembrane movement of protons, occupies a central position in the mitochondrial respiratory chain and in the cyclic electron flow of photosynthetic bacteria. A preliminary X-ray structure has recently been reported for the cytochrome *bc*$_1$ complex of *Rba. capsulatus* (Berry et al. 2004), which reveals that this minimal functional unit, containing only the catalytic cytochromes *b*, *c*$_1$ and the Rieske iron-sulfur proteins, also exists as an intertwined dimer, like its multisubunit mitochondrial counterpart. As for the mitochondrial complex, eight transmembrane α-helices were assigned to cytochrome *b*, with one each assigned to the Rieske protein and cytochrome *c*$_1$; these form their respective *N*- and *C*-terminal membrane anchors, with the bulk of each protein protruding about 35 Å out into the periplasm. The cytochrome *b* subunit, which contains two *b*-type hemes that catalyze the Q-cycle, serves as the central core of the complex, and also provides the quinone binding sites.

Unlike most purple bacteria, which also contain a three-subunit cytochrome *bc*$_1$ complex, an additional polypeptide (subunit IV) is tightly associated with the cytochrome *bc*$_1$ complex of *Rba. sphaeroides*, and is likely to function in a structural role through conformational alterations that favor ubiquinone binding by cytochrome *b* (Tso et al. 2000). Scheuring et al. (2004) have raised the additional possibility that the gap in the LH1 ellipse can act as a docking site for the cytochrome *bc*$_1$ complex via subunit IV, which could also act as a terminator of LH1 oligomerization during the assembly of the core complex.

A puzzling aspect of recent AFM studies of the bacterial photosynthetic apparatus in the native ICM, is the absence of the cytochrome bc_1 complex from any of the images (Scheuring et al. 2005). It is noteworthy that all four subunits of the *Rba. sphaeroides* complex were demonstrated in Western blots of the ICM patches used to obtain the AFM topographs. Although the structure of the three-subunit *Rba. capsulatus* shows a limited topological exposure at the cytoplasmic surface of the membrane, studies of the topology of *Rba. sphaeroides* subunit IV suggest that the *N*-terminal half of this 14-kDa protein projects out into the cytoplasm (Tso et al. 2000), which should make the complex easy to visualize. Some possibilities that have been suggested to explain this paradox are that the LH2 and LH1-RC complexes are densely packed in the flat regions of the ICM imaged by AFM, with the bc_1 complex located close to the CM and or on the edges of invaginations (Scheuring et al. 2005). It would then be necessary to invoke channels within the bilayer for the observed rapid quinone diffusion between distantly dispersed RC and bc_1 complexes. Clearly, considerably more effort will be needed to resolve this apparent quandary.

2.3 Formation of Intracytoplasmic Photosynthetic Membranes

Rba. sphaeroides also provides an ideal experimental system for membrane biogenesis and assembly studies. When grown photoheterotrophically (under anaerobic conditions in the light), levels of LH2 relative to LH1-RC core complexes in the ICM are related inversely to light intensity. Although ICM formation is repressed by high oxygen tension under chemoheterotrophic conditions, lowering the oxygen partial pressure results in a gratuitous induction of ICM assembly in the dark (greening) by invagination of the CM, together with the synthesis and assembly of LH and RC complexes.

Much of the earlier work on the biosynthesis of the ICM and the assembly of photosynthetic complexes has been reviewed by Drews and Golecki (1995), Drews (1996) and Drews and Niederman (2002). Instead, we will mainly focus here on more recent work involving regulatory aspects, the role of invagination sites in ICM formation, assembly in greening cells, and further perspectives on the photosynthetic unit assembly and organization and its relevance.

2.3.1 Regulatory Mechanisms Controlling Intracytoplasmic Photosynthetic Membrane Formation

The developmental changes occurring during ICM formation are under the control of a number of interacting regulatory circuits that act at the transcriptional level to control formation of both the pigment and apoprotein components of the LH and RC complexes in response to oxygen and light (reviewed by Oh and Kaplan 2001; Young and Beatty 2003). Foremost among these regulatory elements is the global two-component RegBA signal transduction system, designated as PrrBA for *Rba. sphaeroides* 2.4.1 (Eraso and Kaplan 1994). These regulatory elements play a central role in oxygen-dependent regulation by sensing redox changes in the *cbb*3-type cytochrome *c* oxidase that activate photosynthesis gene expression under reducing conditions. Crucial structural genes under RegBA control include those of the *puf* operon (*pufBALM*), encoding the LH1-β and -α Polypeptides and the RC-L and -M subunits, respectively, those of the *puc* operon (*pucBA*) encoding the respective LH2-β and -α Polypeptides, as well as respiratory components such as the bc_1 complex, *c*-type cytochromes of cyclic electron flow, and terminal oxidases.

Additional circuits involved in oxygen regulation of *puc*, *puh* (encoding RC-H subunit), *bch* and *crt* operon expression, include the aerobic repressor PpsR, and FnrL, a positive, global anaerobic regulator required for enhanced expression of *puc* and selective structural genes encoding oxidoreductases and enzymes of pigment biosynthesis. PpsR forms part of the AppA-PspR antirepressor–repressor system in which AppA integrates both redox and light signals and serves as a blue-light photoreceptor, in addition to sensing the redox state of the quinone pool (Masuda and Bauer 2002; Braatsch et al. 2004).

Possible scenarios to explain how these and other interacting regulatory components come into play in controlling the development of the photosynthetic apparatus have been proposed by Oh and Kaplan (2001). With the recent discovery of additional expression regulators and the elucidation of their expression profiles (Roh et al. 2004), the interplay of the different regulatory mechanisms in the induction of ICM formation, together with the exact roles of the several complex-specific, assembly factors (Young and Beatty 2003) (see below), remain important topics for future investigation.

2.3.2
Role of Cytoplasmic Membrane Invagination Sites

The ability to isolate apparent sites of membrane invagination from *Rba. sphaeroides* has facilitated a number of definitive biochemical and biophysical studies of the ICM development process. These developing membrane domains are obtained in an upper-pigmented band which sediments more slowly than the ICM-derived chromatophore vesicle fraction during rate-zone sedimentation in sucrose density gradients (Fig. 4a). Pulse-chase (Niederman et al. 1979) and spectroscopic studies (Bowyer et al. 1985) have demonstrated that, in addition to containing respiratory CM, the upper pigmented fraction is enriched in nascent LH1-RC core complexes; these developing centers mature into photosynthetic units after further addition of LH2 (Reilly and

Niederman 1986), which drives vesicularization of the ICM (Hunter et al. 1988).

Recent near-IR fast repetition rate fluorescence measurements have shown that in comparison to chromatophores, the upper pigmented band showed a nearly 40% reduced photochemical yield and an approx. 10-fold slower rate of fluorescence relaxation (~ 100 and 8 ms, respectively), indicating that electron transport was only partially active in the developing centers (these results as well as a description of this new fluorescence technique, and its applications to phototrophic bacteria are detailed in Koblízek et al. 2005). The demonstration of partially active RCs in the upper fraction is in line with earlier redox-controlled, flash-induced absorption measurements (Bowyer et al. 1985), which showed that in addition to insertion of developing LH1-RC core structures into respiratory CM, the attainment of photosynthetic competence

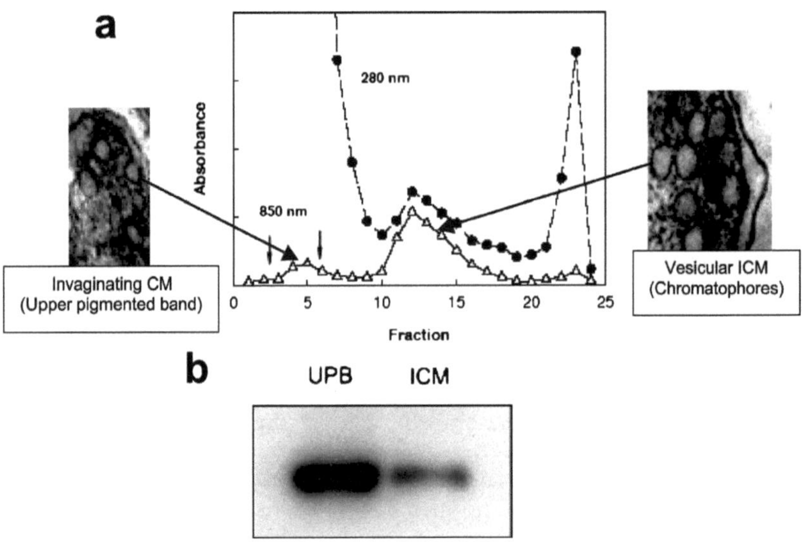

Fig. 4 Isolation and proposed origin of membrane invagination sites of *Rba. sphaeroides*. **a** Rate-zone sedimentation profile of French-press extract of high light-intensity grown *Rba. sphaeroides* on a 5–35% (w/w) sucrose gradient. *Vertical arrows* denote upper-pigmented band containing developing centers; mature ICM (chromatophore) band in fractions 11–15, while fractions 21–24 contain cell envelope material. BChl was localized from absorbance at 850 nm. Transmission electron micrographs of thin-cell sections show intracellular sites from which the upper-pigmented band and chromatophores are believed to be derived. Note that as predicted by the model shown in Fig. 2, the ICM vesicles assume a semispherical shape with a preferential long axis. **b** Role of membrane invagination sites in photosynthetic pigment biosynthesis. Western blot of upper pigmented band and ICM vesicles; gel loadings were normalized on the basis of BChl content (Hunter et al. 2005). The blot was probed with polyclonal antibodies raised against the BChl synthase protein, BchG. UPB, upper pigmented band fraction; ICM, intracytoplasmic membrane vesicle fraction

requires proper organization of fully assembled respiratory components that are active in light-driven cyclic electron flow.

Apart from serving as sites of preferential accumulation of newly-synthesized LH1 and RC complexes, as well as respiratory components in various stages of assembly, the putative membrane invagination sites are also the locale for preferential synthesis of BChl cofactors. This conclusion is based upon radiolabeling studies demonstrating that during photoacclimation to lowered light intensity, BChl cofactors are preferentially synthesized in the membranes isolated in the UPB (Hunter et al. 2005). Moreover, Western blotting with polyclonal antibodies raised against the BChl synthase (BChlG) protein, established that in comparison to chromatophores, the UPB fraction is highly enriched in this terminal enzyme of BChl biosynthesis (Fig. 4b).

2.3.3
Stepwise Assembly of Photosynthetic Units During Induction of Intracytoplasmic Membrane Formation Under Semiaerobic Conditions

Near-IR fast repetition rate fluorescence measurements have recently been applied to a functional analysis of photosynthetic unit assembly in *Rba. sphaeroides* (Koblízek et al. 2005). To gratuitously induce the formation of the ICM (greening), cells grown aerobically both in batch culture and in a carbon-

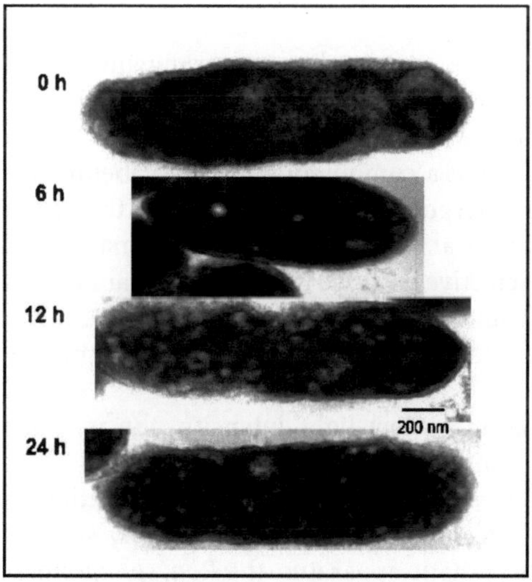

Fig. 5 Transmission electron micrographs of thin sections of *Rba. sphaeroides* cells undergoing ICM induction under 3% oxygen in a chemostat (Koblízek et al. 2005). At 0 h, most of the CM is closely apposed to the outer membrane, typical of cells bleached of photosynthetic pigments; at 6 h, membrane invaginations are seen and the cell has begun to fill up with vesicular ICM, which has reached an apparent steady state at 24 h

limited chemostat were transferred to semiaerobic conditions. Transmission electron micrographs of thin-cell sections (Fig. 5) showed that the kinetics of ICM induction in the chemostat culture could be correlated to morphological changes in the intracellular membranes of the inducing cells. At 0 h, most of the CM is seen at the cell periphery in close apposition to the outer membrane, typical of *Rba. sphaeroides* when grown chemoheterotrophically at high aeration. At 2.5 h (not shown), the CM had begun to invaginate to form vesicular ICM, while at 6 and 12 h, the cells began to fill up with ICM, reflecting an apparent steady state at 24 h.

In both aerobic cultures, low levels of LH1-RC complexes were observed, and in the batch cultures, the RCs were essentially inactive in forward electron transfer. Surprisingly, the chemostat culture displayed small amounts of fully assembled core complexes, containing functional RCs with rather rapid (1–2 ms) electron transfer turnover, as well as high photochemical yields. In both cases, the transfer to semiaerobiosis resulted in rapid induction of BChl biosynthesis that was reflected by both an increase in the number of LH1-RC and peripheral LH2 antenna complexes. These studies established that photosynthetic units are assembled in a sequential manner, where the appearance of the LH1-RC cores is followed by the activation of functional electron transfer, and finally by the accumulation of the LH2 complexes.

2.3.4 Perspectives on Photosynthetic Unit Assembly and Organization

The elucidation of the supramolecular arrangement of PSUs by AFM has prompted a re-evaluation of the mode of ICM assembly, and as a first approach, we have re-investigated the response of *Rba. sphaeroides* to lowered light intensity, through a combination of radiolabeling, membrane fractionation and gentle detergent treatment to isolate the various photosynthetic complexes (Hunter et al. 2005). This photoacclimation process is known to mainly involve a selective increase in the peripheral LH2 antenna (Sturgis and Niederman 1996), and accordingly, the sites of LH2 assembly within the CM-ICM continuum, and their specific localization within the developing PSU were examined.

During adaptation to low light intensity, the increase in LH2 antenna size was mainly confined to the ICM fraction, while the antenna composition of the precursor membrane in the UPB fraction remained relatively constant. In order to investigate the fate of newly synthesized BChls in the developing photosynthetic apparatus, the *Rba. sphaeroides* H5 mutant, which requires the porphyrin precursor δ-aminolevulinate, was radiolabelled with δ-^{3}H aminolevulinate at the onset of the light intensity shift-down and isolated membrane fractions were subjected to limited solubilization in lithium dodecyl sulfate at 4°C in the dark, followed by polyacrylamide gel electrophoresis (Broglie et al. 1980). This procedure results in the separation of the ICM into its constituent LH2, LH1 and RC complexes and is believed to be sufficiently gentle to preserve apparent regions of contact between LH2 and LH1, giving rise to a series of "intermediate" complexes, containing variable LH2 : LH1 stoichiometries.

Two interesting points have emerged from the labeling patterns of the LH complexes isolated from upper-pigmented and chromatophore fractions. First, the highest level of δ-^{3}H aminolevulinate incorporation into BChl after 10 h was seen in the apparent sites of LH1 and LH2 contact in the upper-pigmented band, with the limited LH2 "pool" of this membrane fraction receiving relatively little in the way of new BChl. The second point is that this pattern of new antenna synthesis was reversed in the ICM; the label (mainly into LH2 BChls) was equally incorporated into intermediate and LH2 "pool" complexes at 10 h, but subsequent incorporation and assembly was increasingly concentrated on the LH2 pool over the next 33 h.

These labeling studies, together with the AFM images of the ICM (Fig. 1), provide the basis for a structural model depicting the response of *Rba. sphaeroides* to low intensity illumination, and the increased LH2 antenna size that this provokes in developing photosynthetic units and in the distinct spatial domains of the CM-ICM continuum (Fig. 6). In this proposal: (*i*) new pigment–protein complexes are assembled at membrane invagination sites isolated in the upper-pigmented band (Fig. 6, upper right panel), where enzymes of BChl synthesis are also preferentially localized. LH2 synthesis is favored under these circumstances, and the new LH2 rings cluster between the dimeric rows of LH1-RC core complexes to foster new LH2-LH1 interactions. Since LH2 assembly at this point is insufficient to accumulate large discrete LH2 pools, some disconnected LH2 is still present. (*ii*) Further expansion of the membrane growth initiation sites occurs as a result of lipid biosynthesis and assembly of protein complexes; this results in further membrane invagination, such that upon cell disruption they acquire the characteristics of ICM vesicles (chromatophores) (Fig. 6, lower right panel). New LH2 rings are added as these membranes mature further, but because the sandwich regions between the arrays of core dimers are already fully occupied, the bulk LH2 pool becomes their favored location. The possibility that this latter LH2 domain forms the light-responsive complement of LH2 antenna within ICM is consistent with the above BChl labeling and photosynthetic complex isolation data.

While the structural and functional approaches to ICM development described here have aided in further elucidating events that are involved in the formation of the ordered assemblages of LH1-RC core complexes and their associated LH2 antenna, there remain many important unanswered question relating to ICM assembly in anoxygenic phototrophic bacteria. Foremost among these unknowns are the roles that general and complex-specific assembly factors play in the membrane insertion and assembly of the LH and RC and bc_1 complexes.

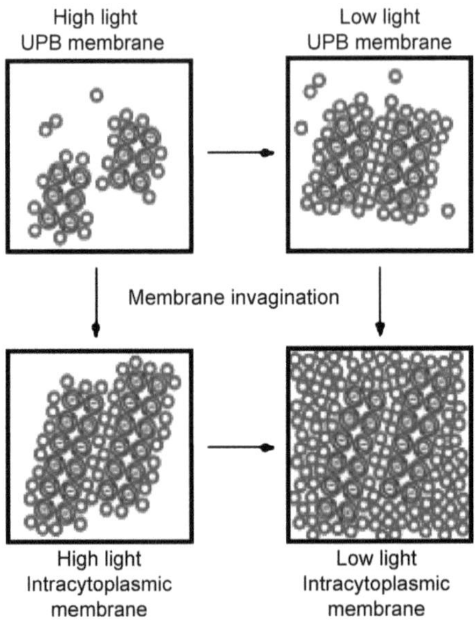

Fig. 6 Structural model for response of individual photosynthetic units within the CM-ICM continuum of *Rba. sphaeroides* to lowered light intensity. The *green circles* represent LH2 complexes and the *dimeric structures* are RC-LH1-PufX complexes. The *panels* present a quantitative model of the adaptation process, based on the known BChl content of LH2 and core complexes, and the physical and functional sizes of the photosynthetic units. UPB, upper pigmented band containing putative membrane invagination sites

The availability of a draft of the annotated *Rba. sphaeroides* genome (Mackenzie et al. 2001) has formed the basis for a recent DNA microarray analysis of oxygen and light-intensity regulation of transcriptome expression, in which transcripts of several complex-specific assembly factor genes, located in the photosynthesis gene cluster, were also identified (Roh et al. 2004). These included *lha*A, encoding the LH1-specific assembly factor LhaA, that along with the LH2 assembly factor PucC, is a member of the putative "chlorophyll delivery branch" of the major facilitator superfamily, and may have a role in BChl cofactor assembly (Young and Beatty 2003). In addition to *puf*X, described above (Sect. 2.1.2), transcripts for factors with apparent roles in photosynthetic complex assembly also included those of the genes encoding PufQ, a putative carrier of BChl and biosynthetic intermediates, PuhB, an apparent dimeric RC assembly factor with a secondary affect on LH1 assembly, and PuhE, a negative modulator of BChl synthesis.

In addition, a search of available *Rba. sphaeroides* genome sequences revealed genes encoding several general membrane assembly factors, among them were: DnaK and GroEL, cytoplasmic chaperonins whose *Rba. capsulatus* homologues were shown in a cell-free, transcription–translation system to play roles in LH1 apoprotein assembly (Drews 1996); SecA, a translocation ATPase, which appears in *Rba. capsulatus* to be involved in the movement of *c*-type apocytochromes to the periplasm for heme attachment (Helde et al. 1997). Genes were also located in *Rba. sphaeroides* for other major general membrane assembly factor homologues that are well described in *Escherichia coli.* Among these are the Sec YEG translocon, the YidC membrane chaperonin, the SecDF/YajC components of a membrane insertion pathway, Fth/FtsY, a signal recognition particle homologue, and Tat, which may be involved in the translocation of the major portion of the Rieske Fe – S protein across the membrane.

This should set the stage for future cell-free, transcription–translation system approaches, and definitive proteomic analyses, to further elucidate the complex assembly mechanism of the functionally important complexes involved in the trapping of light energy, and its conversion into useful chemical potential by intracytoplasmic photosynthetic membranes. In this connection, it is noteworthy that in a proteomic analysis of the *Rps. palustris* chromatophores (Fejes et al. 2003), peptides were detected which arose from the photosynthetic complex assembly factors PuhB and PuhC, and the chaperonin DnaK.

2.3.5 Relevance of Studies on ICM Formation in Anoxygenic Phototrophic Bacteria to Photosystem Assembly in Oxygenic Phototrophs

The demonstration that both photosystem I and II (PSI and PSII) subcomplexes are assembled in the plasma membrane of the cyanobacterium *Synechocystis* 6803, has suggested that this structure, rather than the thylakoid membrane, is the primary site for initial steps of photosystem biogenesis (Zak et al. 2001). The assembled RC cores of both PSI and PSII are subsequently translocated to the thylakoid membrane, where photosystem formation is completed by incorporation of additional regulatory and LH subunits directly into the RC core structures. This is strikingly similar to the stepwise PSU assembly mechanism demonstrated for *Rba. sphaeroides* (Koblízek et al. 2005), in which the appearance of LH1-RC core structures at CM invagination sites is followed by the activation of functional electron transfer, and finally by the accumulation of LH2 within the ICM (see Sect. 2.3.2). A sequential activation of PSII electron transfer has also been reported recently for *Synechocystis* 6803 (Keren et al. 2005), in which partially assembled, plasma-membrane localized PSII complexes exhibit a functional electron acceptor side; assembly of the donor side, together with the insertion of the LH proteins CP43 and CP47, subsequently occurs in the thylakoid membrane. It is noteworthy that purple non-sulfur bacteria such as *Rba. sphaeroides* contain a PSII related type-II RC.

Cyanobacteria are considered to be the progenitors of chloroplasts of green plants, and the chloroplast inner envelope has been proposed to play a role in photosystem biogenesis, similar to that of the cyanobacterial plasma membrane (Vothknecht and Westhoff 2001). In plant proplastids, a continuum is sometimes observed between the inner envelope and developing internal membranes, suggesting that the initial synthesis of thylakoid membranes might involve invagination of the inner envelope. The mechanism by which the partially assembled core complexes are transported from the plasma membrane to the thylakoid membrane is unknown, and may involve either vesicle transport or lateral movement through connecting membranes. Such lateral transfer of developing centers apparently comes into play in *Rba. sphaeroides* during invagination of the CM to form the ICM, and the ability to isolate and characterize the specific membrane growth initiation sites involved in this process, serves as a useful model for crucial aspects of the thylakoid membrane developmental pathway in higher phototrophs.

3
Intracytoplasmic Membranes of Other Proteobacteria

This section covers the ICM structures of the aerobic methanotrophs, an ecologically important group of proteobacteria, that impact upon the global carbon cycle by consuming atmospheric methane. Lastly, we discuss the respiratory components responsible for the oxidation of ammonia and nitrite, believed to be associated with the ICM of chemolithotrophic nitrifying bacteria.

3.1
The Aerobic Methanotrophs

The methanotrophs form a ubiquitous, obligately aerobic subgroup of the methylotrophic proteobacteria, and are the only group capable of utilizing methane as a sole source of carbon and energy under oxic conditions (Hanson and Hanson 1996). In contrast, non-methanotrophic methylotrophs depend upon other one-carbon compounds and methyl groups as sources for their metabolism. As consumers of the greenhouse gas methane, the methanotrophs function as important methane sinks and therefore play a major role in the global carbon cycle; however, recent genomic analyses by Hallam et al. (2004) have suggested that the flux of greenhouse gases between the ocean and the atmosphere is also impacted by methanogen-related, methane-oxidizing Archea in anoxic deep-sea sediments. In deep-sea environments, such as hydrothermal vents, methanotrophs together with chemolithotrophs, exist at the base of a photosynthesis-independent food chain in a symbiotic relationship with invertebrates. The biotechnological potential of the methan-

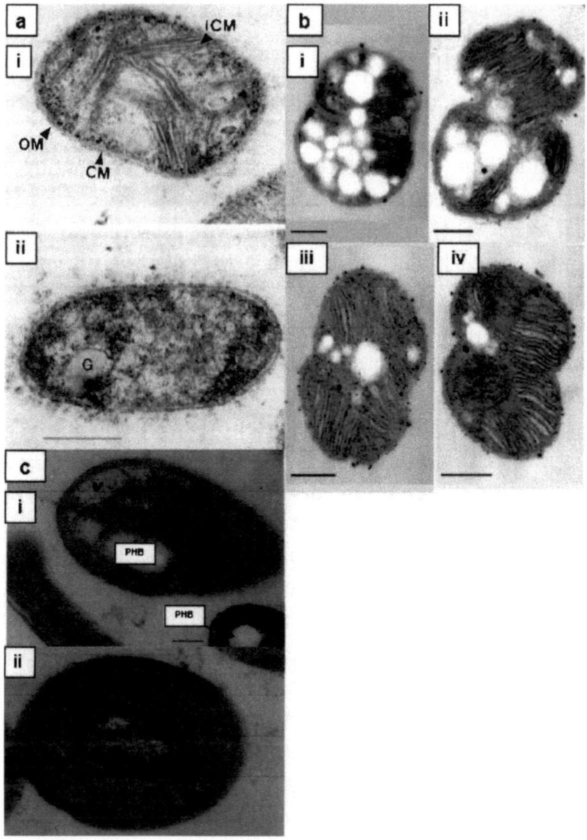

Fig. 7 Transmission electron micrographs of thin-cell sections showing the dependence of ICM formation in methanotrophs upon copper availability and the presence of particulate methane monooxygenase. **a** *Methylmicrobium albus* (type I methanotroph) with (i) and without (ii) copper supplementation in growth medium (Collins et al. 1991). Note that this type I methanotroph exhibits a copper-dependent formation of typical ICM bundles distributed within the cytoplasm. *Scale bar* represents 0.5 μM. OM, outer membrane; G, granule. **b** *Metylococcus capsulatus* (Bath) (type X methanotroph) cultured in chemostats with increasing copper concentrations (Choi et al. 2003). Cells were harvested at copper concentrations of 5 (i), 20 (ii), 40 (iii) and 60 (iv) μM. Maximal ICM development and pMMO concentration on a cellular basis occurred at 60 μM copper. This study demonstrated that in addition to controlling the metabolic switch between sMMO and pMMO, copper regulates the level of pMMO expression and the development of the ICM. Note that *M. capsulatus* also exhibits type-I ICM ultrastructure. *Scale bar* represents 0.2 μM. **c** *Methylosinus* (type II methanotroph) mutant lacking pMMO that constitutively expresses sMMO in the presence of copper (Phelps et al. 1992). (i) Mutant and (ii) revertant cultured in 5 μM copper. In the revertant, note the paired type-II ICM that forms stacks, running along most of the length of the cell periphery. The mutant lacks the peripheral stacked ICM, but contains some vesicular membrane (V). Polyhydroxybutyrate granules (PHB) are also seen. This copper-resistant mutant strain appeared to be altered in the ability to transport copper. Reprinted with permission of the authors

otrophs includes large-scale production as a source of single-cell protein and commercial methanol production, and because of their ability to co-oxidize a wide variety of aliphatic compounds, they have achieved importance in the bioremediation of recalcitrant xenobiotic pollutants. Methanotrophs are also unique among methylotrophs in forming an ICM.

Methanotrophs can be divided into two distinct groups, on the basis of phylogenetic relationships, together with differences in ICM fine structure, fatty acid composition, and formaldehyde assimilation pathways. Type I methanotrophs, which include the genera *Methylomicrobium*, *Methylobacter* and *Methylomonas*, are members of the γ-Proteobacteria and their phospholipids contain a predominance of 16-carbon fatty acids. This group utilizes the ribulose monophosphate cycle for formaldehyde assimilation and forms distinct ICM bundles of stacked vesicular disks, distributed throughout the cytoplasm (Fig. 7a(i)). Type II methanotrophs, consisting of the genera *Methylosinus* and *Methylocystis*, belong to the α-proteobacteria; 18-carbon fatty acids predominate in their membranes, they use the serine pathway, and possess paired ICM aligned at the periphery of the cell (Fig. 7c(ii)). Although the well-studied genus *Metylococcus* shares the characteristics of the type I methanotrophs (Fig. 7b), this organism is usually classified in a separate group (type X), since they grow at higher temperatures and contain low levels of the serine pathway enzyme ribulose-bisphosphate carboxylase (Hanson and Hanson 1996). The question of how the methanol arising from methane monooxygenase (MMO) is metabolized, needs to be reassessed in view of the work of Vorholt et al. (1999) in which a tetrahydromethanopterin-linked C1 transfer pathway of formaldehyde metabolism was found in all three types of methanotrophs (see below).

3.1.1 Enzymes of Methane Oxidation

Methane monooxygenase. The initial step of methane oxidation is catalyzed by the MMOs, enzymes that are unique to methanotrophs. As typical monooxgenases, MMOs utilize two reducing equivalents in the cleavage of the dioxygen bond. One of the oxygen atoms is incorporated into methane to form the methanol hydroxyl group, while the second oxygen undergoes reduction to form H_2O. Distinct soluble (sMMO) and particulate (pMMO) forms of the enzyme exist, which are distinguished by sedimentation of cell extracts at 150 000 xg for 75 min (Dalton 1991). The biosynthesis of pMMO is dependent upon copper availability, and this form of the enzyme is found in virtually all methanotrophs. The soluble form is usually present in type II and X strains; some type I species lack sMMO, and thus serve as useful tools for biochemical and biophysical studies of pMMO in whole cells (Yuan et al. 1999), and as an expression system for molecular genetic studies of the soluble enzyme (Lloyd et al. 1999). Although, NADH serves as the electron donor for sMMO, the physiological reductant for pMMO remains to be established. The possibility that duroquinol may be the source of electrons for pMMO in vivo has been further supported by Choi et al. (2003), and it has been suggested that pMMO is coupled to the electron transport chain at the level of the cytochrome bc_1 complex, rather than to NADH:quinone oxidoreductase.

The sMMO consists of three components: a heterotrimeric hydroxylase; protein B and protein C. Component B is a 15.8-kDa activity regulating protein, while component C is a 38.4-kDa, NADH-dependent reductase, containing FAD and a [2Fe – 2S] ferredoxin. The hydroxylase polypeptide subunits, designated as α, β and γ, with respective masses of ~ 60, 45 and 20 kDa, form the holoenzyme in an $(\alpha\beta\gamma)_2$ configuration (Lieberman and Rosenzweig 2004). The α polypeptide contains the carboxylate-bridged non-heme diiron center where methanol arises from methane and O_2. A high-resolution crystal structure was first reported for the hydroxylase of *Methylococcus capsulatus* by Rosenzweig et al. (1993), in which the dinuclear iron cluster was located in a hydrophobic, active-site cavity, beneath a floor formed by two "canyon regions" of the α- and β-hydroxylase subunits.

The structural genes for the sMMO protein components are found within a single operon consisting of *mmo*X, *mmo*Y (encoding respective α- and β-subunits of the hydroxylase), *mmo*B (encoding component B), *mmo*Z (encoding hydroxylase γ-subunit), *orf*Y and *mmo*C (encoding component C reductase) (Murrell et al. 2000). Merkx and Lippard (2002) have recently demonstrated that *orf*Y encodes a 12-kDa protein, designated as MmoD, that binds to the hydroxylase protein and may function in the assembly of the binuclear iron center. One chromosomal copy of the *mmo*XYBZDC operon is present in the complete genome sequence recently reported for *M. capsulatus* (Bath) (Ward et al. 2004). A transposase gene is interspersed between mmoB and Z, and a hypothetical protein separates these genes from the *mmo*GQSR regulatory locus.

Although pMMO catalyzes essentially the same basic reaction as sMMO, this ICM localized enzyme (Brantner et al. 2002) uses a different catalytic mechanism, involving copper atoms at the active site. The pMMO protomer consists of α, β and γ, subunits, respectively designated as PmoB (α, ~ 47 kDa), PmoA (β, ~ 24 kDa), and PmoC (γ, ~ 22 kDa), encoded by the *pmo*CAB structural gene cluster. Two complete *pmo*CAB copies are present on the *M. capsulatus* (Bath) chromosome, along with a third copy of *pmo*C, which is present with three other apparent genes that may form an operon with a function in methane oxidation (Ward et al. 2004).

Much of the controversy surrounding the pMMO metal centers has been resolved by the recent landmark crystal structure of the holoenzyme from *M. capsulatus* (Bath) determined at 2.8-Å resolution by Lieberman and Rosenzweig (2005) (Fig. 8). Each pMMO subunit contained three metal centers; one was assigned as mononuclear copper coordinated to the δ nitrogen atoms of two linearly aligned His residues and is the apparent source of

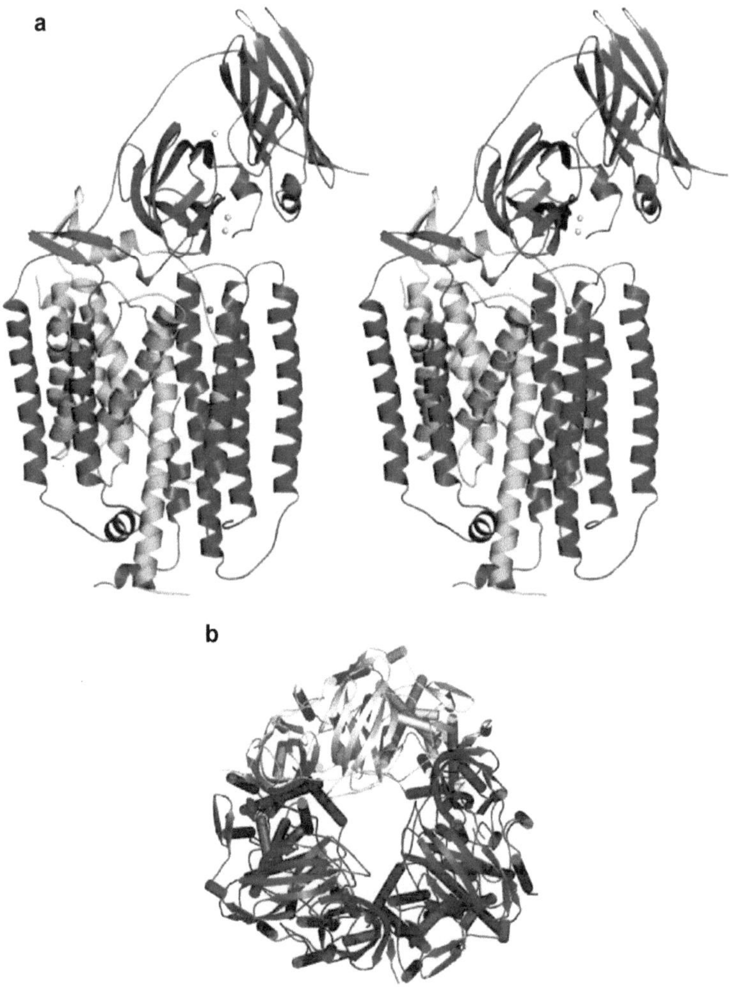

Fig. 8 Structure of the particulate methane monooxygenase of *Metylococcus capsulatus* (Bath) (Lieberman and Rosenzweig 2005). **a** Stereo view of a single pMMO protomer parallel to the membrane normal; PmoA (*yellow*), PmoB (*magenta*) and PmoC (*blue*), copper ions (*cyan spheres*), zinc ion (*grey sphere*). Periplasmic space is at *top*. Note that the protomer contains a total of 14 transmembrane α-helices with seven, two and five assigned to PmoA, PmoB and PmoC, respectively. **b** The three protomers (*dark purple*, *magenta*, *light pink*) forming the trimer are shown at periplasmic surface perpendicular to the membrane normal. Reprinted with permission of the authors

the type 2 (Cu(II)) mononuclear EPR signal. A second metal center was accounted for by dinuclear copper, consistent with the proposed Cu – Cu interacting cluster, and at least one of these copper atoms was assigned as Cu(I). His residues also provide the ligands to each copper atom in the dinuclear cluster. Both copper sites are located in the soluble periplasmic region of the large PmoB subunit (Fig. 8a). A third metal site was assigned to an apparent adventitiously bound zinc atom, located near the periplasmic surface of the membrane in the intramembrane region of the small PmoC subunit; this site may be occupied by iron, or an additional copper in vivo. Interestingly, the structure of the PmoB subunit is similar to that of cytochrome *c* oxidase subunit II, with the PmoB dinuclear copper site located in a position comparable to that of the cytochrome oxidase dinuclear Cu_A site, which has a pivotal role in electron transfer. These structural similarities suggest that the pMMO dinuclear copper cluster may shuttle electrons to the mononuclear copper site, which may then function as the catalytic center. However, it should be noted that the ligand environments of the dinuclear copper clusters in pMMO and subunit II of cytochrome *c* oxidase differ markedly.

The fully oligomerized pMMO forms a cylindrical $(\alpha\beta\gamma)_3$ trimer with an opening in the center lined by Glu, Asp and Arg residues (Fig. 8b). It is also noteworthy that putative docking sites for NADH:quinone oxidoreductase and quinol, the possible electron donor to pMMO, were also revealed in the structure. Using the detailed molecular framework provided by this structural milestone, it is now possible to elucidate the elusive and unique methane oxidation mechanism of the pMMO. In this regard, the possibility that the site occupied by zinc in the crystal structure may represent the catalytic site has also been considered (Lieberman and Rosenzweig 2005). The understanding of this biological methane oxidation mechanism is of considerable industrial importance, since no chemical catalysts have been developed that can efficiently perform the conversion of the methane, present in natural gas, to methanol, a potential alternative, and a less polluting energy source than petroleum or coal.

Methanol dehydrogenase. Methanol is oxidized to formaldehyde in methylotrophs by a periplasmically localized, quinone-containing methanol dehydrogenase (Hanson and Hanson 1996). The holoenzyme exists as an $(\alpha\beta)_2$ tetramer with subunit molecular weights of 66 and 8.5 kDa, respectively, and contains a pyrroloquinoline quinone prosthetic group and a calcium atom. Cytochrome_L, an atypical heme protein, serves as the immediate electron acceptor for methanol dehydrogenase and is subsequently oxidized by a second cytochrome (cytochrome_H). The methanol dehydrogenase and cytochrome components are unique to the methylotrophs. In the *M. capsulatus* (Bath) genome, the structural genes (*mxa*F and *mxa*I) for the large and small methanol dehydrogenase subunits are present within a large gene cluster that, in addition to the gene coding for cytochrome_L (*mxa*G), includes seven additional genes required for optimal catalytic function and assembly of the enzyme (Ward et al. 2004). In addition to a second cluster containing only *mxa*F and a second gene of unknown function, a third locus is found that contains genes necessary for synthesis of the enzyme.

Several high-resolution X-ray structures of methanol dehydrogenase have been determined in two different methylotrophs (reviewed by Anthony and

Williams 2003). The large α-subunit is arranged in a β-propeller fold composed of a superbarrel of eight four-stranded β-sheets, arranged radially to form the "propeller blades". The active site, containing the pyrroloquinoline quinone, is located at the center of this structure, coordinated to the adjacent calcium atom. The β-subunit, which is present only in this quinoprotein, wraps around one side of the α-subunit. A pathway for electrons from the reduced pyrroloquinoline quinone to the cytochrome$_L$ heme has been proposed in which the shortest electron transfer route involves a disulfide ring formed by two Cys close to the Q cofactor, and nearby Asp and Asn residues and water molecules (Anthony and Williams 2003). MxaD, a 17-kDa periplasmic protein encoded in the Mxa operon that increases the rate of electron transfer between methanol dehydrogenase and cytochrome$_L$, may also be involved in this pathway through the stimulation of interactions between the enzyme and the heme protein (Toyama et al. 2003).

Oxidation of formaldehyde and formate. The oxidation of formaldehyde via formate to CO_2 produces much of the reducing power that is essential for methane metabolism. The only protein found in the *M. capsulatus* (Bath) genome that could be annotated as a formaldehyde dehydrogenase (Ward et al. 2004), is the membrane-bound, pyrroloquinoline quinone enzyme produced in a high copper growth medium (Zahn et al. 2001). Genes ascribed to the ribulose monophosphate cycle of formaldehyde assimilation, and the tetrahydromethanopterin-linked pathway, were also identified. Genome analysis also suggested the presence of a complete tetrahydrofolate-linked pathway, thought to function in carbon assimilation from both formate and formaldehyde; the tetrahydromethanopterin-linked route is apparently involved in oxidation of formaldehyde to formate, while the ribulose monophosphate cycle is used for formaldehyde assimilation (Vorholt 2002). This permits tight control of both the metabolism and detoxification of formaldehyde. Three soluble formate dehydrogenase homologues were found in the genome, but the importance of this NAD-dependent enzyme in *M. capsulatus* remains to be determined. In this connection, Christoserdova et al. (2004) have recently shown that in mutants of the serine-cycle methylotroph, *Methylobacterium extorquens*, that lack all three formate dehydrogenase forms, growth on methanol was retained, possibly as a result of using the tetrahydrofolate-linked pathway. A multisubunit formate hydrogenlyase homologue was also found in the *M. capsulatus* (Bath) genome, and the role played by this enzyme, which produces CO_2 and dihydrogen from formate, should also be considered.

3.1.2
Intracytoplasmic Membranes are the Sites of Early Methane Oxidation Steps

The intracellular localization of pMMO in *Methylmicrobium album*, a type I methanotroph lacking sMMO, has recently been established by immunoelectron microscopy (Brantner et al. 2002). Collins et al. (1991) had previously shown that pMMO could account for up to one-half of the membrane protein in *Mm. album* when sufficient copper was present in the growth medium. By employing an indirect immunogold labeling procedure, together with transmission electron microscopy of thin-cell sections, Brantner et al. (2002) localized this integral membrane protein to the ICM, with $\sim 75\%$ of the gold particles residing on pMMO small subunit epitopes exposed at the surface of ICM stacks. In this same study, more than 80% of the methanol dehydrogenase was also observed on the ICM, suggesting that this enzyme is mainly located in the intra-ICM space, in a position facilitating the oxidation of methanol arising from pMMO activity, and for coupling to the subsequent cytochrome electron acceptors. This confirmed previous work in a *Methylomonas* sp. by Fassel et al. (1992), and a role for pMMO in the development and proliferation of the ICM was also proposed. A pMMO-containing membrane vesicle fraction arising from the *Mm. album* ICM has been isolated by rate-zone sedimentation in sucrose density gradients (Brantner et al. 2000), which should facilitate further studies of ICM structure, function and biogenesis in the methanotrophs.

3.1.3 Copper Uptake and Regulatory Mechanisms Initiating Intracytoplasmic Membrane Formation in Methanotrophs

In addition to the crucial roles that copper homeostasis mechanisms play in MMO regulation and ICM development (Fig. 7), they also control the expression of formaldehyde dehydrogenase (Zahn et al. 2001) and proteins that are essential to copper regulation and transport (Choi et al. 2003; Stafford et al. 2003). Central to copper homeostasis in this group of organisms is a specialized copper-trafficking mechanism mediated by methanobactin, a small fluorescent chromopeptide, analogous to the extracellular iron-binding siderophores (Kim et al. 2004). Methanobactin functions as a copper acquisition and sequestration compound, involved in the uptake of the high levels of copper that are required by the methanotrophs, while guarding against their potential toxicity. Choi et al. (2005) have recently demonstrated that methanobactin also interacts with pMMO, resulting in increased catalytic activity and electron flow to the enzyme.

While it has been known for many years that copper availability regulates the biosynthesis of pMMO, together with the formation of the ICM, a detailed understanding of the transcriptional mechanisms that control MMO levels is only now beginning to emerge (recently reviewed by Lieberman and Rosenzweig 2004).

In methanotrophs that form both pMMO and sMMO, a unique cooper ion-mediated metabolic switch regulates the reciprocal expression of their transcripts (Murrell et al. 2000). When the copper-to-biomass ratio is low, sMMO expression is maximal, while conversely, at higher copper-to-biomass ratios, pMMO is expressed and an extensive ICM system develops. The involvement of an unidentified copper-sensing mechanism as part of this copper switch, was suggested from the lowering in sMMO mRNA levels observed during cooper addition to the growth medium, and the correlation between the appearance pMMO transcripts and the presence of copper. The possibility might be considered that methanobactin participates in this copper-sensing role, but the finding of the genes for a variety of P-type cation ATPase, that include putative multiple copper ion pumps (Ward et al. 2004), suggests that other factors are also involved.

In *M. capsulatus* (Bath), the genes of the previously noted regulatory sMMO locus (*mmo*GQSR), located downstream (3′) of the *mmo* operon, have been shown to encode a GroEL-like chaperonin (MmoG), a two-component sensor-regulator system (MmoQ, MmoS), and a σ^{N}-dependent transcriptional activator (*mmo*R) (Csáki et al. 2003). Transcription is initiated upon interaction of MmoR with a σ^{N}-promoter (encoded by *rpo*N), located upstream (5′) of the *mmo*X structural gene. Under high-copper conditions, copper-sensing components are thought to transmit their signal to MmoS, a sensor protein which then undergoes autophosphorylation. The phosphoryl group is transferred to the MmoQ regulatory protein, which, in turn, interacts with the MmoR transcriptional activator to shut down transcription of the sMMO structural genes. However, this may not be the sole mechanism of copper regulation, since previous studies had suggested a role for targeted sMMO degradation as an additional copper response (Fitch et al. 1993).

In contrast to the detailed picture that has emerged for transcriptional regulation of sMMO, the factors that regulate expression of pMMO have yet to be identified. An analysis of the *pmo*CAB operons from *M. capsulatus* (Bath) (Stolyar et al. 1999, 2001) showed that both copies are transcribed from σ^{70}-type promoters located upstream of the *pmo*C structural genes. Although these two gene clusters appear to be functionally redundant, Stolyar et al. (2001) also demonstrated that copper availability alters their relative expression, with copy 2 transcripts predominating at an optimal copper concentration for growth (5 μM), while the levels of copy 1 and copy 2 transcripts were equalized at 10-fold higher copper concentrations. Whether the same copper sensor is involved in the regulation of both pMMO and sMMO remains to be determined, as well as how these signals are transmitted to pMMO operons, and ultimately lead to ICM formation (Fig. 7).

In this connection, when copper levels become sufficiently high and levels of the accumulating pMMO protein are in excess of those that can be inserted into the peripheral CM, newly synthesized pMMO may play a role in the invagination of the CM to form the ICM, as well as in the stacking of the proliferating ICM vesicular disks in type I methanotrophs, and in the growth and pairing of the ICM in the type II group. It can be envisioned that the large cylindrical pMMO trimer (Fig. 8b) may drive the membrane invagination process by acting as a "membrane-bending protein", analogous to the morphogenesis role proposed for the LH2 complex of *Rba. sphaeroides* (Sturgis and Niederman 1996, see Sect. 2.3.1), but how pMMO could participate in the subsequent ICM proliferation and surface association events is less clear. Studies of the ability of reconstituted pigment–protein complexes to promote the adhesion of proteoliposomes (Varga and Staehelin 1985), suggested that for the stacked intracytoplasmic photosynthetic membrane of *Rps. palustris*, the LH1 complex may act as the principal adhesion factor for these thylakoid-like structures; a role for a specific ICM adhesion protein in the methanotrophs could be examined in a similar manner.

The advent of a complete genome sequence for *Methylococcus capsulatus* (Bath) (Ward et al. 2004) now provides the basis for transcriptome expression profiling (Roh et al. 2004) and proteome analyses (Fejes et al. 2003) for an improved understanding of the multiple components that are likely to be involved in the mechanisms of assembly, growth and self-associations of the methanotroph ICM.

3.2 Respiratory Components are associated with Intracytoplasmic Membranes of Aerobic Chemolithotrophic Nitrifying Bacteria

Many of the enzymes and associated cofactors that are necessary for respiratory processes, are believed to be located in the ICM of aerobic chemolithotrophic nitrifying bacteria (Arp et al. 2002). *Nitrosomonas europaea*, one of the best described members of this group, is an ammonia (NH_3)-oxidizing obligate chemolithotroph of the β-subclass Proteobacteria, that derives all of its reducing power from the oxidation of NH_3 to nitrite (NO_2^-); it is also an obligate autotroph, obtaining necessary carbon from CO_2. The resulting NO_2^- is subsequently oxidized to nitrate (NO_3^-) by NO_2^- oxidizing *Nitrobacter* sp., an α-Proteobacterium that also forms an ICM on which particles of NO_2^- oxidoreductase are attached at the cytoplasmic surface. This sequential transformation of NH_3 into NO_3^- is known as nitrification, and forms part of the biogeochemical N cycle (Arp et al. 2002), with implications for the pollution of groundwater associated with croplands that are heavily fertilized with NH_3.

NH_3 is oxidized to NO_2^- in a two-step process: in the first step, the oxidation of NH_3 to hydroxylamine (NH_2OH) is catalyzed by ammonia monooxygenase (AMO), a particulate enzyme localized in the ICM of *N. europaea*; in the second step, NH_2OH is oxidized to NO_2^- by a multiple *c*-type heme containing hydroxylamine oxidase (HAO), located in the periplasm. AMO is similar to the pMMO of methanotrophs in primary structure, apparent subunit composition, and in inhibitor and substrate profiles, and may also use Cu as a cofactor (Arp et al. 2002). Electrons released during NH_2OH oxidation by HAO are transferred to the soluble cytochrome c_{554}, membrane-bound cytochrome c_{m552} and to the ubiquinone pool, which partitions reducing equivalents to the terminal oxidase and back to AMO (Whittaker et al. 2000). The flow of electrons from the ubiquinol to O_2 is thought to involve a cytochrome bc_1 complex, a soluble cytochrome c_{552}, and cytochrome aa_3 oxidase, with a branch directly to NADH dehydrogenase to generate NADH by reverse electron flow. The definitive intracellular localization of these membrane-bound components remains to be determined.

Under NH_3 oxidizing, aerobic conditions, when high levels of AMO are present, the ICM assumes a flattened vesicular structure and is located at the cell periphery. However, ICM morphology is markedly altered when *N. europaea* is grown anaerobically under denitrifying conditions, with hydrogen as the electron donor and NO_2^- as the electron acceptor (Schmidt et al. 2001). As a consequence, only low levels of AMO are formed and the ICM assumes a circular vesicular arrangement. During the recovery of the NH_3 oxidation capacity and the accompanying formation of AMO, these circular structures gradually reappear as flattened peripheral ICM, and it is possible that AMO may also participate in promoting these morphological alterations.

The genome of *N. europaea* has been sequenced and genes necessary for NH_3 catabolism, energy and reductant generation, and NH_3 and CO_2 assimilation, were identified (Chain et al. 2003). With the application of this information in transcriptome and proteome analyses, the basis for the obligate chemolithotrophic and autotrophic lifestyles will become better understood, along with the unique intracellular structures that accompany them.

Acknowledgements The author thanks Chris Chen and Riyesh Menon for conducting the searches of the *Rba. sphaeroides* genome and *Rps. palustris* proteomics data. Profs. Xiche Hu, Mary Lynne Perille Collins, Alan A. Despirito, George Georgiou and Amy C. Rosenzweig are gratefully acknowledged for permission to use figures from their publications.

References

Allen JP, Feher G, Yeates TO, Komiya H, Rees DC (1987) Structure of the reaction center from *Rhodobacter sphaeroides* R26: The protein subunits. Proc Natl Acad Sci USA 84:6162–6166

Anthony C, Williams P (2003) The structure and mechanism of methanol dehydrogenase. Biochim Biophys Acta 1647:18–23

Arp DJ, Sayavedra-Soto LA, Hommes NG (2002) Molecular biology and biochemistry of ammonia oxidation by *Nitrosomonas europaea*. Arch Microbiol 178:250–255

Bahatyrova S, Frese RN, Siebert CA, van der Werf KO, van Grondelle R, Niederman RA, Bullough PA, Otto C, Olsen JD, Hunter CN (2004a) The native architecture of a photosynthetic membrane. Nature 430:1058–1062

Bahatyrova S, Frese RN, van der Werf KO, Otto C, Hunter CN, Olsen JD (2004b) Flexibility and size heterogeneity of the LH1 light harvesting complex revealed by atomic force microscopy: functional significance for bacterial photosynthesis. J Biol Chem 279:21327–21333

Barz WP, Vermeglio A, Francia F, Venturoli G, Melandri BA, Oesterhelt D (1995) Role of the PufX protein in photosynthetic growth of *Rhodobacter sphaeroides*. 2. PufX is required for efficient ubiquinone/ubiquinol exchange between the reaction center Q_B site and the cytochrome bc_1 complex. Biochemistry 34:15248–15258

Berry EA, Huang LS, Saechao LK, Pon NG, Valkova-Valchanova M, Daldal F (2004) X-Ray structure of *Rhodobacter capsulatus* cytochrome bc_1: Comparison with its mitochondrial and chloroplast counterparts. Photosynth Res 81:251–275

Bowyer JR, Hunter CN, Ohnishi T, Niederman RA (1985) Photosynthetic membrane development in *Rhodopseudomonas sphaeroides*. Spectral and kinetic characterization of redox components of light-driven electron flow in apparent photosynthetic membrane growth initiation sites. J Biol Chem 260:3295–3304

Braatsch S, Gomelsky M, Kuphal S, Klug G (2002) A single flavoprotein, AppA, integrates both redox and light signals in *Rhodobacter sphaeroides*. Mol Microbiol 45:827–836

Brantner CA, Buchholz LA, Remsen CC, Collins MLP (2000) Isolation of intracytoplasmic membrane from the methanotrophic bacterium *Methylomicrobium album* BG8. Curr Microbiol 40:132–134

Brantner CA, Remsen CC, Owen HA, Buchholz LA, Perille Collins ML (2002) Intracellular localization of the particulate methane monooxygenase and methanol dehydrogenase in *Methylomicrobium album* BG8. Arch Microbiol 178:59–64

Broglie RM, Hunter CN, Delepelaire P, Niederman RA, Chua N-H, Clayton RK (1980) Isolation and characterization of the pigment–protein complexes of *Rhodopseudomonas sphaeroides* by lithium dodecyl sulfate/polyacrylamide gel electrophoresis. Proc Natl Acad Sci USA 77:87–91

Chain P, Lamerdin J, Larimer F, Regala W, Lao V, Land M, Hauser L, Hooper A, Klotz M, Norton J, Sayavedra-Soto L, Arciero D, Hommes N, Whittaker M, Arp D (2003) Complete genome sequence of the ammonia-oxidizing bacterium and obligate chemolithoautotroph *Nitrosomonas europaea*. J Bacteriol 185:2759–2773

Chistoserdova L, Laukel M, Portais JC, Vorholt JA, Lidstrom ME (2004) Multiple formate dehydrogenase enzymes in the facultative methylotroph *Methylobacterium extorquens* AM1 are dispensable for growth on methanol. J Bacteriol 186:22–28

Choi DW, Antholine WE, Do YS, Semrau JD, Kisting CJ, Kunz RC, Campbell D, Rao V, Hartsel SC, DiSpirito AA (2005) Effect of methanobactin on the activity and electron paramagnetic resonance spectra of the membrane-associated methane monooxygenase in *Methylococcus capsulatus* Bath. Microbiology 151:3417–3426

Choi DW, Kunz RC, Boyd ES, Semrau JD, Antholine WE, Han JI, Zahn JA, Boyd JM, de la Mora AM, DiSpirito AA (2003) The membrane-associated methane monooxygenase (pMMO) and pMMO-NADH:quinone oxidoreductase complex from *Methylococcus capsulatus*. Bath J Bacteriol 185:5755–5764

Cogdell RJ, Gardiner AT, Roszak AW, Law CJ, Southall J, Isaacs NW (2004) Rings, ellipses and horseshoes: how purple bacteria harvest solar energy. Photosynth Res 81:207–214

Collins MLP, Buchholz LA, Remsen CC (1991) Effect of copper on *Methylomonas albus* BG8. Appl Environ Microbiol 57:1261–1264

Csáki R, Bodrossy L, Klem J, Murrell JC, Kovacs KL (2003) Genes involved in the copper-dependent regulation of soluble methane monooxygenase of *Methylococcus capsulatus* (Bath): cloning, sequencing and mutational analysis. Microbiology 149:1785–1795

Dalton H (1991) Structure and mechanism of action of the enzymes involved in methane oxidation. In: Kelley JW (ed) Applications of Enzyme Biotechnology. Plenum Press, New York, pp 55–68

Deisenhofer J, Michel H (1989) The photosynthetic reaction center from the purple bacterium *Rhodopseudomonas vridis*. Science 245:1463–1473

Drews G (1996) Formation of the light-harvesting complex I (B870) of anoxygenic phototrophic purple bacteria. Arch Microbiol 166:151–159

Drews G, Golecki JR (1995) Structure, molecular organization, and biosynthesis of membranes of purple bacteria. In: Blankenship RE, Madigan MT, Bauer CE (eds) Anoxygenic Photosynthetic Bacteria. Kluwer Academic Publishers, Dordrecht, The Netherlands, pp 231–257

Drews G, Niederman RA (2002) Membrane biogenesis in anoxygenic photosynthetic prokaryotes. Historical minireview for History of photosynthesis: A celebration of the millennium. Photosynth Res 73:87–94

Eraso JM, Kaplan S (1994) prrA, A putative response regulator involved in oxygen regulation of photosynthesis gene expression in *Rhodobacter sphaeroides*. J Bacteriol 176:32–43

Fassel TA, Buchholz LA, Collins MLP, Remsen CC (1992) Localization of methanol dehydrogenase in two strains of methylotrophic bacteria detected by immunogold labeling. Appl Environ Microbiol 58:2302–2307

Fejes AP, Yi EC, Goodlett DR, Beatty JT (2003) Shotgun proteomic analysis of a chromatophore-enriched preparation from the purple phototrophic bacterium *Rhodopseudomonas palustris*. Photosynth Res 78:195–203

Fitch MW, Graham DW, Arnold RG, Agarwal SK, Phelps R, Speitel GE, Georgiou G (1993) Phenotypic characterization of cooper-resistant mutants of *Methylosinus trichosporium* OB3b. Appl Envirort Microbiol 159:2771–2776

Fotiadis D, Qian P, Philippsen A, Bullough PA, Engel A, Hunter CN (2004) Structural analysis of the reaction center light-harvesting complex I photosynthetic core complex of *Rhodospirillum rubrum* using atomic force microscopy. J Biol Chem 279:2063–2068

Francia F, Dezi M, Rebecchi A, Mallardi A, Palazzo G, Melandri BA, Venturoli G (2004) Light-harvesting complex 1 stabilizes P^+QB^- charge separation in reaction centers of *Rhodobacter sphaeroides*. Biochemistry 43:14199–14210

Frese RN, Siebert CA, Niederman RA, Hunter CN, Otto C, van Grondelle R (2004) The long-range organization of a native photosynthetic membrane. Proc Natl Acad Sci USA 101:17994–17999

Frese RN, Olsen JD, Branvall R, Westerhuis WH, Hunter CN, van Grondelle R (2000) The long-range supraorganization of the bacterial photosynthetic unit: A key role for PufX. Proc Natl Acad Sci USA 97:5197–5202

Hallam SJ, Putnam N, Preston CM, Detter JC, Rokhsar D, Richardson PM, DeLong EF (2004) Reverse methanogenesis: testing the hypothesis with environmental genomics. Science 305:1457–1462

Hanson RS, Hanson TE (1996) Methanotrophic bacteria. Microbiol Rev 60:439–471

Helde R, Wiesler B, Wachter E, Neubuser A, Hoffschulte HK, Hengelage T, Schimz KL, Stuart RA, Muller M (1997) Comparative characterization of SecA from the alpha-subclass purple bacterium *Rhodobacter capsulatus* and *Escherichia coli* reveals differences in membrane and precursor specificity. J Bacteriol 179:4003–4012

Hu X, Damjanovic A, Ritz T, Schulten K (1998) Architecture and mechanism of the light-harvestin g apparatus of purple bacteria. Proc Natl Acad Sci USA 95:5935–5941

Hunter CN, Tucker JD, Niederman RA (2005) The assembly and organisation of photosynthetic membranes in *Rhodobacter sphaeroides*. Photochem Photobiol Sci 4:1023–1027

Hunter CN, Pennoyer JD, Sturgis JN, Farrelly D, Niederman RA (1988) Oligomerization states and associations of light-harvesting pigment-protein complexes of *Rhodobacter sphaeroides* as analyzed by lithium dodecyl sulfate-polyacrylamide gel electrophoresis. Biochemistry 27:3459–3467

Karrasch S, Bullough PA, Ghosh R (1995) The 8.5 Å projection map of the light-harvesting complex I from *Rhodospirillum rubrum* reveals a ring composed of 16 subunits. EMBO J 14:631–638

Keren N, Liberton M, Pakrasi HB (2005) Photochemical competence of assembled photosystem II core complex in cyanobacterial plasma membrane. J Biol Chem 280:6548–6553

Kiley PJ, Varga A, Kaplan S (1988) Physiological and structural analysis of light-harvesting mutants of *Rhodobacter sphaeroides.* J Bacteriol 170:1103–1115

Kim HJ, Graham DW, DiSpirito AA, Alterman MA, Galeva N, Larive CK, Asunskis D, Sherwood PM (2004) Methanobactin, a copper-acquisition compound from methane-oxidizing bacteria. Science 305:1612–1615

Koblízek M, Shih JD, Breitbart SI, Ratcliffe EC, Kolber ZS, Hunter CN, Niederman RA (2005) Sequential assembly of photosynthetic units in *Rhodobacter sphaeroides* as revealed by fast repetition rate analysis of variable bacteriochlorophyll a fluorescence. Biochim Biophys Acta 1706:220–231

Koepke J, Hu X, Muenke C, Schulten K, Michel H (1996) The crystal structure of the light-harvesting complex II (B800–850) from *Rhodospirillum molischianum.* Structure 4:581–597

Kramer HJM, Pennoyer JD, van Grondelle R, Westerhuis WHJ, Niederman RA, Amesz J (1984) Low-temperature optical properties and pigment organization of the B875 light-harvesting bacteriochlorophyll-protein complex of purple photosynthetic bacteria. Biochim Biophys Acta 767:335–344

Lieberman RL, Rosenzweig AC (2005) Crystal structure of a membrane-bound metalloenzyme that catalyses the biological oxidation of methane. Nature 434:177–182

Lieberman RL, Rosenzweig AC (2004) Biological methane oxidation: regulation, biochemistry, and active site structure of particulate methane monooxygenase. Crit Rev Biochem Mol Biol 39:147–164

Lloyd JS, De Marco P, Dalton H, Murrell JC (1999) Heterologous expression of soluble methane monooxygenase genes in methanotrophs containing only particulate methane monooxygenase. Arch Microbiol 171:364–370

Mackenzie C, Choudhary M, Larimer FW, Predki PF, Stilwagen S, Armitage JP, Barber RD, Donohue TJ, Hosler JP, Newman JE, Shapleigh JP, Sockett RE, Zeilstra-Ryalls J, Kaplan S (2001) The home stretch, a first analysis of the nearly completed genome of *Rhodobacter sphaeroides* 2.4.1. Photosynth Res 70:19–41

Masuda S, Bauer CE (2002) AppA is a blue light photoreceptor that antirepresses photosynthesis gene expression in *Rhodobacter sphaeroides.* Cell 110:613–623

McDermott G, Prince SM, Freer AA, Hawthornthwaite-Lawless AM, Papiz MZ, Cogdell RJ, Isaacs NW (1995) Crystal structure of an integral membrane light-harvesting complex from photosynthetic bacteria. Nature 374:517–521

Merkx M, Lippard SJ (2002) Why OrfY Characterization of *mmoD,* a long overlooked component of the soluble methane monooxygenase from *Melhylococcus capsulatus* (Bath). J Biol Chem 277:5858–5865

Monger TG, Parson WW (1977) Singlet–triplet fusion in *Rhodopseudomonas sphaeroides* chromatophores. A probe of the organization of the photosynthetic apparatus. Biochim Biophys Acta 460:393–407

Murrell JC, Gilbert B, McDonald IR (2000) Molecular biology and regulation of methane monooxygenase. Arch Microbiol 173:325–332

Niederman RA, Mallon DE, Parks LC (1979) Membranes of *Rhodopseudomonas sphaeroides*. VI. Isolation of a fraction enriched in newly synthesized bacteriochlorophyll *a*-protein complexes. Biochim Biophys Acta 555:210–220

Niederman RA, Gibson KD (1978) Isolation and physicochemical properties of membranes from purple photosynthetic bacteria. In: Clayton RK, Sistrom WR (eds) The Photosynthetic Bacteria. Plenum Publishing Corp, New York, NY, pp 79–118

Oh J-I, Kaplan S (2001) Generalized approach to the regulation and integration of gene expression. Mol Microbiol 39:1116–1123

Phelps PA, Agarwal SK, Speitel GE, Georgiou G (1992) *Methylosinus trichosporium* OB3b mutants having constitutive expression of soluble methane monooxygenase in the presence of high levels of copper. Appl Environ Microbiol 58:3701–3708

Qian P, Hunter CN, Bullough PA (2005) The 8.5 Å projection structure of the core RC-LH1-PufX dimer of *Rhodobacter sphaeroides*. J Mol Biol 349:948–960

Reilly PA, Niederman RA (1986) Role of apparent membrane growth initiation sites during photosynthetic membrane development in synchronously dividing *Rhodopseudomonas sphaeroides*. J Bacteriol 167:153–159

Roh JH, Smith WE, Kaplan S (2004) Effects of oxygen and light intensity on transcriptome expression in *Rhodobacter sphaeroides* 2.4.1. Redox active gene expression profile. J Biol Chem 279:9146–9155

Rosenzweig AC, Frederick CA, Lippard SJ, Nordlund P (1993) Crystal structure of a bacterial non-haem iron hydroxylase that catalyses the biological oxidation of methane. Nature 366:537–543

Roszak AW, Howard TD, Southall J, Gardiner AT, Law CJ, Isaacs NW, Cogdell RJ (2003) Crystal structure of the RC-LH1 core complex from *Rhodopseudomonas palustris*. Science 302:1969–1972

Scheuring S, Levy D, Rigaud JL (2005) Watching the components of photosynthetic bacterial membranes and their in situ organisation by atomic force microscopy. Biochim Biophys Acta 1712:109–127

Scheuring S, Francia F, Busselez J, Melandri BA, Rigaud JL, Levy D (2004) Structural role of PufX in the dimerization of the photosynthetic core complex of *Rhodobacter sphaeroides*. J Biol Chem 279:3620–3626

Schmidt I, Zart D, Bock E (2001) Effects of gaseous NO_2 on cells of *Nitrosomonas eutropha* previously incapable of using ammonia as an energy source. Antonie van Leeuwenhoek 79:39–47

Stafford GP, Scanlan J, McDonald IR, Murrell JC (2003) rpoN, mmoR and mmoG, genes involved in regulating the expression of soluble methane monooxygenase in *Methylosinus trichosporium* OB3b. Microbiology 149:1771–1784

Stolyar S, Costello AM, Peeples TL, Lidstrom ME (1999) Role of multiple gene copies in particulate methane monooxygenase activity in the methane-oxidizing bacterium *Methylococcus capsulatus* Bath. Microbiology 145:1235–1244

Stolyar S, Franke M, Lidstrom ME (2001) Expression of individual copies *of Methylococcus capsulatus* Bath particulate methane monooxygenase genes. J Bacteriol 183:1810–1812

Sturgis JN, Niederman RA (1996) The effect of different levels of the B800–850 light-harvesting complex on intracytoplasmic membrane development in *Rhodobacter sphaeroides*. Arch Microbiol 165:235–242

Toyama H, Inagaki H, Matsushita K, Anthony C, Adachi O (2003) The role of the MxaD protein in the respiratory chain of *Methylobacterium extorquens* during growth on methanol. Biochim Biophys Acta 1647:372–375

Tso SC, Shenoy SK, Quinn BN, Yu L (2000) Subunit IV of cytochrome bc_1 complex from *Rhodobacter sphaeroides*. Localization of regions essential for interaction with the three-subunit core complex. J Biol Chem 275:15287–15294

Varga AR, Staehelin LA (1985) Membrane adhesion in photosynthetic bacterial membranes. Light harvesting complex I (LHI) appears to be the main adhesion factor. Arch Microbiol 141:290–296

Vorholt JA (2002) Cofactor-dependent pathways of formaldehyde oxidation in methylotrophic bacteria. Arch Microbiol 178:239–249

Vorholt JA, Chistoserdova L, Stolyar SM, Thauer RK, Lidstrom ME (1999) Distribution of tetrahydromethanopterin-dependent enzymes in methylotrophic bacteria and phylogeny of methenyl tetrahydromethanopterin cyclohydrolases. J Bacteriol 181:5750–5757

Vos M, van Grondelle R, van der Kooij FW, van de Poll D, Amesz J, Duysens LMN (1986) Singlet–singlet annihilation at low temperatures in the antenna of purple bacteria. Biochim Biophys Acta 850:501–512

Vothknecht UC, Westhoff P (2001) Biogenesis and origin of thylakoid membranes. Biochim Biophys Acta 1541:91–101

Walz T, Jamieson SJ, Bowers CM, Bullough PA, Hunter CN (1998) Projection structures of three photosynthetic complexes from *Rhodobacter sphaeroides*: LH2 at 6 Å, LH1 and RC–LH1 at 25 Å. J Mol Biol 282:833–845

Ward N, Larsen Ø, Sakwa J, Bruseth L, Khouri H, Durkin AS, Dimitrov G, Jiang L, Scanlan D, Kang KH, Lewis M, Nelson KE, Methe B, Wu M, Heidelberg JF, Paulsen IT, Fouts D, Ravel J, Tettelin H, Ren Q, Read T, DeBoy RT, Seshadri R, Salzberg SL, Jensen HB, Birkeland NK, Nelson WC, Dodson RJ, Grindhaug SH, Holt I, Eidhammer I, Jonasen I, Vanaken S, Utterback T, Feldblyum TV, Fraser CM, Lillehaug JR, Eisen JA (2004) Genomic insights into methanotrophy: The complete genome sequence of Methylococcus capsulatus (Bath). PLoS Biol 2:e303

Westerhuis WHJ, Sturgis JN, Ratcliffe EC, Hunter CN, Niederman RA (2002) Isolation, size estimates, and spectral heterogeneity of an oligomeric series of light-harvesting 1 complexes from *Rhodobacter sphaeroides*. Biochemistry 41:8698–8707

Westerhuis WHJ, Hunter CN, van Grondelle R, Niederman RA (1999) Modeling of oligomeric-state dependent spectral heterogeneity in the B875 light-harvesting complex of *Rhodobacter sphaeroides* by numerical simulation. J Phys Chem B 103:7733–7742

Whittaker M, Bergmann D, Arciero D, Hooper AB (2000) Electron transfer during the oxidation of ammonia by the chemolithotrophic bacterium *Nitrosomonas europaea*. Biochim Biophys Acta 1459:346–355

Young CS, Beatty JT (2003) Multi-level regulation of purple bacterial light-harvesting complexes. In: Green BR, Parson WW (eds) Light-harvesting Antennas in Photosynthesis. Kluwer Academic Publishers, Dordrecht, The Netherlands, pp 449–470

Yuan H, Collins MLP, Antholine WE (1999) Type 2 Cu^{2+} in pMMO from *Methylomicrobium album* BG8. Biophys J 76:2223–2229

Zahn JA, Bergmann DJ, Boyd JM, Kunz RC, DiSpirito AA (2001) Membrane-associated quinoprotein formaldehyde dehydrogenase from *Methylococcus capsulatus* Bath. J Bacteriol 183:6832–6840

Zak E, Norling B, Maitra R, Huang F, Andersson B, Pakrasi HB (2001) The initial steps of biogenesis of cyanobacterial photosystems occur in plasma membranes. Proc Natl Acad Sci USA 98:13443–13448

Microbiol Monogr (2)
J.M. Shively: Complex Intracellular Structures in Prokaryotes
DOI 10.1007/7171_026/Published online: 20 May 2006

Membrane-bounded Nucleoids and Pirellulosomes of Planctomycetes

John A. Fuerst

School of Molecular and Microbial Sciences, University of Queensland, Queensland 4072, Brisbane, Australia
j.fuerst@uq.edu.au

Abstract Planctomycetes are peptidoglycan-less organisms that form a distinct phylum *Planctomycetes* of domain Bacteria. The cells of all planctomycetes examined share a cell plan involving intracellular membranes, in which the cytoplasm is divided into compartments. Planctomycetes in genera *Pirellula*, *Blastopirellula* and *Rhodopirellula* possess a pirellulosome compartment bounded by a single membrane and containing ribosome-like particles, but all planctomycetes examined possess a topologically equivalent compartment. In all planctomycetes the nucleoid is enclosed by the pirellulosome membrane. The space between the pirellulosome's membrane and the cytoplasmic membrane forms another planctomycete-characteristic compartment, the paryphoplasm. In *Gemmata obscuriglobus*, cells possess a membrane-bounded nucleoid, where the nucleoid is surrounded by an envelope comprising two apposed membranes, forming a 'nuclear body' region within the pirellulosome, analogous to the structure of a eukaryotic nucleus. Planctomycete compartmentalization may have functional roles in anaerobic ammonium-oxidizing anammox planctomycetes, where the anammoxosome harbours specialized enzymes and its envelope possesses unique ladderane lipids (discussed in detail elsewhere in this volume, Fuerst 2006, in this volume). Compartmentalized cells of members of the distinctive phylum *Planctomycetes*, especially those of *Gemmata* strains, form an exception to the prokaryote cell plan and have significant implications for models for origins of the eukaryote nucleus.

1 Introduction

Nucleoids of microorganisms in the Domains Bacteria and Archaea are the defining characteristic of the prokaryotic form of cell organization, with no membranes surrounding their DNA, in contrast to the chromosomal DNA of interphase nuclei of eukaryotes. The DNA of the cryofixed nucleoid of the typical models from the Bacteria, the Gram-negative *E. coli* and Gram-positive *Bacillus subtilis*, extends in coralline fashion throughout the cell cytoplasm (Bohrmann et al. 1991; Kellenberger 1991a). This was an insight only made available by the advent of cryosubstitution preparation methods for examining thin sectioned prokaryote cells via transmission electron microscopy, since earlier chemical fixation methods were subject to artefacts connected with condensation and coagulation of the nucleoid DNA which may have destroyed any connection with internal membranes. Such chemical fixation methods also gave rise in some cases to artefactual internal membrane structures known as mesosomes, demonstrated to be an artefact induced by chemical fixation after the application of cryosubstitution methods (Ebersold et al. 1981; Dubochet et al. 1983).

Cryosubstitution has enabled the relationship of the nucleoid to cell function to be assayed to some degree, so that immunogold labeling has revealed the occurrence of single-stranded DNA at the periphery of the nucleoid, where it might be expected to be prepared for and active in transcription (Hobot et al. 1987). Consistent with the need for a dynamic structure for transcription and DNA replication, distribution of RNA polymerase and DNA topoisomerase is also peripheral to cryosubstituted nucleoids, though histone-like protein HU (presumed to bind to folded nucleoid DNA) unexpectedly also follows this pattern (Durrenberger et al. 1988). Evidence from autoradiography used to localize sites of RNA synthesis in *Bacillus* and *Escherichia* supports a concept of actively transcribing DNA spread throughout the cell as far as the cytoplasmic membrane (Ryter and Chang 1975). The application of green fluorescent protein-based methods may be needed however to yield a more detailed picture of such processes or to confirm that they are indeed accurate at all (Lewis 2004). The distribution of the nucleoid and its functional extensions and molecular associations in cryosubstituted cells is consistent with an "open plan" for the classical bacterial models for prokaryote structure. In this plan, even extension of nucleoid DNA to the cytoplasmic membrane can occur in such a way that DNA is anchored to the cytoplasmic membrane at many points, which is necessary if the assumed coupled transcription-translation typical of prokaryotes is to be combined with co-translational protein secretion, a phenomenon which has been termed co-transcriptional anchoring (Lynch and Wang 1993).

One might assume then that the prokaryotes, both Bacteria and Archaea, might be universally contrasted with members of the domain Eucarya, which can be defined on the basis of their cell organization as in their simplest form nucleated, with a nucleus characteristically surrounded by a double-membrane envelope. However, our knowledge of prokaryote organization in Bacteria and Archaea is based on relatively limited sampling especially that involving modern methods for preparation for electron microscopy such as cryosubstitution. At least one major exception to the concept of prokaryotes with a naked nucleoid not wrapped in a membrane of any type has come to light with the discovery of membrane compartmentalization in the organisms of the phylum *Planctomycetes*.

2 Planctomycetes

The planctomycetes, organisms within the order *Planctomycetales*, are members of the distinct phylum *Planctomycetes* of Domain Bacteria (Garrity et al. 2004), a phylum which represents a deep-branching group within the Bacteria on the basis of 16S rRNA sequence phylogenetics (Schlesner and Stackebrandt 1986; Van De Peer et al. 1994; Fuerst 1995, 2005). A recent important phylogenetic study applying an alignment of only slowly evolving positions to tree generation suggests that this division may be the deepest branching among the Bacteria, rather than hyperthermophiles like the Aquificales (Brochier and Philippe 2002) though there is controversy about this conclusion (Di Giulio 2003).

Planctomycetes are distinctive relative to other bacteria with respect to their peptidoglycan-less cell walls, budding reproduction and their compartmentalized cell structure. These bacteria have been identified in diverse freshwater, marine and soil habitats and even invertebrate animals (Staley et al. 1992; Schlesner 1994; Fuerst 1995; Fuerst et al. 1997; Neef et al. 1998; Wang et al. 2002). They have been isolated as chemoheterotrophs from an equally diverse range of habitats, using for example selective media based on their inherent resistance to antibiotics targeting peptidoglycan synthesis and their predilection for *N*-acetylglucosamine as a carbon substrate (Schlesner 1994), but only representatives of six genera exist in pure culture, *Pirellula, Blastopirellula, Rhodopirellula, Planctomyces, Gemmata* and *Isosphaera. Pirellula*-like species (*Pirellula, Blastopirellula*, and *Rhodopirellula*) tend to have ovoid or elliptical cell shapes, while cells of *Gemmata* and *Isosphaera* are spherical cocci, the latter often occurring in chains in the case of *Isosphaera.* The cell walls of the planctomycetes that have been examined, including *Pirellula staleyi, Blastopirellula marina, Planctomyces maris, Gemmata obscuriglobus* and *Isosphaera pallida*, possess cell walls with a substantially protein composition and no detectable peptidoglycan assayed via muramic acid and diaminopimelic acid, consistent with the resistance of planctomycetes in pure culture to antibiotics based on peptidoglycan synthesis inhibition (Konig et al. 1984; Liesack et al. 1986; Stackebrandt et al. 1986). They thus resemble some Archaea in their possession of protein cell walls, but were placed more definitely within the Bacteria by application of electron microscopy, 16S rRNA phylogenetics and the determination of a bacteria-like reaction to diphtheria toxin (Stackebrandt et al. 1984; Starr and Schmidt 1989). Cell surfaces of all planctomycetes examined characteristically possess pit-like structures termed crateriform structures when examined by negative staining and transmission electron microscopy, apparently reflecting structures of the cell wall or those structures firmly attached to the wall (and preserved after 10% sodium dodecyl sulfate treatment) (Liesack et al. 1986; Stackebrandt et al. 1986), and their presence can be useful in preliminary identification of isolates (Wang et al. 2002). A unique group of planctomycetes, the autotrophic "anammox" planctomycetes, comprising at least three distinct "Candidatus" genera ("Brocadia", "Kuenenia" and "Scalindua") (Schmid et al. 2000; Kuenen and Jetten 2001; Schmid et al. 2003), perform a novel type of autotrophic metabolism based on anaerobic oxidation of ammonium, the "anammox" process; these exist in culture so far only in bioreactor mixed cultures, though these can be quite enriched in anammox planctomycetes as the major component of the microbial community (Strous et al. 1999; Schmid et al. 2003).

Many of the early observations and species designations of planctomycetes were based on natural microbial communities or enrichments (Starr and Schmidt 1989), and even now some of those species such as the rosette-forming *Planctomyces bekefii* (type species of the genus) remain uncultured, and enrichments such as bioreactor cultures remain an important contributor to our knowledge of new planctomycetes.

Planctomycetes are distinctive for their peptidoglycan-less cell walls, budding reproduction, and compartmentalized cell structure of great evolutionary significance discussed below. These bacteria have been identified in diverse freshwater, marine and soil habitats and even invertebrate animals (Staley et al. 1992; Schlesner 1994; Fuerst 1995; Fuerst et al. 1997; Neef et al. 1998; Wang et al. 2002). They have been isolated in culture as chemoheterotrophs from an equally diverse range of habitats, using for example selective media based on their inherent resistance to antibiotics targeting peptidoglycan synthesis and their predilection for *N*-acetylglucosamine as a carbon substrate (Schlesner 1994), but only representatives of 6 genera exist in pure culture, *Pirellula*, *Blastopirellula*, *Rhodopirellula*, *Planctomyces*, *Gemmata* and *Isosphaera*. We know already that members of several more genera remain to be isolated in pure culture, and that their physiological range is wider than axenically cultured strains suggest.

3
The Compartmentalized Cell Plan of Planctomycetes

Cells of all planctomycetes possess compartments within the cytoplasm defined as the region bounded by cytoplasmic membrane. These intracellular compartments are bounded by a membrane or membranes. Although different genera or suprageneric clades of planctomycetes may possess different types of compartment, there are also shared features of compartmentalization of the cytoplasm common to all planctomycetes examined, and which can be said to be a defining characteristic of members of the phylum *Planctomycetes*.

3.1
Shared Features of Cell Plan and Structure Involving Separation of Cytoplasm into Membrane-bounded Compartments in Planctomycetes

All planctomycete species which have been examined possess a shared cell plan including the central elements of 1) an intracytoplasmic membrane (ICM) within the cytoplasm bounding 2) an inner pirellulosome compartment containing ribosome-like particles and the nucleoid DNA and 3) an outer ribosome-free paryphoplasm compartment (derived from the Greek *paryphe* – "a border along a robe") between the intracytoplasmic membrane and the cytoplasmic membrane (Lindsay et al. 2001; Fuerst 2005). Cytoplasm of the pirellulosome containing ribosome-like particles is distinguished from paryphoplasm by use of the term "riboplasm". These shared elements can be understood most clearly by comparison of the diagrams in Fig. 1. The variations in the cell plans of planctomycetes derive either from 1) differences in scale and position of topologically identical compartments, as in the case of *Isosphaera pallida* where the central ribosome-free region is actually occupied by paryphoplasm which has come to be placed centrally due to invagination of the ICM (Fig. 1a), or 2) occurrence of additional membrane-bounded compartments within the pirellulosome. The latter is illustrated by *Gemmata obscuriglobus* and related organisms with a double-membrane bounded nuclear body within the pirellulosome, and the anammox planctomycetes such as "*Candidatus* Brocadia anammoxidans" with a single membrane-bounded organelle, the anammoxosome, within the pirellulosome (Fig. 1b). The re-

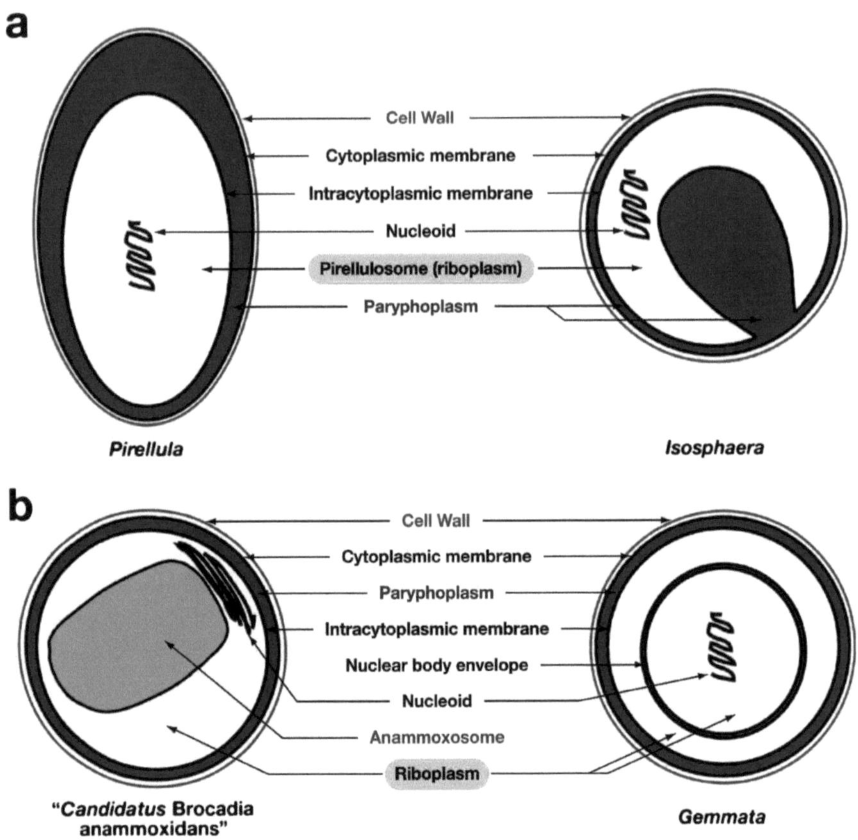

Fig. 1 Diagrams of cell organization and compartmentalization in **a** *Pirellula* (e.g., *Pirellula staleyi*) and *Isosphaera* (e.g., *Isosphaera pallida* – plan also applies to *Planctomyces maris*), and **b** "*Candidatus* Brocadia anammoxidans", and *Gemmata* (e.g., *Gemmata obscuriglobus*). The varieties of cell compartmentalization found in different planctomycetes are shown, as well as the underlying similarities in topology of their internal organization (i.e., possession of paryphoplasm compartment and intracytoplasmic membrane). (Reproduced from Fuerst 2005, with permission)

markable nuclear body of *Gemmata obscuriglobus* is an organelle surrounded by a "double-membrane" nuclear envelope analogous in structure to that of the nucleus of the eukaryote cell.

In summary, all planctomycetes examined, including representatives of all the extant genera, have at least one membrane-bounded cell compartment enclosing the nucleoid, thus providing a clear exception to the classical prokaryote organization of the cell. And in the case of *Gemmata obscuriglobus*, this exceptional organization goes further to mimic eukaryote organization itself in the form of a nuclear compartment surrounded by an envelope comprising two membranes, so that, since these cells also have the

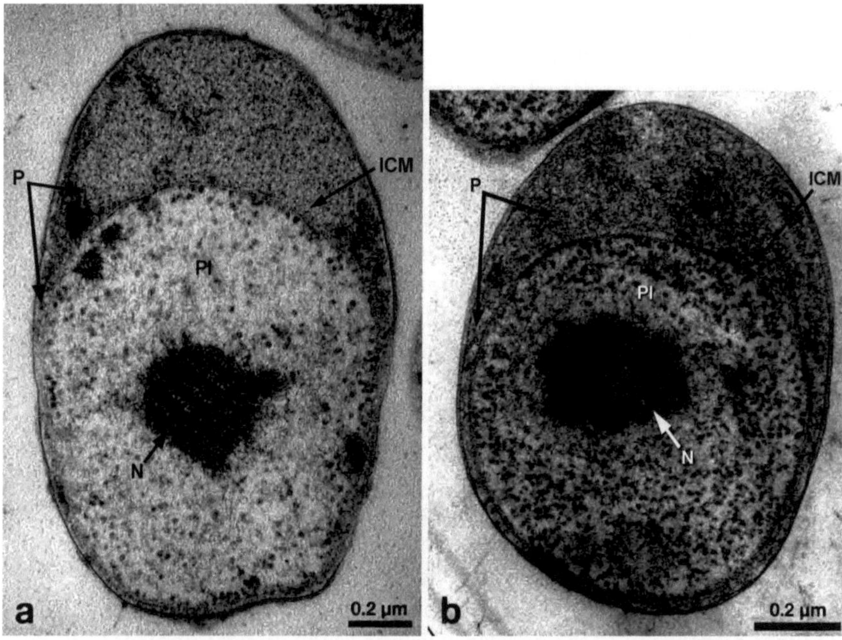

Fig. 2 Thin sections of cryosubstituted *Blastopirellula* and *Pirellula* illustrating a type of planctomycete cell plan involving a pirellulosome as the major intracellular compartment **a** Transmission electron micrograph of thin section of cryosubstituted cell of *Blastopirellula marina* (formerly *Pirellula marina*) displaying compartmentalization into pirellulosome (PI) and paryphoplasm (P) separated by the intracytoplasmic membrane (ICM). The nucleoid (N) is contained within the pirellulosome and is thus compartmentalized and surrounded by the single ICM membrane. Some of the ribosome-like particles in the pirellulosome form a linear array along the inside of the ICM bounding the pirellulosome. *Scale bar* 0.2 μm. Modification of figure in Lindsay et al. 1997. (Reproduced from Fuerst 2005, with permission) **b** Transmission electron micrograph of thin section of cryosubstituted cell of *Pirellula staleyi* displaying compartmentalization into pirellulosome (PI) and paryphoplasm (P) separated by the intracytoplasmic membrane (ICM). The nucleoid (N) is contained within the pirellulosome and is thus compartmentalized and surrounded by the single ICM membrane. Some of the ribosome-like particles in the pirellulosome form a linear array along the inside of the ICM bounding the pirellulosome. *Scale bar* 0.2 μm. Modification of figure of Lindsay et al. 1997. (Reproduced from Fuerst 2005, with permission)

pirellulosome compartment bounded by the single ICM membrane, the nucleoid is bounded by three membranes in this species.

3.2
Other Shared Features of Cell Plan and Structure in Planctomycetes

Condensed nucleoids appear to be a consistent feature of all cryosubstituted planctomycetes. They are distinctly fibrillar and often folded, and in such a way that similarity with the folded chromosomes of eukaryote dinoflagellates (Kellenberger 1991b) has been made (Kellenberger 1991b; Lindsay et al. 2001). This is in complete contrast to the situation in cryosubstituted non-planctomycete bacteria where the nucleoid is diffuse and granular. The relationship of the condensation of nucleoids in planctomyetes to cell compartmentalization in these organisms is not known, but such condensation could conceivably have implications for replication and the location of "replication factories" within membrane-bounded compartments, the distribution of chromosome and pirellulosome during division, and the need for transport mechanisms for the products of transcription. In the case of anammox planctomycetes (see Chapter 10), the nucleoid is not only condensed, but also associated with the anammoxosome membrane, suggesting a possibility of a linked distribution during cell division. Nuecleoids of other planctomycetes do not appear to be associated closely with internal membranes.

Another feature shared by some planctomycetes at least is the occurrence of arrays of ribosome-like particles lining one side of a membrane surface within the cytoplasm in a manner reminiscent of rough endoplasmic reticulum of eukaryotic cells. This can occur on the inner surface of the pirellulosome in *Pirellula* and *Blastopirellula* and in *Planctomyces maris* but also on the inner membrane of the nuclear envelope of the nuclear body of *Gemmata obscuriglobus*.

4 Pirellulosomes

The genera *Pirellula, Blastopirellula, Rhodopirellula* of the *Pirellula* group members of which cluster together in phylogenetic trees, and the genera *Isosphaera* and *Planctomyces* which are separate from this *Pirellula* group, nevertheless share a pirellulosome as the major nucleoid-containing membrane-bounded compartment and also share the feature that no further compartments appear within this pirellulosome. The genera differ in the degree to which the paryphoplasm intrudes into the cell interior or the extent to which it makes up the cytoplasmic area, but topologically their underlying cell plans appear to be identical.

4.1 Pirellulosomes of *Pirellula, Blastopirellula, Rhodopirellula*

Pirellulosomes were first described in the genus *Pirellula*, in the freshwater species *Pirellula staleyi* and the marine species *Pirellula marina* (now reclassified as *Blastopirellula marina*) (Lindsay et al. 1997). These appear via electron microscopy in sections of cryosubstituted cells as large regions of the cell cytoplasm bounded by a single ICM and containing electron-dense ribosome-like particles (9–18 nm diameter) as well as the cell's fibrillar nucleoid (Fig. 2). Ribosome-like particles sometimes appear to be arrayed along the internal side of the ICM (Fig. 2b). The pirellulosome is surrounded by a region of relatively amorphous ribosome-free cytoplasm often more extensive at one pole of the cell, giving rise to a "polar cap" appearance – this material varies considerably in its electron density from cell to cell (Fig. 2a,b). This region lies between the ICM surrounding the pirellulosome and the cytoplasmic membrane, which is difficult to discern due to its close apposition to the cell wall. The region between the pirellulosome and the cytoplasmic membrane was later termed the *paryphoplasm* when both it and the pirellulosome was recognized as a general feature of all planctomycetes examined by cryosubstitution, including those phylogenetically outside the genus *Pirellula* and also those entirely outside the three related "*Pirellula*-group" genera. In *P. staleyi* and *B. marina*, the "polar cap region" paryphoplasm varies both in size and electron density, appearing as small as 9% of the total cell volume or as large as 45% of that volume, and in most cells (57% of the total) the paryphoplasm may be filled with electron-dense material while a minor proportion of the cells has an electron-transparent paryphoplasm. The band of paryphoplasm away from the poles can be as thin as 20 nm wide. *P. staleyi* and *B. marina* possess the simplest cell structures of all planctomycetes, in which a single ICM separates the cytoplasm into two compartments, an inner pirellulosome and an outer paryphoplasm. The phylogenetic significance of this structure is indicated by study of a strain originally classified as a planctomycete on solely morphological grounds, ATCC 35122, later confirmed via 16S rRNA gene sequencing to be closely related to the type strain of *Pirellula staleyi*, ATCC 27377; thin sections of this strain reveal an identical cell plan to that of *Pirellula staleyi* and *B. marina* (Butler et al. 2002).

Cells of *Rhodopirellula baltica* strain SH 1, the first planctomycete for which a completed genome sequence has become available, as well as a closely related strain SH 796, possess a major pirellulosome structure within cryosubstituted cells, but also display several smaller structures apparently similar in appearance to the major pirellulosome but without obvious nucleoid (Schlesner et al. 2004). These smaller structures are apparently within the polar cap region paryphoplasm, but the nucleoid and most ribosomes are held within the major pirellulosome. This is a new variation on the pirellulosome theme within *Pirellula*-like planctomycetes, and differs from *Blastopirellula marina* and *Pirellula staleyi* where the paryphoplasm is homogeneous in structure. It would appear that as in *B. marina* and *P. staleyi*; however, the nucleoid is situated within the major membrane-bounded pirellulosome, and that there is a polar organization to the cell as illustrated by a large polar cap paryphoplasm region, albeit one containing small membrane-bounded nucleoid-less "mini-pirellulosomes". Strain SH796 from a marine sponge *Aplysina aerophoba*, referred to as a *Pirellula* species in one study but clearly clustering within *Rhodopirellula baltica* in the formal taxonomic description, clearly displays this "mini-pirellulosome" variation in cryosubstituted cells and ribosome-like particles can be seen within the mini-pirellulosomes of the polar cap paryphoplasm (Gade et al. 2004).

DNA and RNA has been localized in cells of *B. marina*, using RNase gold and anti-ds/ss DNA immunogold labeling. Most of the DNA labeling in *B. marina* is found co-localized with the fibrillar nucleoid within the pirellulosome, with only a minor amount of labeling within the pirellulosome outside the nucleoid, and all the cell DNA appears confined to the pirellulosome. In contrast, although most of the cell RNA appears within the pirellulosome as expected from the distribution of ribosome-like particles, at least some RNA appears to occur in the apparently ribosome-free paryphoplasm outside the pirellulosome, at levels significantly above background.This unexpected RNA distribution has significant implications both functionally and concerning arguments supporting the true cytoplasmic nature of paryphoplasm. Since ribosome-like particles do not appear in the paryphoplasm but RNA does, at least to some extent, this RNA would not seem to be associated with translation.

The paryophoplasm in *Pirellula*-like species does not appear to be merely an enlarged periplasm. The arguments supporting this interpretation of the electron micrograph data are that cytoplasmic membrane closely appresssed to the cell wall can be demonstrated where it has retracted from the wall, freeze-fracture replicas confirm presence of a cytoplasmic membrane boundary outside the membrane bounding the pirellulososome, and in the case of *Blastopirellula marina*, there is RNA in the paryphoplasm polar cap region of *Pirellula marina* supporting its nature as true cytoplasm (Lindsay et al. 1997). An analogous situation where ICM is distinct from cytoplasmic membrane and the paryphoplasm is clearly different from any possible periplasm applies to other planctomycetes that have been examined, such as *Gemmata obscuriglobus* (Lindsay et al. 2001); paryphoplasm appears to contain some RNA in all planctomycetes examined and freeze-fracture replicas clearly have demonstrated the occurrence of a distinct cytoplasmic membrane in *Blastopirellula marina*, *Gemmata obscuriglobus* and *Isosphaera pallida* (Lindsay et al. 2001). No connection between cytoplasmic membrane and the ICM surrounding the pirellulosome has been seen, so that the pirellulosome membrane does not seem to clearly originate via an invagination of the cytoplasmic membrane. This suggests that the pirellulosome and its surrounding ICM might be distributed intact during cell division.

Examination of chemically fixed cells of *Pirellula staleyi* and *Blastopirellula marina* prepared using a fixative containing ruthenium red results in dense staining of the paryphoplasm (polar cap region) and has revealed electron dense particles larger than ribosome-like particles within the pirellulosome towards the polar cap region (Lindsay et al. 1997). The densely stained pirellulosome in these cells is much less extensive in area than in cryosubstituted cells. The paryphoplasm in these planctomycetes and the large pirellulosome particles are conceivably rich in polysaccharide since ruthenium red stains mainly acid mucopolysaccarides, but the specificity of such a stain is not sufficient to exclude a possible composition of protein, glycoprotein or glycogen. Attempted staining for glycogen or glycoproteins using phosphotungstic acid or tannic acid/uranyl acetate was inconclusive due to resistance of such staining to pre-treatment with proteinase K or periodate. Regardless of its source, the dense staining of the paryphoplasm with ruthenium red does suggest that there may be differences in the molecular composition of the paryphoplasm relative to the pirellulosome in these planctomycetes, perhaps correlated with a functional distinction.

Planctomycetes in the *Pirellula* group are known to possess a life cycle involving budding mother cells and daughter cells released as motile flagellated swarmer cells. This is known for a *Pirellula staleyi*-like isolate from freshwater (Tekniepe et al. 1981) and for *Rhodopirellula baltica* known to possess a pirellulosome (Gade et al. 2005a). In *R. baltica*, crateriform structures are restricted to one cell pole (the broader one, considered for *R. baltica* as the "reproductive" pole from which budding occurs) in "adult" mother cells but are distributed uniformly across buds, and only mother cells appear to possess fimbriae, which are correlated with the crateriform structures (Gade et al. 2005a). In the stationary phase, non-motile cells in rosettes may dominate the population. No clear structural correlation has yet been established between the position of the pirellulosome or polarity of the internal cell structure and the so-called reproductive pole in *Blastopirellula marina* and *Pirellula staleyi*. Although the polar cap region of the paryphoplasm does seem more common at the budding pole, budding is not necessarily connected with such a polar cap, since buds can be observed to originate from either pole. However, the question of a role for the pirellulosome and paryphoplasm in cell division and thus in the organization of the life cycle is one needing examination systematically, and might be made possible if a genetic system for these planctomycetes could be developed so that the experimental model so well developed for *Caulobacter* could be applied. Cell compartmentalization in planctomycetes could conceivably be simply a consequence of the budding reproductive life cycle of these bacteria, and such an approach could test this hypothesis. Interestingly, *R. baltica* does not possess a homolog of the master regulator response protein CtrA which plays a central role in the *Caulobacter* life cycle and correlated protein compartmentalization during that cycle (Glockner et al. 2003; Skerker and Laub 2004).

The cell plan occurring in members of the *Pirellula* group forms the simplest type of planctomycete cell plan, and at the same time the basis of the cell plan of all the other planctomycete genera. Thus all planctomycetes so far examined have a pirellulosome containing riboplasm (cytoplasm containing ribosome-like particles), an ICM bounding the pirellulosome and a paryphoplasm region between intracytoplasmic membrane and cytoplasmic membrane (Lindsay et al. 2001). Members of the genus *Isosphaera* merely have a topologically equivalent cell plan to that of *Pirellula* and *Blastopirellula*, one in which the ICM appears to have invaginated so as to form a ribosome-free central region of the cell continuous with the paryphoplasm (Lindsay et al. 2001). The more complex forms have in addition further membrane-bounded compartments within the pirellulosome, but they all share a pirellulosome, so that the term "pirellulosome" has become much wider than its etymological source genus. Cells of *Gemmata obscuriglobus* and related organisms possess a compartment within their pirellulosome containing the nucleoid and bounded by an envelope consisting of two associated membranes, while cells of the anammox planctomycetes possess a compartment within their pirellulosome bounded by a single membrane and containing an enzyme associated with their unique anaerobic ammonium-oxidizing physiology and called the anammoxosome (see Fuerst 2006, this volume).

4.2
Pirellulosomes of *Isosphaera* and *Planctomyces*

Pirellulosomes of planctomycetes related most closely to the genus *Isosphaera*, including *Isosphaera pallida* and "*Nostocoida limicola*" type III (Liu et al. 2001) display a different superficial appearance to planctomycetes in the *Pirellula* group. In *Isosphaera pallida* itself this is manifested by the occurrence of a central region free of ribosome-like particles and in apparent continuity with the paryphoplasm. This is due to the invagination of the ICM so that a large inclusion comprising up to 60% of the cell interior is formed. This forms a central region of the cell, continuous with the paryphoplasm at the cell rim; the rest of the cell is equivalent to the pirellulosome of *Pirellula* (and *Blastopirellula*) species, and contains ribosome-like particles and the nucleoid (Fig. 3). Two further *Isosphaera*-like non-thermophilic isolates from water have also been found to share a cell plan similar to that of *Isosphaera pallida*. The marine planctomycete *Planctomyces maris*, not phylogenetically a member of the *Isosphaera* group, displays a similar two-compartment structure where a large region of paryphoplasm occupies a substantial but variable area of the cell, and the rest of the cell interior is occupied by a pirellulosome enclosing a condensed nucleoid and ribosome-like particles (Fig. 4). Such particles can appear lining the inner, pirellulosome, side of the ICM (Fig. 4). Other species of *Planctomyces* have more complex compartmentalization which nevertheless conforms to the general shared planctomycete features.

A variation on the *Isosphaera pallida* cell plan is found in the filamentous bacterium "*Nostocoida limicola*" morphotype III (Liu et al. 2001) isolated from activated sludge biomass in pure culture, a planctomycete most closely related to *Isosphaera pallida*. When viewed via electron microscopy after cryosubstitution, isolates possess complex intracellular membrane systems resulting in compartmentalization. These bacteria appear to possess a shared

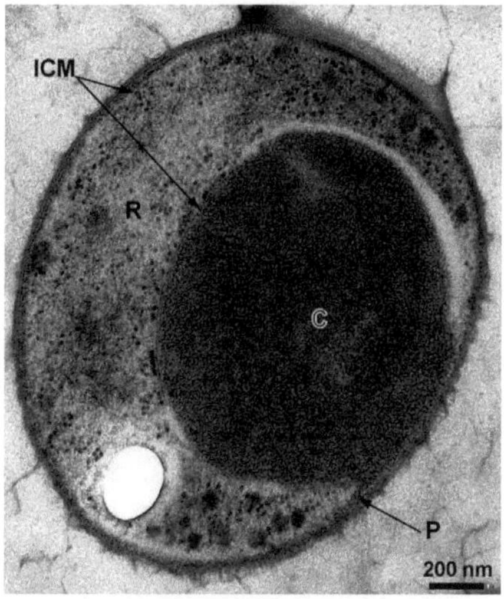

Fig. 3 Transmission electron micrograph of thin section of cryosubstituted cell of *Isosphaera pallida*, displaying a large central region (C) defined by the intracytoplasmic membrane (ICM). This central region is continuous with the outer paryphoplasm (P). The ICM defines a pirellulosome compartment containing ribosome-like particles defining the riboplasm (R). *Scale bar* 200 nm. (Reproduced from Lindsay et al. 2001, with permission)

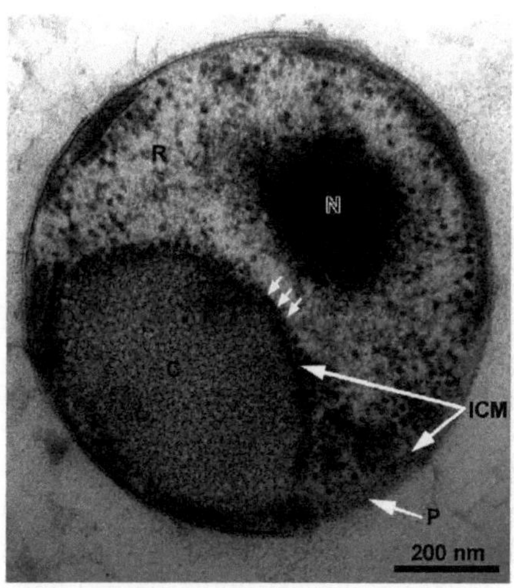

Fig. 4 Transmission electron micrograph of thin section of cryosubstituted cell of *Planctomyces maris*. Paryphoplasm (P) surrounds the outer rim of the cell and is continuous with a large ovoid central region (C). An intracytoplasmic membrane (ICM) separates a pirellulosome containing a riboplasm (R) of ribosome-like particles from the central region, and ribosome-like particles (*arrowheads*) line the pirellulosome side of the ICM bordering the ovoid central region. *Scale bar* 200 nm. (Reproduced from Lindsay et al. 2001, with permission)

planctomycete cell plan, but the ICM separates cell compartments in such a way that regions equivalent to paryphoplasm appear as large membrane-bounded regions ramifying throughout the cytoplasm, sometimes appearing as tubules or irregularly shaped regions. In one strain of the *Nostocoida limicola* III morphotype, Ben 225, distinctive tubules with a diameter of 70–80 nm were present sometimes forming parallel arrays within the cytoplasm, a feature unique within planctomycetes but bearing resemblance to arrays of tubules found in anammoxosomes of anammox planctomycetes described elsewhere in this volume (Fuerst 2006, in this volume). In sections of these isolates prepared by cryosubstitution, nucleoids were typically condensed after cryosubstitution in a way characteristic of all planctomycetes studied so far, and multiple nucleoids were also seen, possibly as a result of multiple sectioning of single folded nucleoids.

In *Pirellula, Blastopirellula, Rhodopirellula, Isosphaera* and *Planctomyces maris*, the ICM bounds a compartment (or compartments) equivalent to the pirellulosome and containing ribosome-like particles, but there are no other compartments within the regions equivalent to the pirellulosome. Only *Gemmata* and anammox planctomycetes (and possibly *Nostocoida limicola* morphotype III) appear to have further membrane-bounded compartments within the pirellulosome.

5
Membrane-bounded Nucleoids of *Gemmata*

The remarkable nuclear body of the freshwater planctomycete *Gemmata obscuriglobus* (Figs. 5 and 6) is an organelle surrounded by a nuclear envelope consisting of two membranes (Fuerst and Webb 1991; Lindsay et al. 2001). The term "nuclear body" is employed in this context only for a double membrane-bounded organelle enclosing a bacterial DNA-containing nucleoid and associated nucleoplasm (material other than the nucleoid but also confined by the double membrane envelope). Because of the double membrane surrounding the nuclear body, it is, thus, analogous in its envelope membrane structure to that of the nucleus of the eukaryote cell. It shares with the riboplasm of the surrounding pirellulosome cytoplasm the occurrence of ribosome-like electron dense particles, which appear within the nuclear body surrounding the fibrillar nucleoid, so that the nuclear body shares these particles with the riboplasm of the pirellulosome cytoplasm surrounding the nuclear body. The term "ribosome-like particle" is used here to refer to these particles since some may be pre-ribosome particles before complete assembly of a ribosome functional in protein synthesis and they cannot be assumed to be all functional ribosomes. The fibrillar nucleoid and thus most of the cell DNA appears confined to the nuclear body (Figs. 5 and 6), though it is not yet

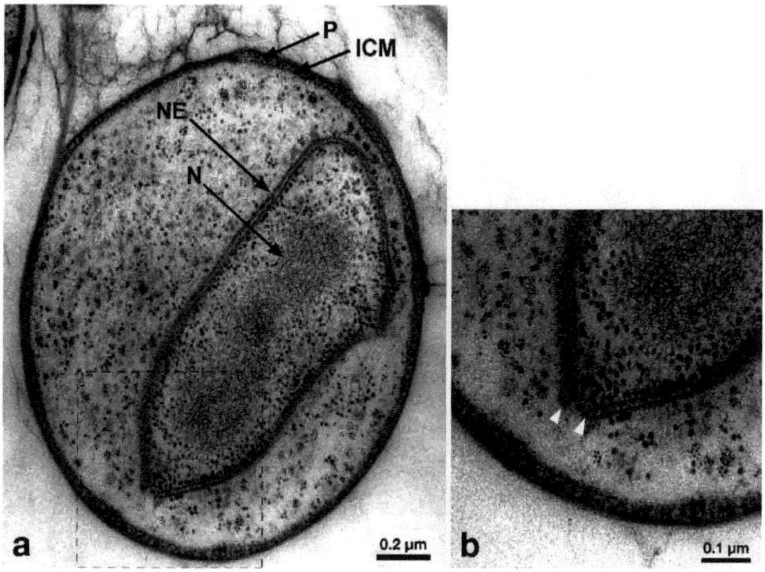

Fig. 5 Thin sections of cryosubstitued *Gemmata obscuriglobus* showing the nuclear body and the folded single membrane nature of its envelope **a** Transmission electron micrograph of thin section of *Gemmata obscuriglobus* fixed and prepared via cryosubstitution, showing the membrane-bounded nuclear body with its nuclear envelope (NE) surrounding the nucleoid (N), as well as the more general features of the planctomycete cell plan including intracytoplasmic membrane (ICM) and paryphoplasm (P). Ribosome-like electron-dense particles appear both inside and outside the nuclear body. *Scale bar* 0.2 μm. **b** Shows an enlarged portion of the nuclear membranes of the same cell as in Fig. 5a where the true continuity of the apparent double membrane is visible as a folded single membrane on both sides of a gap in the nuclear envelope (*arrowheads*). Ribosome-like particles can be seen lining the inner membrane of the nuclear envelope. *Scale bar* 0.1 μm. Modification of figure of Lindsay et al. 2001. (Reproduced from Fuerst 2005, with permission)

clear whether all the cell DNA is confined. Initial qualitative immunogold cytochemistry with antibody against double- and single-stranded DNA was consistent with a view that all the cell DNA is confined within the nuclear body. More detailed quantitative analysis is needed to determine whether the riboplasm outside the nuclear body is completely free of DNA. As expected from the distribution of ribosome-like particles, RNA can be detected in cryosubstituted sectioned cells via RNase-gold immunocytochemistry to be in both the nuclear body and in the riboplasm outside it. There are also indications from rRNA-based oligonucleotide probes that rRNA occurs both inside the nuclear body and in the riboplasm outside it (Margaret Butler, unpublished data). The ICM of *G. obscuriglobus*, thus, encloses a large region containing riboplasm equivalent to the pirellulosome, but within this pirellulosome is the nuclear body which also contains riboplasm and RNA as

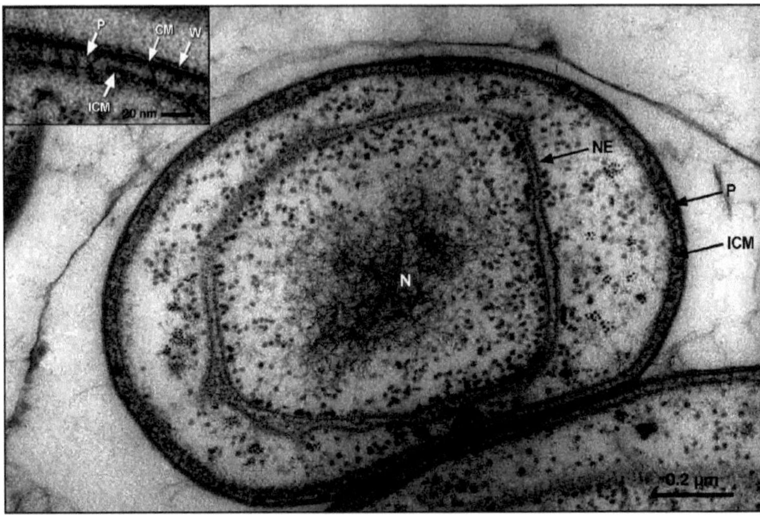

Fig. 6 Thin section of cryosubstituted *Gemmata obscuriglobus* bud cell with large nuclear body containing an extensive fibrillar nucleoid (N) bounded by nuclear envelope (NE) consisting of two membranes between which is a clear electron-transparent space. The nuclear body contains ribosome-like particles as well as nucleoid; some of these can be seen to be arrayed along some regions of the inner membrane of the nuclear envelope. It is surrounded by cytoplasm which contains superficially similar ribosome-like particles and is bounded by a single intracytoplasmic membrane (ICM). The cell external to this ICM is rimmed by a paryphoplasm containing relatively electron-dense cytoplasm (P). *Scale bar* 0.2 µm. *Inset:* enlarged view of envelope of bud cell showing a bilayer cytoplasmic membrane (CM) underneath the cell wall (W) but closely apposed to that wall. Paryphoplasm (P) is bounded peripherally by the cytoplasmic membrane (CM) and internally by the intracytoplasmic membrane (ICM). *Scale bar* 20 nm. Modification of figure of Lindsay 2001. (Reproduced from Fuerst 2005, with permission)

well as the nucleoid and DNA. The riboplasm of the nuclear body is similar in appearance to that of the cytoplasm of the rest of the pirellulosome, but ribosome-like particles can often be seen lining the inner membrane of the nuclear envelope (Figs. 5b and 6). The paryphoplasm between the ICM and the cytoplasmic membrane in *G. obscuriglobus* occurs as an outer rim.

5.1 The Nuclear Body Envelope and its Evolutionary Relevance

The envelope of the *G. obscuriglobus* nuclear body consists of two double-track (trilaminar) membranes, each 5–8 nm wide, separated by an electron-transparent "perinuclear" space 14–21 nm wide (Fuerst and Webb 1991; Lindsay et al. 2001). Each membrane of this nuclear envelope is trilaminar, but there is an asymmetry to the membrane in that the dense layers of the double track comprise a relatively dense and relatively light layer, separated from each other by a much lighter electron transparent layer (Fuerst and Webb 1991). The outer membrane of the nuclear envelope sometimes appears to be connected to and continuous with the ICM (Lindsay et al. 2001), suggesting a potential mechanism for evolution of the double-membrane bounded nuclear body via invagination of the ICM in a *Pirellula*-like cell possessing only an ICM-bounded pirellulosome. Within this perspective, the electron transparent "perinuclear" space between the two nuclear membranes is effectively a continuation of the paryphoplasm. The effective compartmentalization of the cell enabled by the nuclear body is especially clear in freeze-fracture replicas, especially where vesicles occurring in the non-nuclear body riboplasm of the cell make the distinction between nuclear body and the remainder of the cell clear (Fig. 7). Nuclear bodies have also been demonstrated by use of the internal membrane stain $DiOC_6$ applied to whole cells viewed by fluorescence microscopy (Lindsay et al. 2001).

The nuclear envelope superficially appears to be composed of two continuous and separate membranes, but a closer examination reveals that there are regions where continuity between the inner and outer nuclear membranes is clearly demonstrated, as an effective fold joining the two membranes, leaving a gap between the two joins (Fig. 5b). The nuclear envelope membranes are

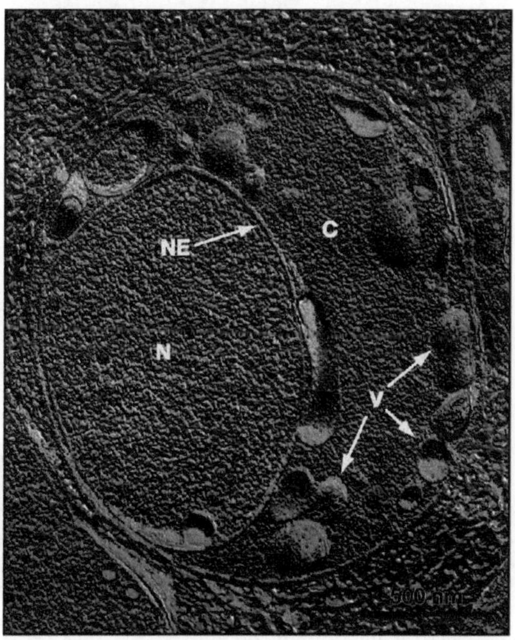

Fig. 7 Transmission electron micrograph of freeze-fracture replica of *Gemmata obscuriglobus* showing cross-fractured nuclear body (N) bounded by nuclear envelope membranes (NE) and resting within a pirellulosome cytoplasm (C) containing vesicles (V). *Scale bar* 500 nm

effectively a series of folded single membranes. It has been pointed out that the eukaryote nuclear envelope also consists of a series of folded single membranes. Major implications for the theories for the origin of the eukaryote nucleus have been drawn from the occurrence of the folded single membrane structure of the eukaryotic nuclear envelope. Endosymbiotic engulfment models for the origin of the nucleus via engulfment of a Gram-negative bacterium by an archaeon become less plausible if the nuclear envelope is not seen as a continuous track of two apposed membranes without areas of join allowing the gaps known to occur in the eukaryote nuclear envelope (Martin 1999; Poole and Penny 2001; Rotte and Martin 2001).

The nuclear body can be seen in newly budded daughter cells (Fig. 6) as well as in the mother cell giving rise to the bud, suggesting that the body is either preserved intact during division or reforms rapidly after any disassembly during division. Continuity of the perinuclear space of the nuclear envelope with paryphoplasm and of the outer nuclear envelope membrane with the ICM in some sectioned cells where the nuclear body is seen in contact with the ICM suggests that nuclear bodies might be able to reform via invagination of the ICM.

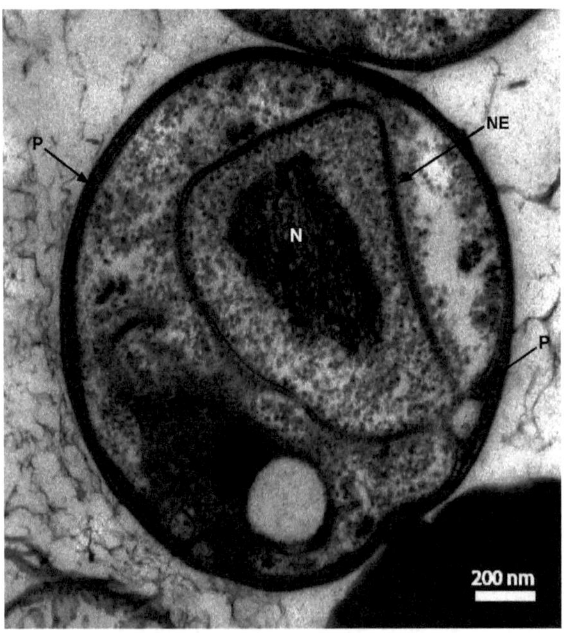

Fig. 8 Transmission electron micrograph of thin section of a cell of cryosubstituted *Gemmata*-related soil planctomycete strain Soil9 possessing a nuclear body containing a nucleoid (N) surrounded by a double-membrane nuclear envelope (NE). Paryphoplasm (P) can be seen at the cell rim. *Scale bar* 200 nm. (Reproduced from Wang-Holmes 2004, with permission)

The nuclear body with envelope composed of two appressed membranes has also been found in 6 additional isolates clustering with *G. obscuriglobus* in a monophyletic *Gemmata* group in phylogenetic trees (Wang et al. 2002), exemplified here by soil strain JW3-8s0 (Fig. 8). The cells of all these strains in addition possess a paryphoplasm and ICM, so that they conform to the shared planctomycete cell plan. Some of these strains are quite deep-branching within the *Gemmata* group. This distribution of the organelle among strains is consistent with a hypothesis that the last common ancestor of the *Gemmata* group also possessed a double membrane-bounded nuclear body. This suggests that a membrane-bounded nuclear body may have been of selective advantage through evolutionary time for members of this group.

5.2 Relevance of Sterols of *Gemmata obscuriglobus*

It is conceivable that the membranes of *Gemmata obscuriglobus* have a special lipid composition correlating with its unusual cell compartmentalization. *Gemmata obscuriglobus* and its close relative *Gemmata* strain Wa1-1 have been found to possess the simple C_{30} sterols lanosterol and its rare isomer parkeol, and homologs of the genes for the sterol synthesis enzymes squalene monooxygenase and oxidosqualene cyclase have been deduced from the genome and lie contiguous to one another. The species has been proposed to possess one of the simplest pathways for biosynthesis of sterols known, perhaps indicating retention of an ancient remnant pathway (Pearson et al. 2003). This is deduced from the absence of any sterols smaller than C_{30}, indicating an absence of enzymes in the sterol synthesis pathway downstream from oxidosqualene cyclase such as α14-demethylase. *Gemmata* strains are unique among the few sterol-synthesizing bacteria in this respect. If sterols are found to be present in the nuclear body envelope as suggested by one brief report (Budin et al. 2004), this may have significant implications for the properties of the nuclear envelope membranes needed for function, and evolutionary implications regarding the nuclear body as a homologous or analogous organelle to the eukaryote nucleus. It has been proposed that genes for sterol synthesis may have been transferred to such bacteria from an ancient eukaryote, but it seems as likely that such genes could perhaps have formed the basis, by gene transfer in the other direction, for the sterol synthesis pathway in ancient eukaryotes. Planctomycetes also synthesize sterol-like hopanoids (Sinninghe Damsté et al. 2004), more typical of other bacteria than sterols, and these are also found in the anaerobic anammox planctomycetes, perhaps forming the basis for sterol biosynthesis evolution in planctomycetes after atmospheric oxygen became available.

6 Functional Implications of Compartmentalization in Planctomycetes

The structural compartmentalization of planctomycetes is unique among Bacteria and Archaea, and one would expect that such internal structural differentiation would be accompanied by functional differences between the cell compartments and potential mechanisms for communication of materials such as macromolecules between compartments. Considering that the distribution of ribosome-like particles and of DNA are correlated with particular membrane-bounded compartments, such functional implications might be expected to have a cell biological nature possibly comparable to the differentiation of sites for transcription and translation and occurrence of such phenomena as RNA splicing and nucleocytoplasmic transport characteristic of eukaryote cells. These possibilities are best considered in the context of the different types of intracellular compartment occurring in different planctomycetes. To solve these problems, the insights from genomics and proteomics will be needed as well as extensions of cell biology experimental approaches to planctomycetes.

6.1 Functional Implications of the Pirellulosome

The simplest form of compartmentalization, the pirellulosome, as it occurs in *Pirellula*-like planctomycetes, involves compartmentalization of one region of the cytoplasm. This region contains all the cell's DNA, most of its RNA and all of its ribosome-like particles, and is bounded by an ICM, separating it from another region, the paryphoplasm, that also contains some RNA (Lindsay et al. 1997). The pirellulosome, thus, appears to contain all the cell's ribosomes and thus newly synthesized proteins, while the paryphoplasm must contain some protein components, at least those needed for insertion into the cytoplasmic membrane and the proteinaceous cell wall. This implies that proteins destined for the paryphoplasm compartment must be transported across the ICM, and thus be targeted via protein secretion pathways and signal peptides. Since RNA is also found in the paryphoplasm it must also be transported across a membrane, but presumably via more specialized mechanisms. The DNA distribution within *Pirellula* and *Blastopirellula* suggests that transcription is compartmentalized so that it occurs only within the pirellulosome. The RNA distribution suggests that at least some transcribed RNA must pass across the ICM from pirellulosome to paryphoplasm, and that thus there must be a mechanism for RNA transport across the ICM, albeit the function of the paryphoplasm RNA is unknown. Since all the ribosome-like particles occur within the pirellulosome in *Pirellula*-like species, it seems likely that all translation occurring in the cell is compartmentalized within the pirellulosome, with the implication that the proteins of the paryphoplasm, cytoplasmic membrane and protein cell wall must be secreted across the ICM.

6.2
Functional Implications of Ribosome-like Particle Distribution

Ribosome-like particles sometimes appear to be arrayed along the internal pirellulosome side of the ICM of *Pirellula*-like planctomycetes (Fig. 2) and *Planctomyces maris* (Fig. 4), suggesting that some transport of material such as newly synthesized protein might occur across the ICM and from the pirellulosome to the paryphoplasm in these cells. If this occurs in these organisms then co-translational protein secretion would occur across an internal membrane rather than the cytoplasmic membrane at the cell surface boundary as in other non-compartmentalized bacteria. Ribosome-like particles are found in linear arrays apparently bound to the inner membrane of the nuclear envelope of the nuclear body of *Gemmata obscuriglobus* (Figs. 5a,b and 6). This suggests that there might be protein translation of a co-translational secretion type across this membrane into the perinuclear space in a manner analogous to that occurring in rough endoplasmic reticulum of eukaryote cells, but if this does occur then this is consistent with a concept that some translation, presumably coupled to transcription, must occur within the nuclear body. Cell biology events within the nuclear body would then be more similar to those occurring in cytoplasm of other bacteria, with the exception that co-translational secretion would occur across an internal membrane rather than the cytoplasmic membrane at the cell surface boundary. Conceivably, however, such linear arrays of particles represent a mechanism for transport of ribosomes or pre-ribosomes to the riboplasm outside the nuclear body and have no direct functional significance for protein synthesis.

6.3
Functional Aspects of the Nuclear Body – Comparison with Eukaryote Cell Biology

If the nuclear body of *Gemmata obscuriglobus* was completely analogous to a eukaryote nucleus in terms of function, then we would expect it to be correlated with features of eukaryote cell biology such as compartmentalization of protein synthesis within the cytoplasm, the uncoupling of transcription from translation, pre-mRNA processing, splicing and transport through the nuclear envelope, and nucleocytoplasmic transport of proteins and RNA facilitated by nuclear pores and a nuclear transport system such as that involving importins and Ran GTP-binding proteins. Establishment of any of these features awaits future work, but some aspects may be examined via the present ultrastructural evidence. If the occurrence of the fibrillar DNA-containing nucleoid within the nuclear body represents a situation in which most or all of the chromosomal DNA is within that compartment, and if we take into account that RNA and ribosome-like particles occur both inside and outside the nuclear body, then this suggests that at least some protein translation may be uncoupled from transcription in *Gemmata obscuriglobus*. This would require some mechanism of mRNA transport through the nuclear envelope to the non-nuclear portion of the riboplasm. Some translation might occur within the nuclear body, and this could be similar to that claimed recently to also occur in nuclei of complex eukaryotic cells (Iborra et al. 2001) (a claim subject to considerable debate), perhaps associated with error-correcting mechanisms such as nonsense-mediated decay. If transcription is confined to the nuclear body compartment as a correlate of the confinement of the fibrillar nucleoid, then RNA must be transported through the nuclear envelope since RNA is found outside that envelope. Of course there is evidence that even in bacteria such as *Bacillus subtilis* without internal membrane compartments, transcription as indicated by RNA polymerase location and translation indicated by ribosomes may occur in separate functional domains (Lewis et al. 2000), and even that some translation may be uncoupled from transcription in such cells, but membranes do not appear to separate these domains.

The folded single membrane nature of the nuclear envelope in *G. obscuriglobus* (Fig. 5b) suggests a possibility that complexes similar to nuclear pore complexes of eukaryote nuclear envelopes might occur between the folds where inner and outer nuclear envelope membranes meet. Combined with a nuclear transport system, RNA export system, and nuclear localization signals on proteins translated in the non-nuclear riboplasm, this would allow the necessary nucleocytoplasmic transport of RNA and protein between nuclear body and the remainder of the riboplasm. It has been suggested that there may be some evidence that pore-like structures may occur in the nuclear envelope of *G. obscuriglobus* (Pennisi 2004) and there is also bioinformatics evidence from the genome of a *Gemmata* strain, Wa1-1, that a gene for a homolog of a nuclear pore protein of a yeast may be present in this planctomycete (Staley et al. 2005).

The occurrence of internal membranes forming compartments in planctomycetes may have implications for the lipid composition enabling their efficient functioning and stability, and the occurrence of sterols and sterol-like hopanoids in planctomycetes may be conceivably related to their functions within the cell. This would be especially so if the suggestion that the nuclear body envelope of *Gemmata obscuriglobus* is specifically enriched in sterols (Budin et al. 2004) is confirmed.

6.4 Implications of Genomics and Proteomics

Bioinformatic analysis of the complete genome sequence for *Rhodopirellula baltica* suggests presence of secretion mechanisms based on *secA*, and the general secretory pathway GSP as indicated by a type II F-domain and GSP type II/III secretion system protein (Glockner et al. 2003). Evidence from the genome of *Rhodopirellula baltica* also supports the view that many proteins of this planctomycete must contain signal peptides regulating their secretion across membranes. *R. baltica* has a very high number of genes with predicted signal peptides (normalized as 178 per megabase), equivalent when other bacterial genomes are examined for the normalized ratio only to *Caulobacter crescentus* with its differentiated cell cycle (Glockner et al. 2003). The *R. baltica* proteome has one the highest number of Tat (twin arginine translocation) signal peptides of any genome (Glockner et al. 2003). One might expect that since the pirellulosome in some form or other is shared by all the planctomycetes that signal peptides and secretion systems will also be found to play a major role in the functional proteome and that additional transport systems might be important in organisms with additional internal membranes such as the membrane-bounded nuclear body of *Gemmata obscuriglobus* and the anammoxosome of the anaerobic ammonium-oxidizing planctomycetes.

Study of the proteome of the first planctomycete for which a complete genome sequence is available, for *Rhodopirellula baltica*, containing a simple pirellulosome compartment and concomitant paryphoplasm, has suggested that a substantial minority of total proteins in the cell may be translocated across membranes (Gade et al. 2005b), confirming the perspective from genome analysis. 26% of the proteins identified from 2D electrophoresis of the soluble protein fraction of cells of this planctomycete possess signal peptides, consistent with but larger than a signal peptide-containing fraction of 16% of proteins predicted from the genome sequence. The procedure used to break open cells for electrophoretic analysis could not have distinguished proteins in the paryphoplasm from those in the riboplasm, so those proteins containing signal peptides may well have been secreted, either across of the ICM into the paryphoplasm or across the cytoplasmic membrane to become cell wall-associated proteins. Housekeeping enzymes (e.g., involved in glycolysis, tricarboxylic acid cycle, amino acid biosynthesis, translation and protein quality control) appear to be confined to the riboplasm as judged by the absence of a signal peptide (but see below for possible exceptions in some translational proteins). Secreted proteins include dehydrogenases, hydrolases for extracellular molecules and signal transduction kinases. Some of the many sulfatases which *R. baltica* possesses as predicted from its genome were detected in the soluble protein map, and possess signal peptides, suggesting they may have been caught in the process of secretion to the external milieu, since they are suggested to function in degradation of sulfated polymers extracellularly. Some problems remain in attribution of secretion via signal peptide presence in *R. baltica* proteins, since two proteins involved in protein synthesis, elongation factor G and lysyl-tRNA synthetase were found to have predicted signal peptides, unexpected since translation presumably occurs in the riboplasm only. Proteomic analysis of individual compartments of *R. baltica* may be needed to resolve such problems, which will await development of methods for isolation of these compartments from cells lysed in such a way as to preserve them intact.

6.5 Protein Miscibility – a Possible Selective Force Driving Evolution of Cell Compartmentalization in Planctomycetes and Eukaryotes?

In an intriguing and entirely novel proposal from theoretical physics, molecular statistics and thermodynamics of physical stability of protein mixtures within the cell cytoplasm have been used to predict that in some microbial species, intracellular compartmentation may be a necessary consequence of certain distributions of expressed protein length (Braun et al. 2005). By distributing different proteins in distinct compartments, a cell may alleviate the protein miscibility constraints. The specific case of the origin of the eukaryote cell and *Gemmata obscuriglobus* were raised as possible examples of such avoidance of the "miscibility" gap – such a proposal awaits a test by assessment of the proteome of planctomycetes for a protein length distribution close to the proposed miscibility limits. While such selective forces may have been active when subcellular compartments first formed, it seems more likely that considerations of molecular cell biology concerned with partitioning of translational abilities, processing of RNA, or segregation of chromosomes were more central to formation of membrane-bounded DNA-containing compartments necessary for the ancestral development of eukaryality. Nevertheless, the proto-eukaryote may have been a cell where protein localization preceded membrane separation of compartments, and the proteomes of planctomycetes might be examined productively with such hypotheses in mind.

7 Evolutionary Implications of Compartmentalization in Planctomycetes

The membrane-bounded compartments of planctomycetes immediately suggest that there may indeed be intermediates between prokaryote and eukaryote forms of cell organization, challenging the concept of a prokaryote-eukaryote dichotomy. The occurrence of membrane-bounded DNA-containing organelles in planctomycetes suggests that models for the origin of the eukaryote nucleus which depend on endosymbiotic engulfment of one ancient lineage cell by another (e.g., a member of the domain Archaea by a domain Bacteria member) (Lopez-Garcia and Moreira 1999; Martin 1999) may be unnecessary. Autogenous models for eukaryote nuclear origins are made more plausible, even if no homology between planctomycete organelles and eukaryote nuclei at the molecular level is found. Continuity of the nuclear body envelope of *G. obscuriglobus* suggests a mechanism for the origin of the nuclear body and its envelope by invagination of the planctomycete ICM. The invention of a pirellulosome and the accompanying structural cell plan may have been a conceivable fundamental step underlying the invention of the double-membrane bounded nucleus, perhaps in a protoeukaryote not necessarily related to contemporary planctomycetes.

The apparent occurrence of a folded single membrane topology of the superficially double membrane nuclear envelope of *G. obscuriglobus* suggests a strong analogy with the eukaryote nuclear envelope, which also has this topology, and is also more consistent with an endogenous origin for the eukaryote nuclear envelope (Fuerst 2005). It has been suggested that endosymbiotic engulfment models for eukaryote nuclear origins predict that the double membrane around the nucleus should be continuous like the cytoplasmic or outer membrane of the engulfed symbiont partner and that existing models for eukaryote nuclear origins do not account for the folded single membrane topology of eukaryote nuclear envelopes (Martin 1999). A mechanism for such origins perhaps involving membrane invagination accompanied by invention of nuclear pores and similar to that which may have given rise to the nuclear body envelope of *G. obscuriglobus* would possibly account for that topology more satisfactorily.

The basic compartmentalized cell plan shared by all planctomycetes so far examined is phylogenetically conserved, and the variations within the plan also plot along phylogenetic lines, e.g., genera related to *Pirellula* have pirellulosomes only and a polar cap region paryphoplasm while all strains isolated clustering in the *Gemmata* monophyletic group have double membrane-bounded nuclear bodies. It seems reasonable to hypothesize that the last common ancestor of the planctomycete phylum must have possessed the shared cell plan. Interestingly, members of a newly discovered phylum of the domain Bacteria, "*Poribacteria*", the cells of which may be compartmentalized in a similar manner to those of *Pirellula*-like species, may be phylogenetically related most closely to the planctomycetes (Fieseler et al. 2004). The possible deep-branching position of the phylum *Planctomycetes* in some phylogenetic trees (Brochier and Philippe 2002) suggests that cell compartmentalization involving membrane-bounded nucleoids and the cell plan exhibited by planctomycetes may have been an ancient feature of cells rather than a necessarily advanced and relatively recent one. Such an evolutionary phenomenon may be consistent with hypotheses suggesting that the Last Universal Common Ancestor of cellular life may have been relatively complex and eukaryote-like (Forterre and Philippe 1999; Glansdorff 2000), formulated from considerations such as comparison of eukaryote cell biology with the theoretical cell biology of the RNA World (Poole et al. 1998), and the concept that a "thermoreduction" stage led to the simplified cell biology of most prokaryotes (Forterre 1995). The occurrence of a relatively high percentage (8%) of genes homologous to those of eukaryotes in the proteome of the planctomycete *Rhodopirellula baltica* (formerly *Pirellula* sp. strain 1) (Glockner et al.

2003) is consistent either with an ancient relationship of planctomycetes with a protoeukaryote or with a lateral gene transfer event of uncertain polarity between eukaryotes and planctomycetes in evolutionary time. Analysis of emerging genome sequence data for *Gemmata* strains and the anammox planctomycetes may contribute to testing of such hypotheses.

Acknowledgements Research on planctomycetes in the author's laboratory is funded by the Australian Research Council. I would like to acknowledge the enthusiastic help of Richard I Webb, my students Margaret Butler, Margaret Lindsay, Cheryl Jenkins and Jenny Wang-Holmes who made important contributions to my lab's planctomycete research program, and Jenny Wang-Holmes for use of an electron micrograph in Fig. 8.

References

Bohrmann B, Villiger W, Johansen R, Kellenberger E (1991) Coralline shape of the bacterial nucleoid after cryofixation. J Bacteriol 173:3149-3158

Braun FN, Paulsen S, Sear RP, Warren PB (2005) Miscibility gap in the microbial fitness landscape. Physical Review Letters 94:178105

Brochier C, Philippe H (2002) Phylogeny: a non-hyperthermophilic ancestor for bacteria. Nature 417:244

Budin M, Jorgenson TL, Pearson A (2004) The evolution of sterol biosynthesis in bacteria: in situ fluorescence localization of sterols in the nucleoid bacterium *Gemmata obscuriglobus*. In: Eos Trans AGU, 85(47) Fall Meet Suppl, Abstract #B31D-05

Butler MK, Wang J, Webb RI, Fuerst JA (2002) Molecular and ultrastructural confirmation of classification of ATCC 35122 as a strain of *Pirellula staleyi*. Int J Syst Evol Microbiol 52:1663–1667

Di Giulio M (2003) The ancestor of the Bacteria domain was a hyperthermophile. J Theor Biol 224:277–283

Dubochet J, Mcdowall AW, Menge B, Schmid EN, Lickfeld KG (1983) Electron-microscopy of frozen-hydrated bacteria. J Bacteriol 155:381–390

Durrenberger M, Bjornsti MA, Uetz T, Hobot JA, Kellenberger E (1988) Intracellular location of the histonelike protein HU in *Escherichia coli*. J Bacteriol 170:4757–4768

Ebersold HR, Cordier JL, Luthy P (1981) Bacterial mesosomes: method dependent artifacts. Arch Microbiol 130:19–22

Fieseler L, Horn M, Wagner M, Hentschel U (2004) Discovery of the novel candidate phylum "*Poribacteria*" in marine sponges. Applied and Environmental Microbiology 70:3724–3732

Forterre P (1995) Thermoreduction, a hypothesis for the origin of prokaryotes. Comptes Rendus De L Academie Des Sciences Serie Iii-Sciences De La Vie – Life Sciences 318:415–422

Forterre P, Philippe H (1999) Where is the root or the universal tree of life? Bioessays 21:871–879

Fuerst JA (1995) The planctomycetes – emerging models for microbial ecology, evolution and cell biology. Microbiology 141:1493–1506

Fuerst JA (2005) Intracellular compartmentation in Planctomycetes. Annual Review of Microbiology 59:299–328

Fuerst JA, Gwilliam HG, Lindsay M, Lichanska A, Belcher C, Vickers JE, Hugenholtz P (1997) Isolation and molecular identification of planctomycete bacteria from postlarvae of the giant tiger prawn, *Penaeus monodon*. Appl Environ Microbiol 63:254–262

Fuerst JA, Webb RI (1991) Membrane-bounded nucleoid in the eubacterium *Gemmata obscuriglobus*. Proc Natl Acad Sci USA 88:8184–8188

Gade D, Schlesner H, Glockner FO, Amann R, Pfeiffer S, Thomm A (2004) Identification of planctomycetes with order-, genus-, and strain-specific 16S rRNA-targeted probes. Microbial Ecology 47:243–251

Gade D, Stuhrmann T, Reinhardt R, Rabus R (2005a) Growth phase dependent regulation of protein composition in *Rhodopirellula baltica*. Environ Microbiol 7:1074–1084

Gade D, Theiss D, Lange D, Mirgorodskaya E, Lombardot T, Glöckner FO, Kube M, Reinhardt R, Amann R, Lehrach H, Rabus R, Gobom J (2005b) Towards the proteome of the marine bacterium *Rhodopirellula baltica*: Mapping the soluble proteins. PROTEOMICS 9999:NA

Garrity GM, Bell JA, Lilburn TG (2004) Taxonomic Outline of the Prokaryotes. In: Bergey's Manual of Systematic Bacteriology, Second Edition Release 5.0 May 2004 edn. Springer, New York

Glansdorff N (2000) About the last common ancestor, the universal life-tree and lateral gene transfer: a reappraisal. Mol Microbiol 38:177–185

Glockner FO, Kube M, Bauer M, Teeling H, Lombardot T, Ludwig W, Gade D, Beck A, Borzym K, Heitmann K, Rabus R, Schlesner H, Amann R, Reinhardt R (2003) Complete genome sequence of the marine planctomycete *Pirellula* sp. strain 1. Proc Natl Acad Sci USA 100:8298–8303

Hobot JA, Bjornsti MA, Kellenberger E (1987) Use of on-section immunolabeling and cryosubstitution for studies of bacterial DNA distribution. J Bacteriol 169:2055–2062

Iborra FJ, Jackson DA, Cook PR (2001) Coupled transcription and translation within nuclei of mammalian cells. Science 293:1139–1142

Kellenberger E (1991a) Functional consequences of improved structural information on bacterial nucleoids. Res Microbiol 142:229–238

Kellenberger E (1991b) The potential of cryofixation and freeze substitution: observations and theoretical considerations. J Microsc 161(Pt2):183–203

Konig E, Schlesner H, Hirsch P (1984) Cell wall studies on budding bacteria of the *Planctomyces/Pasteuria* group and on a *Prosthecomicrobium* sp. Archives of Microbiology 138:200–205

Kuenen G, Jetten M (2001) Extraordinary anaerobic ammonium-oxidizing bacteria. ASM News 67:456–463

Lewis PJ (2004) Bacterial subcellular architecture: recent advances and future prospects. Molecular Microbiology 54:1135–1150

Lewis PJ, Thaker SD, Errington J (2000) Compartmentalization of transcription and translation in *Bacillus subtilis*. EMBO J 19:710–718

Liesack W, Konig H, Schlesner H, Hirsch P (1986) Chemical composition of the peptidoglycan-free cell envelopes of budding bacteria of the *Pirella/Planctomyces* group. Arch Microbiol 145:361–366

Lindsay MR, Webb RI, Fuerst JA (1997) Pirellulosomes: A new type of membrane-bounded cell compartment in planctomycete bacteria of the genus *Pirellula*. Microbiology-Uk 143:739–748

Lindsay MR, Webb RI, Strous M, Jetten MS, Butler MK, Forde RJ, Fuerst JA (2001) Cell compartmentalisation in planctomycetes: novel types of structural organisation for the bacterial cell. Arch Microbiol 175:413–429

Liu JR, McKenzie CA, Seviour EM, Webb RI, Blackall LL, Saint CP, Seviour RJ (2001) Phylogeny of the filamentous bacterium "*Nostocoida limicola*" III from activated sludge. Int J Syst Evol Microbiol 51:195–202

Lopez-Garcia P, Moreira D (1999) Metabolic symbiosis at the origin of eukaryotes. Trends Biochem Sci 24:88–93

Lynch A, Wang J (1993) Anchoring of DNA to the bacterial cytoplasmic membrane through cotranscriptional synthesis of polypeptides encoding membrane proteins or proteins for export: a mechanism of plasmid hypernegative supercoiling in mutants deficient in DNA topoisomerase I. J Bacteriol 175:1645–1655

Martin W (1999) A briefly argued case that mitochondria and plastids are descendants of endosymbionts, but that the nuclear compartment is not. Proceedings of the Royal Society of London Series B Biological Sciences 266:1387–1395

Neef A, Amann R, Schlesner H, Schleifer K (1998) Monitoring a widespread bacterial group: in situ detection of planctomycetes with 16S rRNA-targeted probes. Microbiology 144:3257–3266

Pearson A, Budin M, Brocks JJ (2003) Phylogenetic and biochemical evidence for sterol synthesis in the bacterium *Gemmata obscuriglobus*. Proc Natl Acad Sci USA 100:15352–15357

Pennisi E (2004) EVOLUTIONARY BIOLOGY: The Birth of the Nucleus. Science 305:766–768

Poole A, Penny D (2001) Does endosymbiosis explain the origin of the nucleus? Nat Cell Biol 3:E173-E173

Poole AM, Jeffares DC, Penny D (1998) The path from the RNA world. J Mol Evol 46:1–17

Rotte C, Martin W (2001) Does endosymbiosis explain the origin of the nucleus? Nat Cell Biol 3:E173-E173

Ryter A, Chang A (1975) Localization of transcribing genes in the bacterial cell by means of high resolution autoradiography. J Mol Biol 98:797

Schlesner H (1994) The development of media suitable for the microorganisms morphologically resembling *Planctomyces* spp., *Pirellula* spp., and other *Planctomycetales* from various aquatic habitats using dilute media. Syst Appl Microbiol 17:135–145

Schlesner H, Rensmann C, Tindall BJ, Gade D, Rabus R, Pfeiffer S, Hirsch P (2004) Taxonomic heterogeneity within the *Planctomycetales* as derived by DNA-DNA hybridization, description of *Rhodopirellula baltica* gen. nov., sp nov., transfer of *Pirellula marina* to the genus *Blastopirellula* gen. nov. as *Blastopirellula marina* comb. nov. and emended description of the genus *Pirellula*. Int J Syst Evol Microbiol 54:1567–1580

Schlesner H, Stackebrandt E (1986) Assignment of the genera *Planctomyces* and *Pirella* to a new family *Planctomycetaceae* fam.nov. and description of the order *Planctomycetales* ord.nov. Syst Appl Microbiol 8:174–176

Schmid M, Twachtmann U, Klein M, Strous M, Juretschko S, Jetten M, Metzger JW, Schleifer KH, Wagner M (2000) Molecular evidence for genus level diversity of bacteria capable of catalyzing anaerobic ammonium oxidation. Syst Appl Microbiol 23:93–106

Schmid M, Walsh K, Webb R, Rijpstra WIC, Van De Pas-Schoonen K, Verbruggen M, Hill T, Moffett B, Fuerst J, Schouten S, Sinninghe Damsté JS, Harris J, Shaw P, Jetten M, Strous M (2003) *Candidatus* "Scalindua brodae", sp. nov., *Candidatus* "Scalindua wagnei", sp. nov., two new species of anaerobic ammonium oxidizing bacteria. Syst Appl Microbiol 26:529–538

Sinninghe Damsté JS, Rijpstra WIC, Schouten S, Fuerst JA, Jetten MSM, Strous M (2004) The occurrence of hopanoids in planctomycetes: implications for the sedimentary biomarker record. Org Geochem 35:561–566

Skerker JM, Laub MT (2004) Cell-cycle progression and the generation of asymmetry in *Caulobacter crescentus*. Nat Rev Microbiol 2:325–337

Stackebrandt E, Ludwig W, Schubert W, Klink F, Schlesner H, Roggentin T, Hirsch P (1984) Molecular genetic evidence for early evolutionary origin of budding peptidoglycan-less eubacteria. Nature 307:735–737

Stackebrandt E, Wehmeyer U, Liesack W (1986) 16S ribosomal RNA- and cell wall analysis of *Gemmata obscuriglobus*, a new member of the Order Planctomycetales. Fems Microbiol Lett 37:289–292

Staley JT, Bouzek H, Jenkins C (2005) Eukaryotic signature proteins of *Prosthecobacter dejongeii* and *Gemmata* sp. Wa-1 as revealed by in silico analysis. FEMS Microbiol Lett 243:9–14

Staley JT, Fuerst JA, Giovannoni S, Schlesner H (1992) The order *Planctomycetales* and the genera *Planctomyces*, *Pirellula*, *Gemmata* and *Isosphaera*. In: Balows A, Truper H, Dworkin M, Harder W, Schleifer K (eds) The Prokaryotes: a Handbook on the Biology of Bacteria: Ecophysiology, Isolation, Identification, Applications. Springer-Verlag, New York, pp 3710–3731

Starr MP, Schmidt JM (1989) Genus *Planctomyces Gimesi* 1924. In: Staley JT, Bryant MP, Pfennig N, Holt JG (eds) Bergey's Manual of Systematic Bacteriology. Williams and Wilkins, pp 1946–1958

Strous M, Fuerst JA, Kramer EH, Logemann S, Muyzer G, van de Pas-Schoonen KT, Webb R, Kuenen JG, Jetten MS (1999) Missing lithotroph identified as new planctomycete. Nature 400:446–449

Tekniepe BL, Schmidt JM, Starr MP (1981) Life-cycle of a budding and appendaged bacterium belonging to morphotype-IV of the *Blastocaulis-Planctomyces* Group. Curr Microbiol 5:1–6

Van De Peer Y, Neefs J-M, De Rijk P, De Vos P, De Wachter R (1994) About the order of divergence of the major bacterial taxa during evolution. Syst Appl Microbiol 17:32–38

Wang-Holmes J (2004) Molecular systematics and biodiversity of planctomycetes. In: Department of Microbiology and Parasitology. University of Queensland, Brisbane, p 219

Wang J, Jenkins C, Webb RI, Fuerst JA (2002) Isolation of *Gemmata*-like and *Isosphaera*-like planctomycete bacteria from soil and freshwater. Appl Environ Microbiol 68:417–422

Microbiol Monogr (2)
J.M. Shively: Complex Intracellular Structures in Prokaryotes
DOI 10.1007/7171_028/Published online: 25 May 2006

Anammoxosomes of Anaerobic Ammonium-oxidizing Planctomycetes

John A. Fuerst[1] (✉) · Richard I. Webb[1,2] · Laura van Niftrik[3] · Mike S. M. Jetten[3] · Marc Strous[3]

[1]School of Molecular and Microbial Sciences, University of Queensland, 4072 Brisbane, Queensland, Australia
j.fuerst@mailbox.uq.edu.au

[2]Centre for Microscopy and Microanalysis, University of Queensland, 4072 Brisbane, Queensland, Australia

[3]Department of Microbiology, Faculty of Science, Radboud University Nijmegen, Toernooiveld 1, 6525 ED Nijmegen, The Netherlands

Abstract Anammoxosomes are unique metabolically significant compartments of planctomycetes performing the **anammox** process. These bacteria carry out **Anaerobic Ammonium Oxidation**, a chemolithotrophic and autotrophic metabolism. They comprise *Candidatus* genera "Brocadia", "Kuenenia" and "Scalindua", mostly from wastewater treatment bioreactors or marine anaerobic habitats and none of which are yet in pure culture. Like cells of other planctomycetes, anammox species possess a shared planctomycete cell plan involving a single-membrane-bounded pirellulosome compartment containing a nucleoid as well as paryphoplasm surrounding the outer rim of the cell. Within the pirellulosome they possess another compartment, the anammoxosome, unique to anammox planctomycetes. The anammoxosome harbours an enzyme, hydrazine oxidoreductase, important for models of anammox. The anammoxosome is wrapped in an envelope possessing cyclobutane-containing ladderane lipids which may confer impermeability to anammoxosome membranes. The ladderanes occur in both ether-linked and ester-linked forms. This envelope may have specialized functions, for example, protection from toxic intermediates of anammox metabolism and proton gradient formation for bioenergetics. The anammoxosome is important for models of anaerobic ammonium oxidation. It has other unusual features related to cell biology, since within the anammoxosome tubules occur which may form distinct arrangements suggesting a cytoskeletal function, and the nucleoid is often in contact with the anammoxosome envelope.

1
Introduction

The planctomycetes form a distinctive phylum within the domain Bacteria, and possibly a deep-branching and thus ancient one (see Fuerst 2006, in this volume for an introduction to the group). All cultivated members of the planctomycetes share a unique type of cell organization in which the intracellular cytoplasm is divided into a number of compartments by internal membranes, in some cases such as that of *Gemmata obscuriglobus*, giving rise to what appears to be a double membrane-bounded nuclear structure analogous to the eukaryote nucleus (Fuerst 2005). The anammox planctomycetes are a divergent and deeply-branching group within the planctomycete phylum, discovered initially in wastewater-processing bioreactors and with the remarkable physiology of anaerobic ammonium oxidation, one unique among known Bacteria and Archaea. With their discovery it was thus clear that the diversity of the planctomycetes must encompass chemoautotrophs as well as the chemoheterotrophs known from isolates in pure culture. Their cell organization is also unique even among the unusual planctomycetes, with the occurrence of an anammoxosome organelle found only in anammox planctomycetes.

2
Anammox Planctomycetes and Their Physiology

The first anammox planctomycete, a new deep-branching member of the phylum *Planctomycetes*, "*Candidatus* Brocadia anammoxidans", was discovered to be the dominant member of the microbial community of an anaerobic wastewater treatment bioreactor performing a novel anaerobic ammonium oxidation (the Anammox process) (Strous et al. 1999a). A second genus and species, "*Candidatus* Kuenenia stuttgartiensis" (Schmid et al. 2000) has been discovered, and a further probable distinct species of this genus, the Kolliken anammox planctomycete, has been claimed (Schmid et al. 2000; Egli et al. 2001). A third genus, "*Candidatus* Scalindua" has been described which would appear to have a marine representative "*Candidatus* Scalindua sorokinii" from the anoxic Black Sea habitat and two other species "*Candidatus* Scalindua brodae" and "*Candidatus* Scalindua wagneri" from a wastewater plant treating landfill leachate (Jetten et al. 2003; Schmid et al. 2003). The known metabolic and phylogenetic diversity of the group will undoubtedly result in further genus descriptions. Anammox morphotypes have not yet been obtained in pure culture, and appear to be bacteria with some of the slowest generation times in the laboratory [up to 3 weeks (Strous et al. 1998) and with even the shortest recorded under optimum conditions as long as 9–11 days (Strous et al. 1999b; Jetten et al. 2005)]—enrichment cultures normally mature only in 100–200 days. Our knowledge of their diversity, physiology, applied significance in waste remediation, and global ecological significance has been reviewed extensively (Jetten et al. 2001, 2003; Sinninghe Damsté et al. 2002; Op den Camp et al. 2006). The anammox planctomycetes may form only a subgroup of an even larger monophyletic group of organisms—group VI planctomyetes in the scheme of Chouari et al.—branching deeply and divergently within the phylum *Planctomycetes* (Chouari et al. 2003) to the extent that some arguments have been advanced for a separate phylum from the planctomycetes for this clade.

The anammox reaction mediated by anammox planctomycetes is most simply defined as the oxidation of ammonium under anaerobic (anoxic) conditions using nitrite as the electron acceptor and with dinitrogen gas as the final product. The process can be summarized in the equation: $NH_4^+ + NO_2^- \rightarrow N_2 + 2H_2O$ [$\Delta G^\circ = -358$ kJ (mol of $NH_4^+)^{-1}$]. The toxic compound hydrazine is an important intermediate in this process, and addition of toxic and reactive hydrazine (a component of some rocket fuels!) can activate purified cells to perform the process. In one early model for the reaction, hydroxylamine generated from nitrite reduction oxidizes ammonium to yield hydrazine which is then oxidized to nitrogen, generating electrons which can be used for the reduction of the nitrite (vandeGraaf et al. 1997; Schalk et al. 2000). The role of hydroxylamine may not now be as clear. In a later model, nitric oxide, also proposed to be generated from nitrite reduction, may take the role hydroxylamine played in earlier models (Strous and Jetten 2004). Later models try to also account for generation of ATP and will be described below. An ^{15}N isotope labelling technique can be used to directly demonstrate anammox, by incorporation of ammonia labelled with ^{15}N and unlabelled nitrite with ^{14}N (vandeGraaf et al. 1997). Anammox results in dinitrogen labelled as $^{14}N\ ^{15}N$. Such experiments have been combined with other evidence to demonstrate that the anammox process is of potentially major global significance for the nitrogen cycle, for example as the only sink for fixed nitrogen in the water column of the Benguela upwelling in the southern Atlantic Ocean, one of the most primary productive bodies of water on Earth (Kuypers et al. 2005).

Anammox planctomycetes are true autotrophs and chemolithoautotrophs, since they have been shown to fix CO_2 when growing anaerobically on ammonium (Strous and Jetten 2004). Some nitrite is oxidized anaerobically to nitrate and this may supply electrons for CO_2 fixation by reverse electron transport. Thus anammox planctomycetes only need to be supplied with ammonium, nitrite and carbon dioxide as the major substrates for their metabolism and growth. Because of the dependence on nitrite they may be found in association with aerobic ammonium-oxidizing nitrifiers that produce nitrite from ammonium and consume inhibitory O_2 (Sliekers et al. 2002), or can derive their nitrite from that produced by anaerobic nitrate reducers and denitrifiers, which gave rise originally to a view that nitrate was the electron acceptor for anammox (Mulder et al. 1995). Alternatively, anammox bacteria can also reduce nitrate to nitrite themselves with organic compounds as the electron donor (Guven et al. 2005). Completely autotrophic systems for ammonium removal can be developed involving anammox planctomycetes and aerobic nitrifiers with no need for added carbon other than CO_2 (Sliekers et al. 2002).

Cells of these anammox planctomycetes possess a unique membrane-bounded compartment, the anammoxosome, specific to anammox planctomycetes (Strous et al. 1999a; Lindsay et al. 2001), and a unique membrane lipid, a concatenated cyclobutane chain lipid termed a "ladderane", appears to be unique to anammox planctomycetes and present in the anammoxosome envelope of the "Dokhaven" strain of "*Candidatus* Brocadia anammoxidans" (Sinninghe Damsté et al. 2002); such ladderanes have also been detected in cell biomass from waste treatment plants containing "*Candidatus* Scalindua brodae" and "*Candidatus* Scalindua wagneri" (Schmid et al. 2003) and in marine waters and sediments with anammox activity (Kuypers et al. 2003). Anammoxosomes and the unique anammox lipids which appear to be associated with them will be described in detail below.

3 Anammox Planctomycete Cell Plans

One of the major characteristic features of planctomycete cells is their internal division into compartments separated by membranes (see Fuerst 2006, in this volume). This gives rise to a cell organizational or structural plan defined by such compartmentalization. Planctomycetes share a common cell plan in major respects, for example in the universal presence of pirellulosome and paryphoplasm compartments of the cytoplasm within the cytoplasmic membrane, but individual genera or groups of genera exhibit significant variations within this shared plan. Anammox planctomycetes display one of the most significant of these variations, only matched in extent and potential functional significance by that of the nucleated *Gemmata obscuriglobus* and related isolates.

3.1 Features of the Anammox Cell Plan Shared with and Differing from Other Planctomycetes

Anammox planctomycetes conform to the shared cell organizational plan common to all planctomycetes examined by cryosubstitution followed by thin-sectioning —that is, the cytoplasm is divided into a pirellulosome compartment bounded by a single intracytoplasmic membrane, and the paryphoplasm compartment bounded at its outer edge by the cytoplasmic membrane and at its inner edge by the intracytoplasmic membrane (Lindsay et al. 2001) (see Fuerst 2006, in this volume). This plan, illustrated diagrammatically in Fig. 1, is clearly seen in the electron micrograph of a cryosubstituted cell of "*Candidatus* Brocadia anammoxidans" in Fig. 2a. The fact that the anammox planctomycetes share features of the cell plan common to all other planctomycetes examined even though they are deep-branching and diverge from other planctomycetes phylogenetically is consistent with the view that that cell plan might have been shared by the last common ancestor of the phylum *Planctomycetes*. Together with molecular sequence data, the characteristic planctomycete ultrastructure of anammox planctomycetes was central to the identification of the bacterium performing anammox metabolism as a member of the planctomycete phylum, and also had provided an initial incentive for continuing to pursue the quest for the bacterium responsible for anammox, since this morphotype was found to dominate the multispecies biofilms performing anammox.

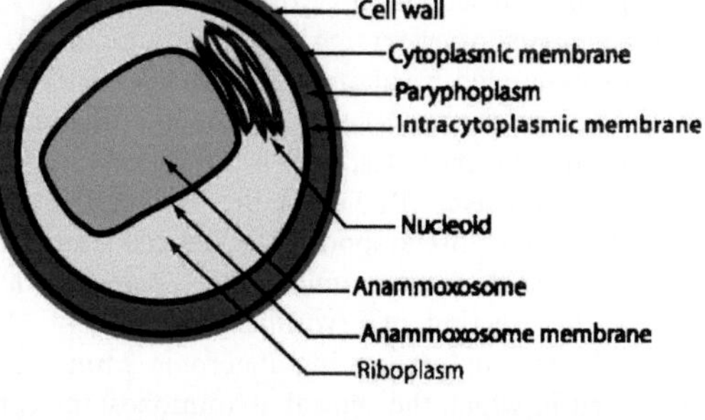

Fig. 1 Diagram of cell plan of anammox planctomycete, showing features shared by all planctomycetes (e.g. paryphoplasm, intracytoplasmic membrane and riboplasm) and features unique to anammox planctomycetes (anammoxosome, anammoxosome membrane and adherence of nucleoid to anammoxosome membrane)

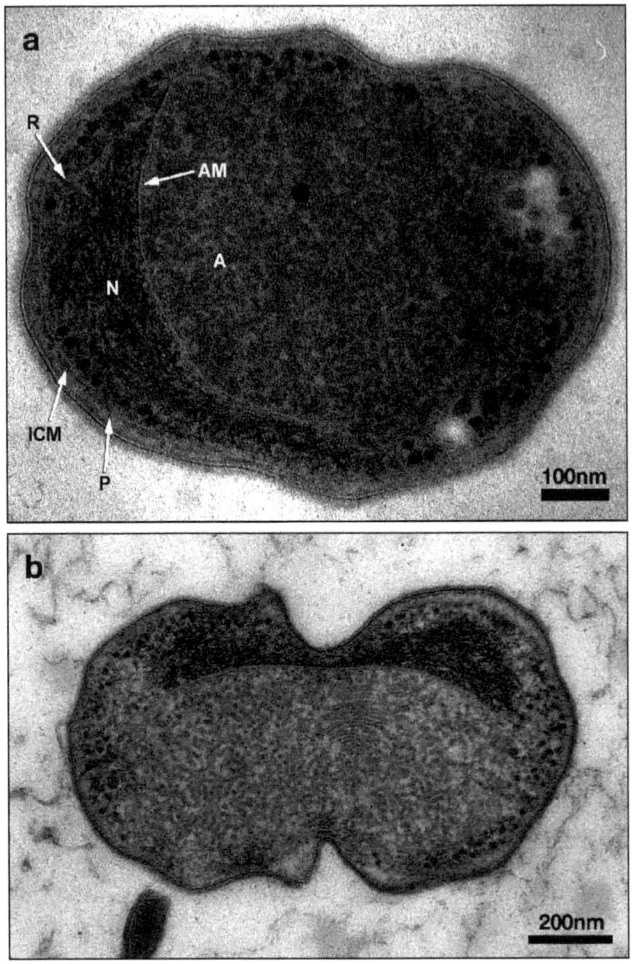

Fig. 2 **a** (reproduced with permission of Annual Review of Microbiology) Transmission electron micrograph of thin section of cryosubstituted cell of "*Candidatus* Brocadia anammoxidans" from a bioreactor performing anaerobic ammonium oxidation. A central anammoxosome (A) is surrounded by a single membrane (AM). The cytoplasm external to the AM contains the nucleoid (N) adhering to the anammoxosome membrane, and also contains ribosome-like particles, and is surrounded by the intracytoplasmic membrane (ICM). A paryphoplasm (P) lies at the rim of the cell between the ICM and the cytoplasmic membrane closely apposed to the cell wall. Transversely sectioned tubule structures are visible inside the anammoxosome. *Scale bar* 100 nm. **b** Transmission electron micrograph of thin section of cryosubstituted cell of "*Candidatus* Brocadia anammoxidans" from a bioreactor performing anaerobic ammonium oxidation. This is an apparently dividing cell in which the central anammoxosome contains tubules, some of which are seen in longitudinal section oriented perpendicular to the plane of division; a nucleoid is seen in each half of the dividing cell accompanying the corresponding anammoxosome half section. Courtesy of J.A. Fuerst, R.I. Webb and M. Strous. *Scale bar* 200 nm

However, anammox planctomycetes display a significant variation of the shared planctomycete plan —there are a total of three cell compartments each separated from the other by a single bilayer membrane (Fig. 1)— (1) the outermost paryphoplasm, separated by an intracytoplasmic membrane (ICM) from (2) a pirellulosome consisting of riboplasm cytoplasm with electron-dense ribosome-like particles and fibrillar, condensed nucleoid, and (3) an innermost and ribosome-free compartment, the anammoxosome, bounded by a single membrane (see Fig. 2a and Figs. 3–6) and only found in anammox planctomycetes (Lindsay et al. 2001). The anammoxosome organelle is thus separated by a membrane from the rest of the cell, and is surrounded by riboplasm. The anammoxosome membrane is not identical in composition to the other cell membranes (see below). In anammox planctomycetes, the pirellulosome is thus similar to those found in other planctomycete species in that it contains electron-dense ribosome-like particles and the cell's nucleoid DNA. The anammoxosome is an example of an additional membrane-bounded compartment occurring within the pirellulosome compartment common to all planctomycetes, the other example being the double membrane-bounded nuclear body of *Gemmata obscuriglobus.*

In cryosubstituted preparations of "*Candidatus* Brocadia anammoxidans" at least, the nucleoid characteristically appears bound to the anammoxosome membrane and is also often crescent-shaped (Fig. 2a). This is also the case in "*Candidatus* Scalindua brodae" and "*Candidatus* Scalindua wagneri" (Schmid et al. 2003). As in cells of other cryosubstituted plancto-

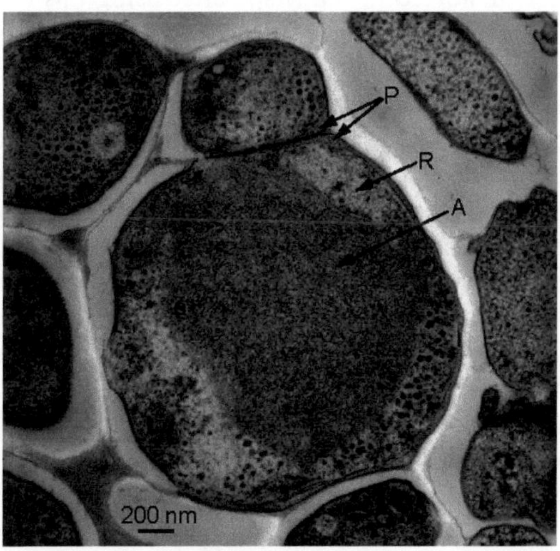

Fig. 3 Transmission electron micrograph of thin section of cryosubstituted cell of "*Candidatus* Kuenenia stuttgartiensis" from a bioreactor performing anaerobic ammonium oxidation. The cell contains a large anammoxosome (A) within the riboplasm (R) of the pirellulosome which is separated from the thin paryphoplasm (P). *Scale bar* 200 nm

mycete species, in anammox planctomycetes the nucleoid appears condensed, an unexpected appearance for nucleoids of non-planctomycete bacteria. In "*Candidatus* Brocadia anammoxidans" at least, simultaneous immunogold labelling for both DNA and RNA (using anti-DNA antibody or RNase-gold, respectively) shows that most of the cell DNA appears associated with the nucleoid (Lindsay et al. 2001). However, from labelling, there is some indication of a very small amount of DNA in the anammoxosome itself, even though there are no nucleoid fibrils apparent in the anammoxosome. This small amount of anammoxosomal DNA may be consistent with a potential role for the anammoxosome in chromosome segregation during cell division suggested by the close binding of the nucleoid outside the anammoxosome to the anammoxosome membrane, and with a potential role of cytoskeletal tubules within the anammoxosome in chromosome segregation or cell division.

3.2
Anammoxosomes

As explained above, the anammoxosome is an organelle of anammox planctomycetes bound by a single membrane (van Niftrik et al. 2004). As seen in cells prepared for electron microscopy via the cryofixation method of cryosubstitution, it usually occupies a central position in the anammox planctomycete cell, and may occupy more than 30% of the cell volume. Chemical fixation may also preserve the anammoxosome, but cells of anammox planctomycetes are usually highly distorted in cell shape after using this method. The occupation of the central region of the cell by the anammoxosome gives rise to the characteristic ring staining pattern when fluorescent 16S rRNA oligonucleotide probes are applied to anammox planctomycetes. The ring staining has been proposed to be due to the staining only of the rRNA of the riboplasm surrounding the central anammoxosome which is devoid of any RNA or ribosome-like particles. Such ring staining has also been observed when FISH probes are applied to other compartmentalized planctomycetes (Neef et al. 1998), and thus may not be exclusive to anammox planctomycetes and may be a consequence of internal compartments as such rather than anammoxosomes. It may not even be exclusive to planctomycetes, since a central nucleoid might give rise to a similar appearance in non-compartmentalized non-planctomycete bacteria.

The anammoxosome is completely distinct from the central intrusion of paryphoplasm seen in the non-anammox planctomycete *Isosphaera pallida* (Lindsay et al. 2001); it is completely separate from the paryphoplasm and published evidence does not suggest continuity between the anammoxosome membrane and any other cell membrane, including the intracytoplasmic membrane (invagination of which forms the central paryphoplasm in *Isosphaera*). This is clearly apparent in Figs. 2a, 3, and 5. Anammoxosomes

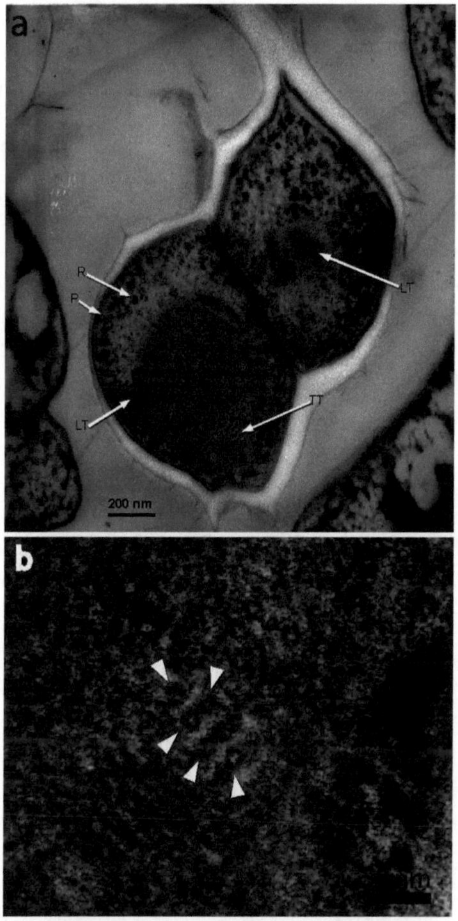

Fig. 4 **a** Transmission electron micrograph of thin section of cryosubstituted cell of "*Candidatus* Kuenenia stuttgartiensis" from a bioreactor performing anaerobic ammonium oxidation. A dividing cell is seen; the mother cell contains a large anammoxosome (A) in which both longitudinally sectioned tubules (LT) and transversely sectioned tubules (TT) are seen, while the daughter cell contains only a rudimentary anammoxosome with longitudinally sectioned tubules apparent (LT). P, paryphoplasm; R, riboplasm. *Scale bar* 200 nm. **b** An enlarged view of a portion of the cell in 4a where the tubular nature of the transversely sectioned tubules indicated by *arrowheads* can be clearly discerned as a central electron-transparent region surrounded by an electron dense ring. *Scale bar* 50 nm

can be seen in both members of pairs of dividing cells in at least a mature stage of cell division (Figs. 2b and 6).

Most significantly from the perspective of understanding the mechanism of anammox metabolism, a key enzyme in this metabolism, hydrazine oxidoreductase (originally termed hydroxylamine oxidoreductase or HAO due to the only activity for it initially demonstrated) has been shown via immunogold labelling of "*Candidatus* Brocadia anammoxidans" using an anti-HAO antibody to be localized only within the anammoxosome—no labelling appears outside this organelle (Lindsay et al. 2001). This result has formed the foundation of the extant models for the biochemistry of anammox metabolism, which will be dealt with below. However, the HAO appears to be inside and throughout the anammoxosome, rather than at or near the anammoxosome envelope.

Anammoxosomes have been found in all the cultured genera of planctomycetes, including the species "*Candidatus* Brocadia anammoxidans", "*Candidatus* Kuenenia stuttgartiensis", "*Candidatus* Scalindua wagneri" and "*Candidatus* Scalindua brodae". It thus appears to be a phylogenetically

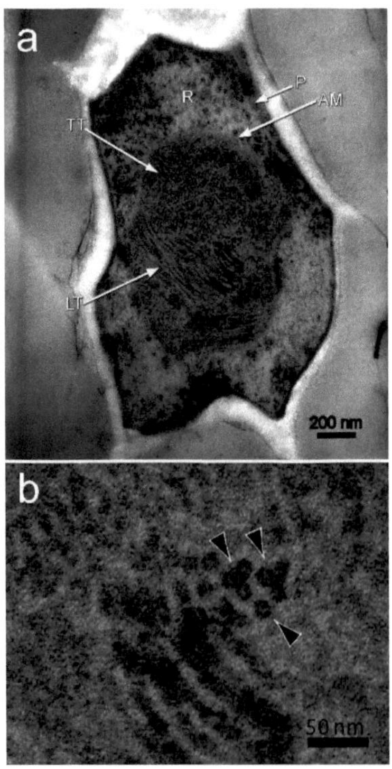

Fig. 5 **a** Transmission electron micrograph of thin section of cryosubstituted cell of "*Candidatus* Kuenenia stuttgartiensis" from a bioreactor performing anaerobic ammonium oxidation. A large anammoxosome is seen with a clear anammoxosome membrane (AM); both longitudinally sectioned tubules (LT) and transversely sectioned tubules (TT) can be seen. P, paryphoplasm; R, riboplasm. *Scale bar* 200 nm. **b** An enlarged view of a portion of the cell in 5a where the tubular nature of the transversely sectioned tubules indicated by *arrowheads* can be clearly discerned as a central electron-transparent region surrounded by an electron dense ring in each tubule. *Scale bar* 50 nm

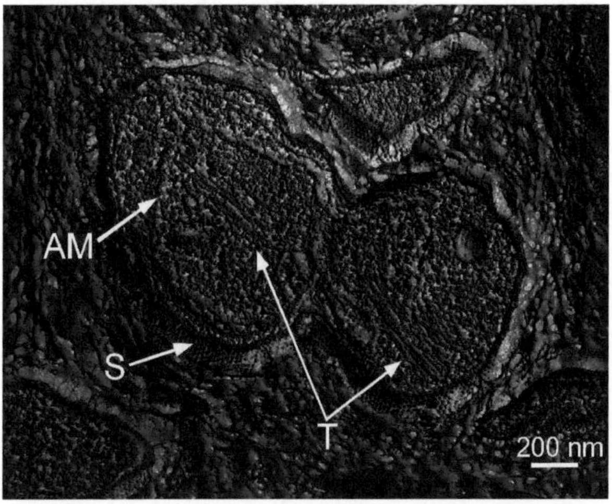

Fig. 6 Transmission electron micrograph of a freeze-fracture replica of "*Candidatus* Kuenenia stuttgartiensis" from a bioreactor performing anaerobic ammonium oxidation. Longitudinally arrayed tubules (T) are seen in both halves of an apparently dividing cell, one of which displays an anammoxosome delineated by a cross-fractured membrane (AM). A surface lattice (S) of regularly arranged subunits covers the surface of the cell. *Scale bar* 200 nm

conserved ultrastructural character, which was presumably present in the common ancestor of the monophyletic anammox planctomycetes.

Ultrastructurally, the anammoxosome appears to be completely filled with small tubules closely packed within the anammoxosome space (Fig. 1a,b). In "*Candidatus* Brocadia anammoxidans", the clearest tubules are 13 nm wide in their outer diameter. These cytoskeleton-like tubules are sometimes arranged in organized patterns (Lindsay et al. 2001) suggesting their assembly for some function in cell organization (Fig. 2b). Their tubular nature is revealed not only in cross section (Figs. 4b and 5b) but also in longitudinal section and by the variation in appearance with the plane of section. They appear to have a clearly tubular structure with an electron-transparent core surrounded by an electron dense ring (Figs. 4b and 5b). This is most clear in freeze-fractured and in sectioned cryosubstituted cells of the anammox species "*Candidatus* Kuenenia stuttgartiensis" (Fig. 3a,b). The longitudinally arranged tubules especially can appear in an organized arrangement (Figs. 5 and 6).

4
Ladderane and Hopanoid Lipids of Anammox Planctomycetes

The correlation of the anammoxosome organelle with the ability of planctomycetes possessing them to oxidize ammonium anaerobically, a unique metabolism unknown in other bacteria, combined with the presence of an enzyme needed for hydrazine oxidation located inside the anammoxosome, is consistent with the hypothesis that the membrane surrounding the anammoxosome may possess special properties, and perhaps even a specific composition. Anammox planctomycetes have been found to contain two classes of candidate membrane lipids, ladderanes and hopanoids. The ladderanes appear to occur only in the anammox planctomycetes, and both ladderane and hopanoid lipids may be relevant to properties of internal anammox membranes.

4.1 Ladderanes

The unique status of anammox planctomycetes and anammoxosomes is emphasized by their possession of unique ladderane lipids with four-membered aliphatic cyclobutane rings with ether linkages to the glycerol backbone in some cases, and which confer much greater density to the anammoxosome membrane than ordinary membranes (Sinninghe Damsté et al. 2002).

These were first identified in a 99.5% pure suspension of the Dokhaven strain of "*Candidatus* Brocadia anammoxidans" purified from modified anammox sequencing batch reactors, and also, in lower amounts, in biomass dominated by "*Candidatus* Kuenenia stuttgartiensis". These first ladderane lipids occurred in two major forms, ether-linked or ester linked to a glycerol. The first type (with an "X" ring system) was an *sn*-2-glycerol monoether bonded to a C_{20} alkyl chain composed of a ring system of three condensed cyclobutane rings and one cyclohexane ring substituted with an octyl chain, while the second (with a "Y" ring system) was an ester of a C_{20} fatty acid containing five cyclobutane rings substituted with a heptyl chain [though at first thought to be a methyl ester, it was later determined that this methylation was an artefact induced by the analytical protocol—see Sinninghe Damsté et al. (2005)]. The cyclobutane rings are fused by *cis*-ring junctions in a staircase-like arrangement termed a ladderane (Fig. 7). Within lipids containing these X and Y ladderanes however, there was even greater diversity. The ladderanes occurred not only as fatty acid methyl esters and *sn*-2-alkyl glycerol monoethers but also as alcohols, *sn*-1,2-dialkyl glycerol diethers, and as an *sn*-2-O-alkyl,*sn*-1-acyl glycerol. Both X and Y rings can be present in one ladderane, giving a remarkable mixed glycerol ether/ester type of lipid (Fig. 7).

A novel mixed ladderane/*n*-alkyl glycerol diether lipid combining the ladderane moiety with the more common bacterial membrane trait of *n*-alkyl chains has been identified in association with biomass containing "*Candidatus* Scalindua" species of anammox planctomycetes growing in a UK wastewater treatment plant (Sinninghe Damsté et al. 2004).

The initial work showing ladderane lipids in the Dokhaven strain of "*Candidatus* Brocadia anammoxidans" has largely been confirmed for a 99.5%

I

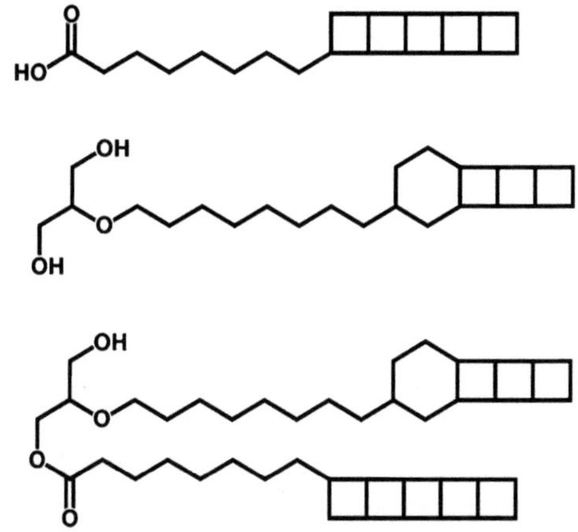

II

III

Fig. 7 Structures of characteristic ladderane lipids of anammox planctomycetes. I Ladderane fatty-acid-containing ring system Y with ester link; II ladderane monoalkyl glycerol ether-containing ring system X; and III a ladderane glycerol ether/ester containing both ring systems X and Y. Modified from van Niftrik et al. (2004); reproduced with permission from Annual Reviews of Microbiology

pure suspension of the Delft strain of "*Candidatus* Brocadia anammoxidans" (Sinninghe Damsté et al. 2005). As in bioreactors where "*Candidatus* Scalindua" was present, the mixed type of glycerol diether with a ladderane moiety and an alkyl group was also found for this species at its higher cell concentration in the purified preparation.

In addition to ladderane lipids, anammox planctomycetes cells also contain more conventional saturated and unsaturated straight chain and branched fatty acids such as 14-methylpentadecanoic acid, 10-methylhexadecanoic acid and 9,14-dimethylpentadecanoic acid in the Delft strain of "*Candidatus* Brocadia anammoxidans" (Sinninghe Damsté et al. 2005) and the branched fatty acids i-C_{16}, 9-methyl hexadecanoic acid and 9,14-dimethyl pentadecanoic acid in the Dokhaven strain of "*Candidatus* Brocadia anammoxidans". At least some of these types are not exclusive to anammox planctomycetes unlike the unique ladderane lipids.

Ladderane lipids appear to be enriched in the anammoxosome organelle relative to the rest of the cell, at least in the case of the Dokhaven strain of "*Candidatus* Brocadia anammoxidans". When anammoxosomes were isolated from the lysed anammox planctomycete (Sinninghe Damsté et al. 2002), the enrichment of anammoxosomes in the purified fraction was matched by an enrichment of ladderane lipid content (53% of total lipids of the anammoxosome are ladderanes relative to 34% ladderanes as a fraction of the total lipids of the intact cell). However, the branched fatty acids i-C_{16}, 9-methyl hexadecanoic acid and 9,14-dimethyl pentadecanoic acid were absent from the anammoxosome membranes, suggesting these are components of other membranes of the anammox planctomycete cell, either the intracytoplasmic membrane or the cytoplasmic membrane. The high ladderane component of the total lipids of "*Candidatus* Brocadia anammoxidans" (Sinninghe Damsté et al. 2002) suggests they are a substantial investment by the cell biosynthetic machinery and likely to be of functional significance.

4.2 Ladderanes and Anammox Physiology

There are several features of ladderane lipids that suggest that they may be essential to the functioning of the anammoxosome in the anammox process and to the ability of the cell to perform such a process using its characteristic and bizarre biochemistry. Molecular modelling assuming the presence of only one type of ladderane lipid suggests that these ladderane lipids confer unusually high density on the anammoxosome membrane (1.5 kg/l in the ladderane part compared to up to only 1.0 kg/l for a conventional membrane). This is consistent with an apparent impermeability of the anammoxosome to fluorophores able to penetrate the more peripheral cytoplasmic membrane (and which presumably penetrate the intracytoplasmic membrane around the riboplasm as well) (Sinninghe Damsté et al. 2002). The density due to the rigidity of ladderane lipids, which must have a lower degree of rotational freedom than ordinary lipids, may result in a limited ability of the anammoxosome membrane to allow diffusion across the membrane of protons and/or toxic and mutagenic intermediates of the anammox reaction such as hydrazine.

Limitation of proton diffusion may be needed since because of the slow rate of anammox catabolism and the translocation of protons across the membrane as a consequence of such catabolism, prevention of leakage of protons across the membrane by diffusion may be of critical importance to the success of energy generation by ammonium oxidation in these bacteria. According to the major (but still unconfirmed) model of anammox metabolism such energy generation depends on generation and maintenance of a proton motive force across the anammoxosome membrane (see below). The anammox intermediate hydrazine is expected to be able to diffuse readily through classical lipid membranes. However, the ladderane-rich and dense anammoxosome membrane may prevent such intermediates, which include mutagenic DNA-damaging hydrazine generated by the anammox process, from contacting the genomic DNA. The possible contributions of the impermeability conferred by ladderane lipids will be dealt with below after consideration of the major model for anammox metabolism dependent on the anammoxosome and its membrane. The cell effectively divides membrane tasks over two or perhaps even three different types of membrane, which releases the anammoxosome membrane to be optimized for prevention of proton leakage and toxic intermediate leakage. Specialization of the anammoxosome membrane leaves the cytoplasmic membrane to be optimized for the permeability and flexibility properties needed for transport and intracellular ion control, and the intracytoplasmic membrane to be optimized for internal protein transport or secretion to the paryphoplasm and transport of ions and nutrients from the paryphoplasm to riboplasm. However, since at least one key enzyme of anammox metabolism, hydrazine oxidoreductase, has been found inside the anammoxosome, there must either be a mechanism for transport of that enzyme into the anammoxosome from the ribosome-rich riboplasm outside it where protein synthesis must occur or the anammoxosome must form around the enzymes after such proteins are sequestered into an anammoxosome compartment region—there is no evidence of any stage in which the anammoxosome membrane is disassembled however, so a transport mechanism seems more likely. Synthesis of the protein within the anammoxosome is very improbable since this is a ribosome-free space.

4.3 Ladderanes and Planctomycete Evolution

Ladderane lipids appear to be confined to anammox planctomycetes, and at least *Blastopirellula marina* and *Gemmata obscuriglobus* among the other planctomycetes do not synthesize them (Sinninghe Damsté et al. 2002). They are thus clearly correlated with the exclusive occurrence of anammoxosomes only in anammox planctomycetes, and this is consistent with their apparent confinement to the anammoxosome membrane in the anammox planctomycete species (Dokhaven strain) where ladderane location within the cell has been determined. The occurrence of both ether- and ester-linkages in ladderane lipids is of special evolutionary interest since ether lipids are often considered characteristic of Archaea. The deep-branching position of the planctomycete phylum (Brochier and Philippe 2002) could be consistent with retention of an early form of lipid, perhaps of the mixed glycerol ether/ester or mixed ladderane/*n*-alkyl glycerol diether form found in some ladderanes. Such lipids could have been synthesized by ancestors of anammox planctomycetes with features of both domains Archaea and Bacteria, perhaps before the annealing of the domain members fully from the progenote.

4.4 Hopanoids

Anammox planctomycetes—"*Candidatus* Brocadia anammoxidans" Delft strain and "*Candidatus* Kuenenia stuttgartiensis" from gas lift reactors, and "*Candidatus* Scalindua brodae" and "*Candidatus* Scalindua wagneri" from a UK waste water treatment plant, have all been found to produce a variety of another class of membrane rigidifier, the hopanoids (Sinninghe Damsté et al. 2004), a class of compounds previously known only from aerobic bacteria. These hopanoids included bacteriohopanetetrol, diplotene and C_{27} hopanoid ketone. Consistent with this, the gene for squalene hopene cyclase, an important enzyme for synthesis of hopanoids, was also present in the genome of "*Candidatus* Kuenenia stuttgartiensis", the only anammox planctomycete for which the genome has been partially sequenced. Occurrence of this gene in *Gemmata obscuriglobus*, and occurrence of hopanoids themselves in *Blastopirellula marina* argues against a specific location of hopanoids in the anammoxosome membrane, but this will wait upon analysis of anammoxosomes for hopanoids. Hopanoids might conceivably play a rigidifying role in cell membranes of planctomycetes analogous to that of sterols in eukaryote membranes (interestingly both hopanoids and sterols may well turn out to be both used in *G. obscuriglobus* membranes). Hopanoids are known useful biomarkers for palaeobiogeochemistry, so may prove useful for detection of anammox planctomycetes in modern sediments and recent or ancient rocks of sedimentary origin (it is not yet known how useful ladderanes will prove as a biomarker in older sedimentary rocks).

5
Anammoxosomes and Models for Anammox Physiology

Models for anammox physiology place the anammoxosome and its membrane envelope in a central position concerning the generation of energy from ammonium oxidation. Within these models, the anammoxosome envelope is crucial for a vectorial process within a chemiosmotic context. The models thus propose a unique situation where ATP generation may occur as part of a chemolithotrophic process involving anaerobic oxidation of an inorganic compound and associated with a membrane-bounded organelle inside a prokaryotic cell. The special chemical composition of the anammoxosome membrane may be also relevant to these mechanisms and to how the peculiar compounds generated by the anammox reaction can be compatible with a viable cell.

5.1
Models for Anammox Physiology

The discovery of the anammoxosome has been central to the formulation of models for anammox metabolism. The enzyme hydrazine oxidoreductase —originally considered only as a hydroxylamine oxidoreductase, HAO — is one of the major enzymes specific to the anaerobic ammonium-oxidizing ability of these organisms, and can form up to 9% of the soluble protein of anammox biomass (Schalk et al. 2000). HAO has been shown by immunogold labelling of thin-sectioned cryosubstituted cells to be localized entirely within the anammoxosome compartment (Lindsay et al. 2001). This data, when combined with the demonstration that the anammoxosome compartment is surrounded by a membrane, has been central to the development of a model for the mechanism of anaerobic ammonium oxidation in these bacteria (Fig. 8). This model involves a pivotal role for knowledge of location of the HAO enzyme (Jetten et al. 2001). All versions of the various models developed involve the concept of a proton motive force being generated across the anammoxosome membrane via a proton gradient as a result of electron transport associated with the oxidation of ammonium to nitrogen via hydrazine oxidation occurring via enzymes associated with the anammoxosome membrane. The anammoxosome membrane plays a central role in the proton gradient generation proposed by these models to occur across this membrane. In such models, the anammoxosome and the riboplasm of the anammox pirellulosome are considered as two separate compartments demarcated by the anammoxosome membrane. Via proton consumption in the riboplasm and proton generation inside the anammoxosome, charge separation results in an electrochemical proton gradient directed from the anammoxosome to the riboplasm, so that the riboplasm is alkaline and negatively charged relative to anammoxosome, thus giving a proton motive force (pmf)

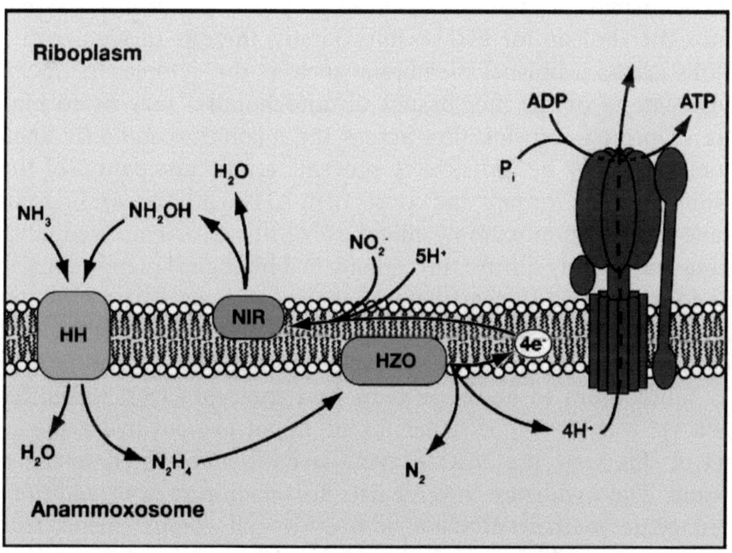

Fig. 8 Model for anaerobic ammonium oxidation coupled to the anammoxosome membrane in anammox bacteria, resulting in a proton motive force and subsequent ATP synthesis via membrane-bound ATPases. HH: hydrazine hydrolase, the hydrazine-forming enzyme; HZO: hydrazine-oxidizing enzyme; NIR: nitrite-reducing enzyme. Note that in one alternative model (Strous and Jetten 2004) nitric oxide can substitute for the place of hydroxylamine in the model shown here. Modified from van Niftrik et al. (2004); reproduced with permission of Annual Review of Microbiology

composed of ΔpH and $\Delta\Psi$. This pmf could theoretically be used to drive ATP synthesis. The models thus predict the presence of an ATPase/synthase embedded within the anammoxosome membrane. Such membrane-bound ATP synthases would be perhaps located with their hydrophilic globular ATP-synthesizing domain in the riboplasm and their hydrophobic proton-translocating domain in the anammoxosome membrane, with synthesized ATP released to the riboplasm (van Niftrik et al. 2004), constituting strong but hard-to-test predictions of the models. However, this ATP synthase has yet to be demonstrated and localized. The models remain stimulating for generation of testable hypotheses concerning the role of the anammoxosome in anaerobic ammonium oxidation. If an ATP synthase were to be found in the membrane and the models confirmed, the anammoxosome would constitute the first example of a mitochondrion-like organelle within a bacterial cell, but one involved in a peculiar and unique anaerobic phosphorylation rather than oxidative phosphorylation.

5.2 Membrane Impermeability and Models for Anammox Physiology

Ladderane lipids impart a property of relative impermeability to the anammoxosome membrane, and this has been proposed as a property essential to anammox metabolism for two reasons. Firstly, the rate of passive diffusion of protons across a normal membrane such as the cytoplasmic membrane of other bacteria or the membranes of mitochondria may be so high that the rate of proton translocation across the anammoxosome by anammox planctomycetes may be too slow to prevent serious dissipation of the proton gradient needed for pmf and consequent ATP synthesis. An impermeable ladderane-rich anammoxosome might solve this problem. Secondly, since hydrazine may readily diffuse through normal biological membranes, the hydrazine formed as an intermediate of anammox metabolism might be lost to an extent with catastrophic consequences to viability (since 1 molecule of hydrazine requires 15 catabolic cycles, consequent on the need for 15 mole of ammonium to be oxidized to fix 1 mole of CO_2). Again, limiting diffusion via a ladderane-rich membrane might prevent hydrazine loss. In support of this view the HZO enzyme is entirely confined to the anammoxosome. The hydrazine intermediate for anammox is in addition toxic and mutagenic so their diffusion into other cell compartments from the anammoxosome could be inimical to cell survival. It has been pointed out that this function would be analogous to the way lysosomes of eukaryotic cells compartmentalize digestive enzymes to protect the rest of the cell from damage (DeLong 2002). It is of interest that "*Candidatus* Kuenenia" and "*Candidatus* Scalindua" species produce hydrazine from exogenously supplied hydroxylamine, just like "*Candidatus* Brocadia anammoxidans", so their anammoxosomes may also function to partition this toxic intermediate from the rest of the cytoplasm and to conserve it from energetically costly diffusion loss, and there is thus biochemical as well as structural evidence that this may apply to all genera of anammox planctomycetes. There are thus at least three arguments why an impermeability-conferring ladderane-rich anammoxosome envelope would be an advantage for survival of anammox planctomycete cells—(1) proton leakage limitation, (2) compartmentation of hydrazine for purposes of energy conservation, and (3) compartmentation of toxic hydrazine thus preventing damage to the riboplasm and nucleoid. However, in order to function in anammox metabolism the anammoxosome membrane cannot be completely impermeable and the membrane must provide transport mechanisms for proteins such as the HZO enzyme and even ions and some low molecular weight compounds additional to anammox intermediates.

6
Potential Cell Biology Functions of Anammoxosome Tubules

The nature of the tubule structures in anammoxosomes is unknown. The occurrence of their apparent organization in arrangements in some cases suggests a possible structural function, and occasionally they are also seen in arrays in dividing cells suggesting a possible role in distribution of the anammoxosome to the daughter cell or a correlation with such distribution. The potential occurrence of some small amount of DNA within anammoxosomes is consistent with a possibility that such tubules could be involved in chromosome or plasmid segregation during division. In addition, the close association of the anammox planctomycete nucleoid with the anammoxosome membrane in at least "*Candidatus* Brocadia" and "*Candidatus* Scalindua" species is consistent with a co-distribution of the anammoxosome with the chromosomes during division. The discovery of bacterial tubulin in *Prosthecobacter* species (Jenkins 2002), members of the phylum *Verrucomicrobia* which according to some analyses may be most closely related to phylum *Planctomycetes* (Fieseler et al. 2004) makes it conceivable that such tubules are also composed of a cytoskeletal protein homologous with tubulin-like proteins. Actual tubule structures as well as reactivity with monoclonal anti-tubulin antibodies have been found in the verrucomicrobial symbiont of the marine ciliate protist *Euplotidium*, the "epixenosome", though it is not yet known whether tubulin protein homologs occur in epixenosome genomes (Petroni et al. 2000). The absence of the cell division protein FtsZ in other planctomycetes for which genome data is available is consistent with a protein other than FtsZ being involved in cell division in both non-anammox and anammox planctomycetes.

7 Anammox Genomics and the Anammoxosome

Several genome projects on anammox planctomycetes are being planned and one of those is in progress at this time, such is the intense interest they have generated. The genome project for completion of the sequencing of the genome of "*Candidatus* Kuenenia stuttgartiensis", a joint project of Genoscope (France), the Department of Microbiology, Radboud University of Nijmegen, the Department of Microbial Ecology, University of Vienna, and the Department für Genom-orientierte Bioinformatik, Technische Universität München, is described at http://www.microbiol.sci.kun.nl/tech/genomics.html. It is conceivable that the whole genome and deduced proteome of "*Candidatus* Kuenenia stuttgartiensis" will yield new insights into the genetic basis for the anammoxosome and ladderane lipid synthesis. In particular, clues to the possible molecular nature of the tubules within the anammoxosome might be found if homologs of cytoskeletal proteins known to occur in Bacteria and Eukaryotes, for example FtsZ and tubulin, were to be discovered. So far analysis of the "*Candidatus* Kuenenia stuttgartiensis" genome has not yielded evidence of any gene for an FtsZ homolog. The nature of the hydrazine oxidoreductase central to anammox metabolism and its relations to hydroxylamine oxidoreductases of other bacteria might be clarified by gene sequencing and the relation of its structure to its function within the anammoxosome could be elucidated. The existence of several compartments within the anammox planctomycete cell between which proteins and perhaps other molecules must be transported implies a prediction of many proteins with signal peptides as in *Rhodopirellula baltica* (see Fuerst 2006, in this volume) and perhaps proteins for specialized secretion systems may be found in the proteome.

The genome sequence of "*Candidatus* K. stuttgartiensis" has revealed 108 ORFs with a classical heme *c* binding motif CXXCH and the unusual heme *c* binding motif CXXX(X)CH, and 62 of these 108 could be annotated as cytochromes (Cirpus et al. 2005). Twelve of the deduced *c*-type cytochromes were thought to correspond to hydroxylamine oxidoreductase-like proteins (and thus could harbour the hydrazine oxidoreductase itself), containing eight heme groups per subunit. Nitrite reductase, nitrate reductase and bc1 complex cytochrome proteins were also identified in the deduced proteome. HAO-like proteins, nitrite reductase and cytochromes of the nitrate reductase cluster were identified by experimental proteomic analysis of 2D gels of "*Candidatus* K. stuttgartiensis" biomass. The occurrence of plasmids or other extrachromosomal DNA elements might be elucidated by a complete genome sequence and this may throw light on the nature and function of the small amount of DNA found within the anammoxosome.

8 Conclusions

The anammoxosome is a unique organelle within known prokaryotes, and is perfectly correlated with the occurrence of a unique type of catabolism among prokaryotes, that of anaerobic autotrophic oxidation of ammonium. Anammoxosomes are only known to occur in anammox planctomycetes, and form yet another type of membrane-bounded compartment as a variation within the fundamental shared cell plan of planctomycetes. They appear to be central to anammox metabolism, since in at least one model of anammox planctomycete they contain a central enzyme in this metabolism, hydrazine oxidoreductase. The anammoxosome membrane and the relationship between the anammoxosome and the riboplasm of the pirellulosome external to it forms a central basis for all models proposed for anammox energy generation. The vectorial and chemiosmotic nature of the energy generation via proton transport across the membrane and consequent proton gradient generation is proposed to enable the ATP synthesis correlated with ammonium oxidation. The occurrence of ladderane lipids in this membrane imparting high density may be needed for preventing proton leakage and loss of hydrazine intermediates as well as sequestering of toxic anammox metabolism intermediates away from other cell compartments. Yet the anammoxosome is just as interesting to consider from the perspective of the cell biology of anammox planctomycetes as a compartment that might function in roles additional to metabolism.

One of the most intriguing features of the anammoxosome is the occurrence of tubule structures packed within the compartment, of as yet undetermined nature and function, and the association of nucleoids with the anammoxosome surface. The anammoxosome may turn out to be a multifunctional organelle of a truly novel type, with functions in metabolism, chromosome segregation and cell division. It appears to have a definite role as a site for the enzymes of a specialized ammonium oxidation catabolism, probably acts as an energy generator and a barrier to diffusion of toxic intermediates in that catabolism, and finally is an organelle associated closely with the nucleoid and may possibly play a role in cell division or anammoxosome distribution to daughter cells.

Anammox planctomycetes may be an example of organisms that have adapted the shared planctomycete cell plan to a specialized role in a unique biochemistry or alternatively represent something close to an ancestral form of planctomycete from which forms with less specialized physiology were derived. It occurs only in anammox planctomycetes, a phylogenetically monophyletic deep branch within the planctomycetes, which themselves appear to be an ancient lineage close to the base of the Bacterial tree. Thus, like anammox metabolism itself it may well have evolved only once and very early within cellular life. It may prove to be significant beyond anammox planctomycete studies, for understanding the evolution of cell organization and function and how eukaryote-like division mechanisms may have evolved, as well as its relevance to understanding a unique physiology pivotal for explaining all the actors in the global nitrogen cycle and how that cycle operated in the Earth's ancient past. The anammox planctomycetes can account for up to 67% of the total N_2 production in continental shelf sediments (Thamdrup and Dalsgaard 2002), may account for up to 15% of the loss of fixed nitrogen from the world's oceans because of their extensive occurrence in the oxygen minimum zones (OMZ) of suboxic ocean waters (Dalsgaard et al. 2003), and may comprise up to 2×10^7 ml^{-1} or 1% of the total bacterial count in the Benguela OMZ of the South Atlantic ocean (Kuypers et al. 2005). It is thus conceivable that anammoxosomes might be one of the world's most numerous and ecologically significant prokaryote compartments.

Note Added in Proof

A near-complete 4.2 Mb genome sequence for "*Candidatus* Kuenenia stuttgartiensis" is now available (Strous et al. 2006) and the annotation indicates major new insights into anammox and other planctomycetes: 1) paralogs of respiratory complex proteins suggest respiration using different electron acceptors is possible, and it has now been shown that iron and manganese oxides can be respired with formate as electron donor, and iron can be oxidized with nitrate as electron acceptor 2) a new model for anammox metabolism has been derived involving NO as an intermediate and is described in detail 3) the complete acetyl-CoA pathway for carbon dioxide fixation is present 4) candidate genes for ladderane biosynthesis can be identified 5) as derived from phylogenetic analysis of a concatenated 5S-16S-23S rRNA dataset, planctomycete and chlamydia phyla group together and this is marked by the signature of a shared 60 kDa outer membrane protein, confirming the concept of a superphylum joining these phyla and 7) most genes needed for peptidoglycan biosynthesis are present, again connecting planctomycetes and chlamydia and suggesting a shared ancestor with a cell wall similar to those in peptidoglycan-synthesizing Bacteria.

Acknowledgements Research on anammox planctomycetes in Fuerst's laboratory is supported by the Australian Research Council. Research in Jetten's and Strous's laboratories on anammox has been supported by the European Union EESD EVK1-CT-2000 000054, the Foundation for Applied Sciences (STW), the Foundation of Applied Water Research (STOWA), the Netherlands Foundation for Earth and Life Sciences (NWO-ALW), the Royal Netherlands Academy of Arts and Sciences (KNAW), DSM-Gist, and Paques BV.

References

Brochier C, Philippe H (2002) Phylogeny: a non-hyperthermophilic ancestor for bacteria. Nature 417:244

Chouari R, Le Paslier D, Daegelen P, Ginestet P, Weissenbach J, Sghir A (2003) Molecular evidence for novel planctomycete diversity in a municipal wastewater treatment plant. Appl Environ Microbiol 69:7354–7363

Cirpus I, Op den Camp H, Kuenen G, Strous M, Le Paslier D, Pluk W, Lasonder E, Allen JWA, Jetten MSM (2005) Importance of cytochromes in the metabolism of the anammox bacterium *Kuenenia stuttgartiensis*. In: 1st Joint German/British Bioenergetics Conference in Cooperation with the GBM study group Bioenergetics "Mechanisms of Bioenergetic Membrane Proteins: Structures and Beyond". German Society for Biochemistry and Molecular Biology (GBM), Naurod (Wiesbaden)

Dalsgaard T, Canfield DE, Petersen J, Thamdrup B, Acuna-Gonzalez J (2003) N-2 production by the anammox reaction in the anoxic water column of Golfo Dulce, Costa Rica. Nature 422:606–608

Egli K, Fanger U, Alvarez PJJ, Siegrist H, van der Meer JR, Zehnder AJB (2001) Enrichment and characterization of an anammox bacterium from a rotating biological contactor treating ammonium-rich leachate. Archives of Microbiology 175:198–207

Fieseler L, Horn M, Wagner M, Hentschel U (2004) Discovery of the novel candidate phylum *Poribacteria* in marine sponges. Appl Environ Microbiol 70:3724–3732

Fuerst JA (2005) Intracellular compartmentation in Planctomycetes. Annual Review of Microbiology 59:299–328

Guven D, Dapena A, Kartal B, Schmid MC, Maas B, van de Pas-Schoonen K, Sozen S, Mendez R, Op den Camp HJ, Jetten MS, Strous M, Schmidt I (2005) Propionate oxidation by and methanol inhibition of anaerobic ammonium-oxidizing bacteria. Appl Environ Microbiol 71:1066–1071

Jenkins C, Samudrala R, Anderson I, Hedlund BP, Petroni G, Michailova N, Pinel N, Overbeek R, Rosati G, Staley JT (2002) Genes for the cytoskeletal protein tubulin in the bacterial genus *Prosthecobacter*. Proc Natl Acad Sci 99:17049–17054

Jetten MS, Cirpus I, Kartal B, van Niftrik L, van de Pas-Schoonen KT, Sliekers O, Haaijer S, van der Star W, Schmid M, van de Vossenberg J, Schmidt I, Harhangi H, van Loosdrecht M, Gijs Kuenen J, Op den Camp H, Strous M (2005) 1994–2004: 10 years of research on the anaerobic oxidation of ammonium. Biochem Soc Trans 33:119–123

Jetten MSM, Sliekers O, Kuypers M, Dalsgaard T, van Niftrik L, Cirpus I, van de Pas-Schoonen K, Lavik G, Thamdrup B, Le Paslier D, Op den Camp HJM, Hulth S, Nielsen LP, Abma W, Third K, Engstrom P, Kuenen JG, Jorgensen BB, Canfield DE, Sinninghe Damsté JS, Revsbech NP, Fuerst J, Weissenbach J, Wagner M, Schmidt I, Schmid M, Strous M (2003) Anaerobic ammonium oxidation by marine and freshwater planctomycete-like bacteria. Applied Microbiology and Biotechnology 63:107–114

Jetten MSM, Wagner M, Fuerst J, van Loosdrecht M, Kuenen G, Strous M (2001) Microbiology and application of the anaerobic ammonium oxidation ("anammox") process. Current Opinion in Biotechnology 12:283–288

Kuypers MMM, Lavik G, Woebken D, Schmid M, Fuchs BM, Amann R, Jorgensen BB, Jetten MSM (2005) Massive nitrogen loss from the Benguela upwelling system through anaerobic ammonium oxidation. Proceedings of the National Academy of Sciences of the United States of America 102:6478–6483

Kuypers MMM, Sliekers AO, Lavik G, Schmid M, Jorgensen BB, Kuenen JG, Sinninghe Damsté JS, Strous M, Jetten MSM (2003) Anaerobic ammonium oxidation by anammox bacteria in the Black Sea. Nature 422:608–611

Lindsay MR, Webb RI, Strous M, Jetten MS, Butler MK, Forde RJ, Fuerst JA (2001) Cell compartmentalisation in planctomycetes: novel types of structural organisation for the bacterial cell. Archives of Microbiology 175:413–429

Mulder A, Vandegraaf AA, Robertson LA, Kuenen JG (1995) Anaerobic ammonium oxidation discovered in a denitrifying fluidized-bed reactor. Fems Microbiology Ecology 16:177–183

Neef A, Amann R, Schlesner H, Schleifer K (1998) Monitoring a widespread bacterial group: in situ detection of planctomycetes with 16S rRNA-targeted probes. Microbiology 144:3257–3266

Op den Camp HJ, Kartal B, Guven D, van Niftrik LA, Haaijer SC, van der Star WR, van de Pas-Schoonen KT, Cabezas A, Ying Z, Schmid MC, Kuypers MM, van de Vossenberg J, Harhangi HR, Picioreanu C, van Loosdrecht MC, Kuenen JG, Strous M, Jetten MS (2006) Global impact and application of the anaerobic ammonium-oxidizing (anammox) bacteria. Biochem Soc Trans 34:174–178

Petroni G, Spring S, Schleifer KH, Verni F, Rosati G (2000) Defensive extrusive ectosymbionts of *Euplotidium* (*Ciliophora*) that contain microtubule-like structures are bacteria related to Verrucomicrobia. Proc Natl Acad Sci USA 97:1813–1817

Schalk J, de Vries S, Kuenen JG, Jetten MSM (2000) Involvement of a novel hydroxylamine oxidoreductase in anaerobic ammonium oxidation. Biochemistry 39:5405–5412

Schmid M, Twachtmann U, Klein M, Strous M, Juretschko S, Jetten M, Metzger JW, Schleifer KH, Wagner M (2000) Molecular evidence for genus level diversity of bacteria capable of catalyzing anaerobic ammonium oxidation. Syst Appl Microbiol 23:93–106

Schmid M, Walsh K, Webb R, Rijpstra WIC, Van De Pas-Schoonen K, Verbruggen M, Hill T, Moffett B, Fuerst J, Schouten S, Sinninghe Damsté JS, Harris J, Shaw P, Jetten M, Strous M (2003) *Candidatus* Scalindua brodae, sp. nov., *Candidatus* Scalindua wagnei, sp. nov., two new species of anaerobic ammonium oxidizing bacteria. Systematic and Applied Microbiology 26:529–538

Sinninghe Damsté JS, Rijpstra WI, Geenevasen JA, Strous M, Jetten MS (2005) Structural identification of ladderane and other membrane lipids of planctomycetes capable of anaerobic ammonium oxidation (anammox). FEBS J 272:4270–4283

Sinninghe Damsté JS, Rijpstra WI, Strous M, Jetten MS, David OR, Geenevasen JA, van Maarseveen JH (2004) A mixed ladderane/n-alkyl glycerol diether membrane lipid in an anaerobic ammonium-oxidizing bacterium. Chem Commun (Camb):2590–2591

Sinninghe Damsté JS, Rijpstra WIC, Schouten S, Fuerst JA, Jetten MSM, Strous M (2004) The occurrence of hopanoids in planctomycetes: implications for the sedimentary biomarker record. Organic Geochemistry 35:561–566

Sinninghe Damsté JS, Strous M, Rijpstra WI, Hopmans EC, Geenevasen JA, van Duin AC, van Niftrik LA, Jetten MS (2002) Linearly concatenated cyclobutane lipids form a dense bacterial membrane. Nature 419:708–712

Sliekers AO, Derwort N, Gomez JL, Strous M, Kuenen JG, Jetten MS (2002) Completely autotrophic nitrogen removal over nitrite in one single reactor. Water Res 36:2475–2482

Strous M, Fuerst JA, Kramer EH, Logemann S, Muyzer G, van de Pas-Schoonen KT, Webb R, Kuenen JG, Jetten MS (1999a) Missing lithotroph identified as new planctomycete. Nature 400:446–449

Strous M, Heijnen JJ, Kuenen JG, Jetten MSM (1998) The sequencing batch reactor as a powerful tool for the study of slowly growing anaerobic ammonium-oxidizing microorganisms. Applied Microbiology and Biotechnology 50:589–596

Strous M, Jetten MSM (2004) Anaerobic oxidation of methane and ammonium. Annual Review of Microbiology 58:99–117

Strous M, Kuenen JG, Jetten MSM (1999b) Key physiology of anaerobic ammonium oxidation. Applied and Environmental Microbiology 65:3248–3250

Strous M, Pelletier E, Mangenot S, Rattei T, Lehner A, Taylor MW, Horn M, Daims H, Bartol-Mavel D, Wincker P, Barbe V, Fonknechten N, Vallenet D, Segurens B, Schenowitz-Truong C, Medigue C, Collingro A, Snel B, Dutilh BE, Op den Camp HJ, van der Drift C, Cirpus I, van de Pas-Schoonen KT, Harhangi HR, van Niftrik L, Schmid M, Keltjens J, van de Vossenberg J, Kartal B, Meier H, Frishman D, Huynen MA, Mewes HW, Weissenbach J, Jetten MS, Wagner M, Le Paslier D (2006) Deciphering the evolution and metabolism of an anammox bacterium from a community genome. Nature 440:790–794

Thamdrup B, Dalsgaard T (2002) Production of N-2 through anaerobic ammonium oxidation coupled to nitrate reduction in marine sediments. Applied and Environmental Microbiology 68:1312–1318

van Niftrik LA, Fuerst JA, Sinninghe Damsté JS, Kuenen JG, Jetten MSM, Strous M (2004) The anammoxosome: an intracytoplasmic compartment in anammox bacteria. Fems Microbiology Letters 233:7–13

vandeGraaf AA, deBruijn P, Robertson LA, Jetten MSM, Kuenen JG (1997) Metabolic pathway of anaerobic ammonium oxidation on the basis of N-15 studies in a fluidized bed reactor. Microbiology-Uk 143:2415–2421

Microbiol Monogr (2)
J.M. Shively: Complex Intracellular Structures in Prokaryotes
DOI 10.1007/7171_027/Published online: 8 June 2006

The Enigmatic Cytoarchitecture of *Epulopiscium* spp.

Esther R. Angert

Department of Microbiology, Cornell University, Wing Hall, Ithaca, 14853, USA
era23@cornell.edu

Abstract *Epulopiscium* spp. and related bacteria are large, intestinal symbionts of surgeonfish. The biggest of these bacteria are cigar-shaped, reaching lengths of 600 μm or more. The architecture of these enormous cells is unusual and it appears that basic bacterial cell components, such as the flagella, cytoplasmic membrane and nucleoids, are present in excess and may serve novel functions. Comparative studies of members of this morphologically diverse group of intestinal symbionts are providing insight into the role of cytoarchitecture in supporting large cytoplasmic volumes as well as the development of unusual reproductive strategies.

1 Introduction

There was a time, in the not so distant past, when microbiologists considered prokaryotes as simple cells with little subcellular organization. These structurally basic cells rely on diffusion to exchange molecules with their immediate environment. Likewise, movement of any biomolecule within a prokaryotic cell is dictated by its diffusion coefficient. Over short distances, diffusive transport is very fast and efficient, but over distances of greater than a few microns, it becomes unreliable and time consuming. To support an active metabolism, prokaryotes need to maintain a high surface-to-volume ratio or ensure that no part of the cytoplasm is very far away from the external environment thus facilitating rapid exchange of metabolites. Hence, most prokaryotes are small. Reports of large bacteria have appeared in the literature for over one-hundred-seventy years [take for example, Ehrenberg's early descriptions of *Spirochaeta plicatus* in 1835 and 1838 (Blakemore and Canale-Parola 1973)], but these and other large bacteria seemed to fit the functional model of the prokaryotic cell. All of these bacteria are extremely long and thin, or if spherical they contain an enormous intracellular vacuole to press the active cytoplasm into a thin layer just under the cell surface (Jannasch et al. 1989). In addition, mineral inclusions further reduce the volume of cytoplasm of many large bacteria. It seemed that only the increased structural complexity of the eukaryotic cell could overcome diffusion-dictated constraints allowing cells to become larger. It was against this backdrop that *Epulopiscium* was discovered.

Epulopiscium spp., and related intestinal symbionts of surgeonfish (Family Acanthuridae), comprise a morphologically diverse group of low G+C Gram-positive bacteria. As yet, none is available in culture. The first described member of this group, *Epulopiscium fishelsoni*, was discovered in the brown surgeonfish *Acanthurus nigrofuscus* taken from the Red Sea (Fishelson et al. 1985). It was classified as a novel eukaryotic protist, based primarily on its size and its unusual ability to produce intracellular offspring (Montgomery and Pollak 1988). The first report suggesting that similar large, intestinal symbionts of surgeonfish were prokaryotes (Clements and Bullivant 1991) was met with skepticism. However, molecular phylogenetic analyses, based on small subunit ribosomal RNA sequence comparisons, placed *Epulopiscium* spp. unequivocally within the Bacteria (Angert et al. 1993). This discovery coincided with the accumulation of evidence that bacteria possess the means of directing the subcellular localization of proteins. Since then, further discoveries demonstrating the ability of bacteria to dictate the position of cellular DNA, to find and mark particular cellular locations, and the existence of cytoskeletal elements have shown that prokaryotic cells are complex and as organized as eukaryotic cells (Losick and Shapiro 1999; Norris et al. 1996; Shapiro and Losick 1997).

2 Diversity of *Epulopiscium*-like Surgeonfish Symbionts

Spurred on by the description of *E. fishelsoni* in surgeonfish in the Red Sea, Clements and colleagues (1989) performed an extensive survey of intestinal biota of tropical reef fish of the South Pacific. They demonstrated that *Epulopiscium*-like symbionts are prevalent in a variety of surgeonfish species. To describe the distribution among fish species, these conspicuous symbionts were classified based on cell shape, size and their reproductive biology (Clements et al. 1989). Many of the morphotypes produce internal offspring. Binary fission, inferred from the observation of cell pairs that have not yet completely separated, is common in several morphotypes. In other morphotypes, evidence for binary fission has not been found. Clements et al. (1989) observed that these symbionts tend to be associated with herbivorous and detritivorous surgeonfish, and are not in species that feed primarily on zooplankton. They confirmed the earlier observation that these symbionts appear specifically associated with certain surgeonfish and are absent in other species of herbivorous fish that regularly feed side-by-side with surgeonfish (Fishelson et al. 1985). The following is a brief description of some notable morphotypes.

The largest cells are members of the A morphotype, cigar-shaped, 70 to > 600 μm in length (see Fig. 1, for example), although cells greater than 400 μm are rare. This morphotype is found in eight herbivorous *Acanthurus* species (Clements et al. 1989) and includes populations described as *Epulopiscium fishelsoni.* These cells produce one, two, or occasionally three, internal offspring. Direct observations of binary fission are rare although some studies indicate that these cells regularly use binary fission for reproduction (Montgomery and Pollak 1988). Phylogenetic surveys and fluorescence in situ hybridization (FISH) studies using small subunit rRNA-targeted probes revealed that A morphotype populations in an individual fish are comprised of at least two distinct but related clades, referred to as A1 and A2 (Angert et al. 1993). In pairwise sequence comparisons, the A1 and A2 clades share approximately 91% sequence identity over 1440 nucleotides compared. Each clade consists of sequence types that vary by less than 2%. Nearly identical A1 and A2 rRNA gene sequences have been obtained from three host species, *A. nigrofuscus*, *A. lineatus* and *A. triostegus*, taken from geographically distant

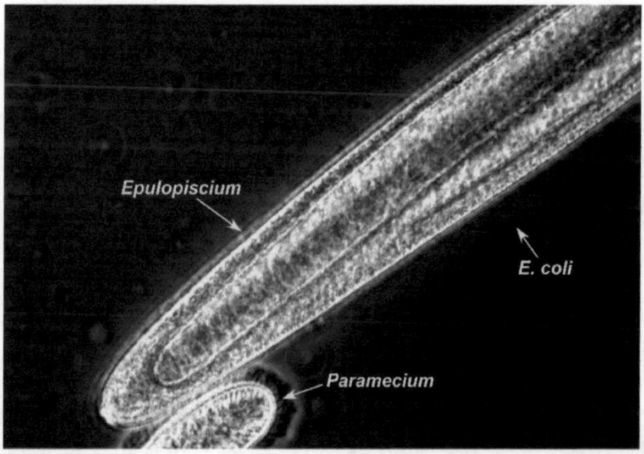

Fig. 1 Phase contrast micrograph showing a portion of a large, morphotype A *Epulopiscium*. For comparison, the cell was mixed with *Paramecium tetraurelia* and *E. coli*. The large, cigar-shaped bodies within the *Epulopiscium* cell, are offspring

sites (Angert 1995; Angert, Clements and Montgomery, unpublished results). These results indicate that the largest *Epulopiscium* morphotype may be confined to *Acanthurus* spp. and that closely related A morphotype cells can be found in a variety of related hosts.

The *Epulopiscium* B morphotype (Fig. 2), from the unicornfish *Naso tonganus* has been used as a model for cell biological studies (Angert and Clements 2004). Natural type B populations in this host tend to be much more homogenous, morphologically and genetically, than *Epulopiscium* populations found in other hosts. Type B cells can reach lengths of 200 to 300 μm when large internal offspring are present. This morphotype generally produces two intracellular offspring but can contain as many as five per mother cell (Clements et al. 1989). No direct evidence of binary fission has been observed in this morphotype.

The slightly smaller, cigar-shaped C morphotype, 40 to 130 μm in length, is found in the herbivorous fish species *Naso lituratus* and *N. unicornis*. Populations from *N. lituratus* have been extensively characterized (Flint et al. 2005). The F morphotype, cigar-shaped cells 15 to 60 μm long, is found in several *Acanthurus* and *Ctenochaetus* species that feed on detritus or graze algae over sand or rubble substrates (Clements et al. 1989). These bacteria have the exceptional ability to produce as many as seven intracellular offspring.

The H morphotype, cigar-shaped, 10 to 100 μm, encompasses a diverse assemblage of symbionts from 20 fish species (Clements et al. 1989). These bacteria harbor intracellular inclusions that may be offspring. Binary fission can occur in some lineages but the reproductive biology of these bacteria is not well characterized.

Another notable variety is the J morphotype. This filamentous morphotype ranges from 10 to upwards of 400 μm in length but is only a few microns in diameter. It is one of the more widely distributed morphotypes, found

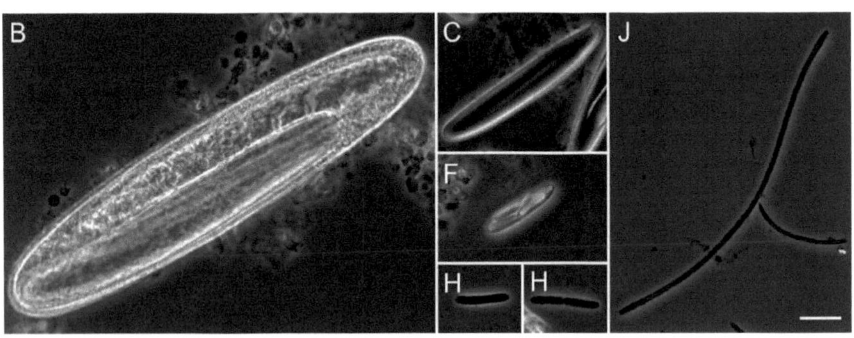

Fig. 2 The morphological diversity of large intestinal symbionts of surgeonfish. Morphotype designations are shown in upper left corner of each panel, see text for details. All micrographs are shown at the same magnification. *Scale bar* represents 20 μm

in nine species of surgeonfish. Some of these cells produce intracellular offspring and they regularly undergo binary fission (Clements et al. 1989).

On the basis of FISH surveys of surgeonfish intestinal contents using an *Epulopiscium* group probe, and phylogenetic analyses, all of these diverse morphotypes are related (Angert 1995; Angert et al. 1993). This situation provides an opportunity for comparative studies to address how changes in cell architecture, symbiotic associations and reproductive strategies may influence the diversification of this group of unusual bacteria (Angert 2005).

3 An Overview of the Cell Architecture of Large *Epulopiscium*

Published transmission electron micrographs (TEMs) of thin sections of large type A *Epulopiscium* cells have originated from intestinal samples of three *Acanthurus* spp. (Clements and Bullivant 1991; Clements et al. 1989; Fishelson et al. 1985; Montgomery and Pollak 1988; Robinow and Angert 1998). On the basis of these studies, a general model of the cytoarchitecture of large *Epu-*

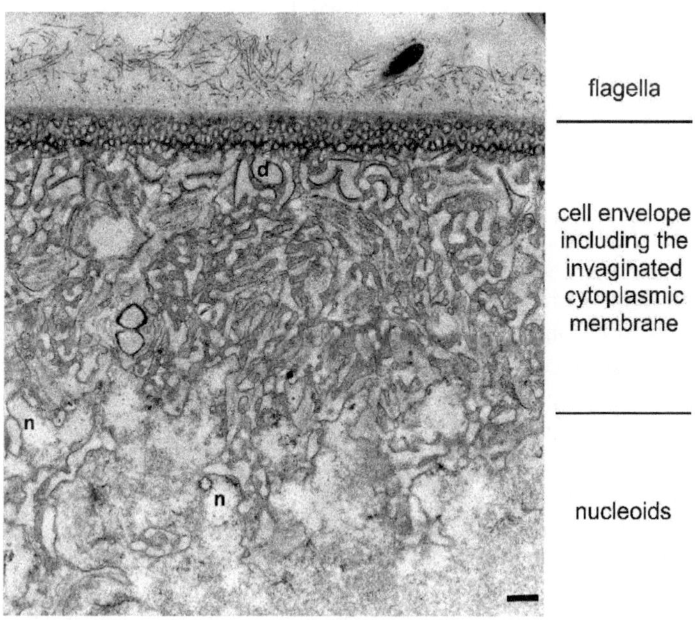

Fig. 3 Transmission electron micrograph of a thin section through the outermost layers of a type A *Epulopiscium* cell. Each structural layer is labeled, see text for details. Note the small bacterium (dark body) trapped within the flagellar mat on the *Epulopiscium* cell surface. *Scale bar* represents 1 μm. Reprinted with permission from Robinow and Angert (1998)

lopiscium cells is emerging (Fig. 3). Starting at the surface, *Epulopiscium* cells are covered with a mat of bacterial flagella, identified on the basis of ultrastructure (Clements and Bullivant 1991). The cell envelope is unlike that of the stereotypical Gram-positive bacterium. Although the envelope contains some peptidoglycan, a thick outermost layer of peptidoglycan is not seen in thin sections of A morphotype cells. Likewise, cells do not stain Gram-positive. The envelope varies in thickness from one morphotype to another. The envelope of A morphotype cells is thick but flexible and retains its shape even when the contents of the cytoplasm have been removed (Montgomery and Pollak 1988). Enigmatic structures within the envelope suggest vesicle-mediated transport but this has yet to be confirmed experimentally (Robinow and Angert 1998). The B morphotype envelope is thinner and more uniform in thickness. The cell membrane of large symbionts is elaborate; it appears either highly invaginated or alternatively membrane tubules extend well into the cytoplasm (Montgomery and Pollak 1988). The function of this internal membrane system is not yet known. Closely associated with the membrane system are the nucleoids. Large *Epulopiscium* cells contain a tremendous amount of DNA, the composition of which is currently under investigation. Within the rather uniform cytoplasm, offspring develop. Structurally, the offspring cell appears as a reiteration of the mother cell with some notable differences. A layer of highly condensed DNA is situated beneath the offspring cell membrane. The cytoplasm of the offspring contains ribosomes but also often holds numerous small inclusions of unknown composition. Each of these structures, and their impact on *Epulopiscium* biology, are explored in more detail below.

3.1 Flagella and Motility

Although flagella are not internal structures, a note on their functional significance is in keeping with an underlying theme of adaptations to large cell size. *Epulopiscium* cells are motile. Although their motility has not been systematically studied, their behavior undoubtedly has a significant impact on their acquisition of nutrients and the maintenance of a productive symbiotic association.

When removed from the fish and placed on a microscope slide with gut fluid, motile type A cells have an average speed of approximately 400 μm/sec (Montgomery and Pollak 1988). Large *Epulopiscium* cells are covered with fine filaments (Fishelson et al. 1985; Montgomery and Pollak 1988) that appear to be typical bacterial flagella (Clements and Bullivant 1991). Flagellin genes have been recovered from type B *Epulopiscium* cells (Flint and Angert, unpublished results) and based on sequence comparisons these genes would produce proteins that resemble other characterized flagellins (Macnab 2003). *Epulopiscium* cells do not exhibit flagella-mediated "run and tumble" motility. As an *Epulopiscium* moves in a given direction, it rotates on its long axis and proceeds in a fairly straight path (Fishelson et al. 1985). A reverse in direction is accompanied by a reversal of axial rotation and a change in which end of the cell leads. Large, cigar-shaped *Epulopiscium* cells have no discernible preferred anterior, although some morphotypes do show a marked preference for one end over the other. These latter cells have a prevalent helical swimming pattern and less axial rotation, similar to other highly motile bacteria and eukaryotic ciliates (Fenchel 1994; Thar and Fenchel 2005).

For *Epulopiscium*, motility may serve several functions. One is maintaining position within the host intestinal tract. Herbivorous and detritivorous surgeonfish hosts have a long, coiled and fairly undifferentiated intestinal tract (Clements 1997; Stevens and Hume 1995). During the day, the highest densities of A or B morphotype populations are found in a region about midway along the length of the intestine (Clements et al. 1989; Fishelson et al. 1985). Few *Epulopiscium* are found in the anterior third or the posterior quarter of the intestinal tract. However, type A populations drift posteriorly at night, when the fish is less active (Montgomery and Pollak 1988). It appears that the symbionts move with the bolus of partially digested algae, as the upper intestinal tract empties while the fish is not feeding. At some time in the early morning, resident populations of symbionts must swim upstream to regain their position in the midgut. It is not known if B populations migrate in a similar pattern but these populations are likewise found in a narrow region within the gut during daylight hours (Clements et al. 1989). The chemical signals or "landmarks" that large *Epulopiscium* cells use to assess their position in the gut are not known.

Flagella may play an additional role in circulating and refreshing the immediate environment to enhance exchange of molecules. Under the microscope, stationary *Epulopiscium* cells sometimes continue to rotate their flagella, as debris can be seen moving over the cell surface. In this way, bacterial flagella may serve a role analogous to the cilia of ciliated epithelia or the ciliated corona of rotifers or other protists. Flagella-mediated motility is clearly important for *Epulopiscium* in navigating within the intestinal lumen, and it is possible that flagella-driven currents may enhance diffusive exchange with the environment. So far, observations of motility in *Epulopiscium* cells have occurred under artificial circumstances, when cells have been removed from the fish and placed on a microscope slide. Therefore the behavior of these intestinal symbionts and the role of flagella in circulation may be very different inside the fish and warrants further study.

3.2
Internal Membrane System

The first published TEMs of large *Epulopiscium* cells revealed a conspicuous and elaborate internal membrane system (Fishelson et al. 1985; Montgomery and Pollak 1988). At this resolution, the membranes are featureless but appear contiguous with the plasma membrane. The invagination of the cell membrane effectively increases the *Epulopiscium* cell surface area relative to its volume (Clements and Bullivant 1991). This cellular modification may accommodate large numbers of transport proteins facilitating exchange with the environment. As such, the membrane system may serve as a transport organelle. When surveying the few published micrographs that represent the morphological diversity of *Epulopiscium* and related surgeonfish symbionts, it appears that the presence of invaginations correlates with cell diameter. Thin cells appear to have little to no internal membrane system, see for example Clements et al. (1989) and Flint et al. (2005). Thick cells (A morphotype cells) have extensive invaginations extending ten microns or more deep into the cytoplasm (Montgomery and Pollak 1988; Robinow and Angert 1998). However, a comprehensive study comparing the development or extent of membrane penetration in related *Epulopiscium* cells of varying diameters has yet to be accomplished.

Membrane-bound compartments and organelles are considered a hallmark of the eukaryotic cell. Internal membrane systems function to organize cellular processes and chemistries. They can be highly specialized in composition and can establish microenvironments within the cell. This is true as well for bacteria that possess intracellular membrane systems or membrane-bound compartments (van Niftrik et al. 2004). These generally serve two known purposes; they (1) provide specialized storage facilities or (2) are sites for the insertion of proteins that enhance a physiological function. Large bacteria like *Thiomargarita* contain a centralized vacuole for nitrate storage (Schulz et al. 1999). The intracytoplasmic membrane systems of phototrophic, nitrifying and methanotrophic bacteria compartmentalize the protein complexes used in these distinct physiological activities. The lipid composition and membrane protein complement of intracellular membranes generally differ from the cytoplasmic membrane. The roles of some membrane systems, such as the membranes that subdivide the cytoplasm of Planctomycetes, are still under investigation (Fuerst 2005). Likewise, the biochemical composition and function of the membranes of large *Epulopiscium* cells have yet to be determined.

3.3 Nucleoids

Large *Epulopiscium* cells contain a tremendous amount of DNA of unknown composition (Fig. 4). The first comprehensive, high-resolution analyses of DNA in large *Epulopiscium* cells demonstrated that in fixed cells, DNA is located primarily in a peripheral layer within the cytoplasm beneath, and in close proximity to, the cell membrane (Robinow and Angert 1998; Robinow and Kellenberger 1994). In addition, highly condensed DNA is located inside

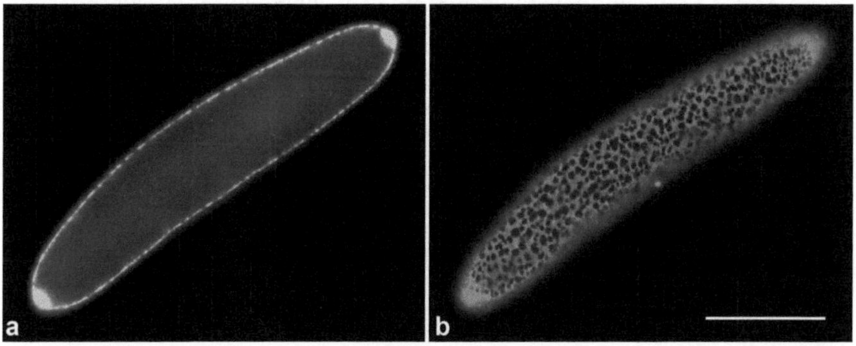

Fig. 4 Two focal planes of a DAPI-stained type B *Epulopiscium* cell. DAPI is a DNA-specific fluorescent dye. **a** A medial focal plane. **b** A focal plane just under the surface of the cell. The bright polar structures in **a** are offspring. Note that only a small proportion of the total DNA is partitioned into the two polar offspring. *Scale bar* represents 50 μm. Reprinted with permission from Robinow and Angert (1998)

developing internal offspring cells. The condensed offspring nucleoids take up fluorescent nucleic acid stains well. When viewed at low magnifications, the brilliant fluorescence of the condensed offspring DNA can overwhelm the more diffuse signal of the mother cell DNA and provide a misleading picture of the distribution of cellular DNA. The relatively intense staining of offspring nucleoids may have led to the original descriptions of offspring as nuclei (Montgomery and Pollak 1988) and subsequent analyses suggesting that parental DNA is equally partitioned into developing offspring cells (Bresler et al. 1998).

In the A morphotype, nucleoids and strands of DNA can extend to deep layers within the cytoplasm of a mother cell; while in the B morphotype the DNA appears superficially to be more confined to a discrete peripheral cytoplasmic stratum (Robinow and Angert 1998). At first glance this may seem like an unusual arrangement of DNA. When stained with fluorescent DNA dyes, most bacteria appear to have a confined centralized nucleoid comprised of DNA and its associated proteins (Robinow and Kellenberger 1994). Through the application of high-resolution imaging, time-lapse microscopy of living cells and classical genetic analyses, a model of the dynamic and highly organized bacterial nucleoid is emerging (Dame 2005; Lewis 2004; Sherratt 2003; Wu 2004). Three-dimensional reconstructions using cryofixed and freeze-substituted cells have mapped the fine tendrils or loops of DNA that extend outward from the compacted bulk of the chromosome (Bohrmann et al. 1991). These relaxed segments of DNA may accommodate transertion, the temporally linked processes of transcription, translation and insertion of proteins into the cytoplasmic membrane (Norris et al. 1996). The vital association of genomic DNA with the cell membrane is even more evident in severe deletion mutants of *Escherichia coli* where nearly 30% of the genome has been eliminated (Hashimoto et al. 2005). In these cells, nucleoids show a marked penchant for the periphery of the cytoplasm. Treatments of either wild type or deletion mutant bacterial cells with translation inhibitors cause the collapse of the bacterial nucleoid into a condensed structure in the middle of the cytoplasm (Bohrmann and Kellenberger 2001; Hashimoto et al. 2005; Zimmerman 2002). Although peripheral localization of DNA in large *Epulopiscium* cells may be a consequence of transertion, it also accommodates the development of intracellular offspring (Angert and Clements 2004; Clements and Bullivant 1991).

We have begun to investigate the composition of the DNA of large *Epulopiscium* cells and envision three possible conformations. The massive amount of DNA may represent one or a few copies of an enormous genome. Alternatively, *Epulopiscium* may have a more typical bacterial size genome (0.5 to 13 Mb) that is regionally amplified or simply multi-copy. It is unlikely that *Epulopiscium* has only a few copies of a huge genome since only a tiny portion of mother cell DNA is partitioned into offspring (Angert and Clements 2004). At this time we cannot rule out either regional or whole genome amplification. Both phenomena have been documented in a number of eukaryotic and prokaryotic cells. In eukaryotes, amplification is important for cell specialization (Tower 2004; Zufall et al. 2005). Whole genome amplification has been described in the bacterial endosymbiont *Buchnera* (Komaki and Ishikawa 1999). In this system, symbiont genome copy number correlates with the physiological status and development of its aphid host. This adaptation may be important for supporting a balanced metabolism in the host (Komaki and Ishikawa 2000). In bacteria, regional amplification of the chromosome allows for an additional level of gene regulation (Reams and Neidle 2004). Amplification in bacteria may play a role in DNA repair, may allow for rapid growth rates and may enhance resistance (Makarova et al. 2001). The distribution of repetitive genetic material throughout the *Epulopiscium* cell could have a significant impact on regional gene expression and response time to environmental signals.

4
Intracellular Offspring Development

Morphologically diverse *Epulopiscium* and related intestinal symbionts produce offspring cells internally (Clements et al. 1989; Fishelson et al. 1985). Surveys of type A populations, taken from fish sacrificed periodically throughout the day, indicate that internal offspring formation follows a circadian cycle (Montgomery and Pollak 1988). Tiny, pole-associated offspring are seen in populations taken from fish early in the morning. As the day progresses, intracellular offspring size, relative to mother cell size, increases steadily, reaching a maximum late in the afternoon. At this time, the active offspring emerge through a tear in the mother cell envelope, and the mother cell is destroyed (Montgomery and Pollak 1988). This process has been referred to as a form of viviparity to distinguish it from the production of dormant internal offspring such as endospores (Angert 2005). Interestingly, in type A populations the average cell size decreases during the night, which may indicate an interval of binary fission (Montgomery and Pollak 1988).

Intracellular offspring formation in *Epulopiscium* spp. is likely related to bacterial endospore formation (Angert 2005; Angert et al. 1996; Angert and Clements 2004). This is based in part on the phylogenetic position of *Epulopiscium* cells within the low G+C Gram-positive, endospore-forming clostridia. The closest known relatives of *Epulopiscium* spp. are a group of gastrointestinal symbionts referred to as *Metabacterium polyspora*. These large (12–35 μm long) symbionts of guinea pigs look a lot like *Epulopiscium*, particularly with respect to the formation of multiple intracellular offspring (Kunstyr et al. 1988). For most endospore-forming, low G+C Gram-positive bacteria, an endospore is produced in response to nutrient deprivation and cell crowding (Errington 2003; Hilbert and Piggot 2004; Stragier and Losick 1996). Sporulation in *M. polyspora*, however, is a normal part of the life cycle (Chatton and Pérard 1913). An individual mother cell can produce as many as nine endospores (Robinow 1957) and sporulation is coordinated with passage of the bacterium through the gastrointestinal tract of its natural host (Angert and Losick 1998). Binary fission occurs rarely and appears to be limited to a brief period in the life cycle, just after a spore germinates. The *M. polyspora* life cycle provides a novel paradigm for the role of endospore formation in the life of a bacterium. Instead of a provisional developmental program called upon in times of stress, sporulation is an important means of reproduction, which may help reinforce the symbiotic association of *M. polyspora* with its host.

Initial development of internal offspring in a large *Epulopiscium* cell bears a striking resemblance to the earliest stages of endospore formation. We have used immunolocalization of the key division protein FtsZ in conjunction with fluorescence microscopy to describe the earliest stages of offspring formation in type B *Epulopiscium* cells (Angert and Clements 2004). Cellular DNA remodeling occurs to oblige the development of intracellular offspring (Fig. 5). DNA accumulates at the tips of a cell and FtsZ localizes in ring-like structures at both poles. The cell divides forming a septum at each pole, trapping DNA in each of the tiny polar partitions formed. The large cell envelops the two, tiny polar cells. Published TEMs of developing type A offspring (Bresler et al. 1998) show membrane forms virtually identical to those seen in forespore engulfment in *Bacillus* or *Clostridium* spp. (see for comparison Fitz-James and Young 1969; Fitz-James 1962; Mackey and Morris 1971). Eventually the growing internal offspring in *Epulopiscium* become free protoplasts within the larger mother cell. These results illustrate mechanisms (the reorganization of cellular DNA, asymmetric division, DNA packaging and engulfment) that are common to endospore formation in *Bacillus subtilis* and intracellular

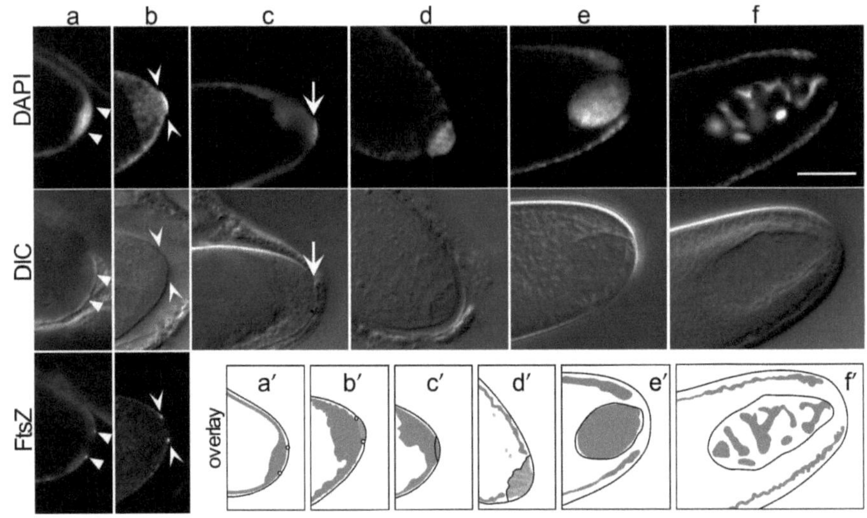

Fig. 5 Early stages of intracellular offspring formation in type B *Epulopiscium*. *Panels* show a close-up view of the tip of a cell where offspring development occurs and represent a progression of developmental stages. Each column **a–f** shows different images of the same cell, illustrating the location of DNA (DAPI fluorescence), a Nomarski differential interference contrast micrograph (DIC) and the immunolocalization of rings of a cell division protein (FtsZ), which marks the incipient division site. In the FtsZ panels, cross sections through FtsZ rings are seen as bright spots near the cell pole. The relative position of FtsZ in the cells, in columns **a** and **b**, is indicated in each *panel* with *arrowheads*. *Panels* **a–d** show a cell still contained within its mother cell while **e** and **f** show cells after release. **a** In anticipation of division, DNA accumulates at the cell pole and FtsZ assembles in a ring. **b** The division septum appears to exert force on the pole-associated DNA. **c** Extreme polar division traps some DNA in a newly formed partition. The division septum is indicated with an *arrow*. **d** As the polar cell is engulfed by the large mother cell, DNA accumulates in the offspring primordium. **e** Engulfment is complete, note the void in the mother cell DNA layer. **f** As the offspring grows, its DNA takes on a peripheral arrangement. *Scale bar* represents 10 μm. *Panels* **a′** through **f′** illustrate the relative positions of DNA (*gray*), cell surfaces (*black lines*) and rings of FtsZ (*small circles*). Reprinted with permission from Angert and Clements (2004)

offspring formation in *Epulopiscium*. One key difference between these processes is that unlike most endospore formers, *Epulopiscium* partitions only a small proportion of mother-cell DNA into the developing offspring (Angert and Clements 2004).

5
Endospores in *Epulopiscium*-like Surgeonfish Symbionts

Early descriptions of *Epulopiscium* and related surgeonfish symbionts suggested that some of these microorganisms produce dormant, and in some cases, phase-bright offspring cells (Clements et al. 1989; Fishelson et al. 1985). As a follow-up on these observations, a recent analysis has shown that several *Epulopiscium*-like morphotypes produce endospores as part of their daily life cycle (Flint et al. 2005). These endospores are best characterized in symbiont populations found in the surgeonfish *Naso lituratus* taken from the waters around Hawaii. *Naso lituratus* harbors a variety of *Epulopiscium*-like morphotypes (Clements et al. 1989), many of which appear to sporulate. Like offspring formation in A and B morphotype cells, endospore production in C and J morphotypes follows a circadian cycle. Small forespores are seen in cells collected in the morning, large forespores are found in cells collected late in the afternoon. Phase-bright endospores are only seen in samples taken from fish at night. Germination begins in the early morning hours.

The most remarkable characteristic of these populations is that endospores are produced every night by a high percentage of cells in a given population of C or J morphotypes. While J morphotype cells regularly reproduce by binary fission, C morphotype cells appear to rely solely on the production of two endospores as their means of cellular reproduction.

The endospores of *Epulopiscium*-like symbionts share features with the endospores of *Bacillus* and *Clostridium* spp. Surgeonfish gut contents with endospores harbor large quantities of dipicolinic acid, a compound that is found in high concentrations in true endospores (Powell and Strange 1953, 1956). Symbiont endospores also appear to have the thick outermost protective layers of cortex and coat (Fig. 6), which are characteristic of mature endospores (Fitz-James and Young 1969; Robinow 1953).

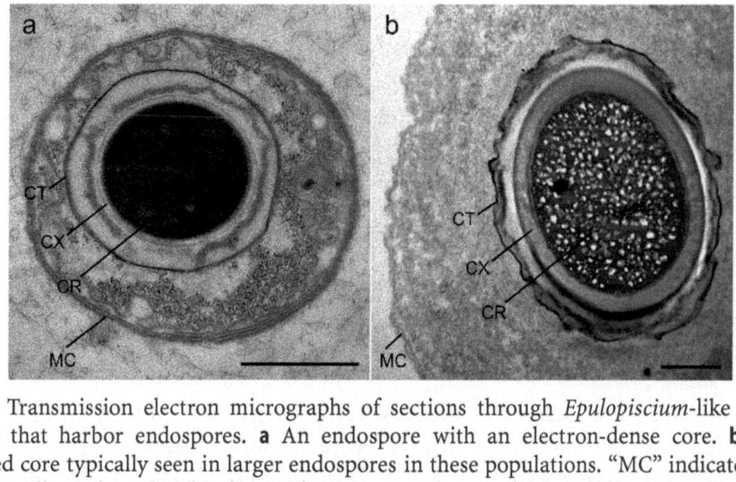

Fig. 6 Transmission electron micrographs of sections through *Epulopiscium*-like symbionts that harbor endospores. **a** An endospore with an electron-dense core. **b** The mottled core typically seen in larger endospores in these populations. "MC" indicates the mother cell envelope, "CT" indicates the outermost layers of the endospore coat(s), "CX" cortex and "CR" is the core. *Scale bar* represents 1 μm. Reprinted with permission from Flint et al. (2005)

Phylogenetic comparisons show that C and J morphotypes fall among the previously described A and B morphotypes (Flint et al. 2005). This raises the possibility that endospore formation is a characteristic of all *Epulopiscium* and related surgeonfish symbionts. Currently, the A and B morphotypes are considered viviparous or live offspring-bearing and we have no evidence that these largest *Epulopiscium* morphotypes regularly produce endospores. It is possible that these cells can produce spores but only under special circumstances or that these cells alternate between sporulation and viviparity, as seen in other low G+C Gram-positive bacteria (Angert 2005). Further analyses are needed to resolve these possibilities but surely reproductive strategy will have a significant impact on the acquisition and maintenance of intestinal symbiont populations.

6
Prepared for Feast and Famine?

Frequently, other inclusions are observed in *Epulopiscium* cells. Published TEMs of thin sections of surgeonfish symbionts show the presence of multiple small inclusions harbored within the cytoplasm of internal offspring or possibly forespores (Clements et al. 1989; Montgomery and Pollak 1988). It is difficult to determine from these micrographs if the inclusions are membrane bound (Shively 1974). Although the composition of these bodies has yet to be characterized they probably consist of some kind of storage compound, such as polyphosphate, polysaccharide or polyhydroxybutryate. Surgeonfish symbionts, particularly those populations that are stably maintained in an individual host, would be subjected to daily fluctuations of available nutrients. In endospore-forming symbionts, these extremes correlate with the production of dormant cells, as endospores are produced when the fish host is not actively feeding (Flint et al. 2005). It is possible that some of these symbionts store abundant nutrients available to them during the day in inclusions that can be used when nutrient levels drop. Clearly more information about the composition of these inclusions is needed to determine their role in population maintenance.

7
Concluding Remarks, There's Much to be Done

Ultrastructural studies of these mystifying surgeonfish intestinal symbionts have identified a variety of unusual cellular modifications. The impact of each of these adaptations on *Epulopiscium* biology remains to be fully elucidated. Flagella are important for maintaining the position of the cell within the host intestinal tract but may provide an additional benefit in enhancing diffusive exchange. There appears to be a correlation between increased cell volume and the development of an extensive internal membrane system and amplified genetic material. The production of inclusions within the cytoplasm of offspring or forespores, as well as an unusual circadian reproductive cycle, may further enhance the ability of *Epulopiscium* to endure predictable changes in nutrient availability. Future high-resolution studies and comparative analyses will provide additional insight into the roles of complex cytoarchitecture in the maintenance of large cell size and unusual reproductive strategies in maintaining *Epulopiscium* populations.

Acknowledgements Research in the author's laboratory is supported by the National Science Foundation (USA). The author is grateful to the students and collaborators who have made this work possible.

References

Angert ER (1995) Molecular Phylogenetic Analyses of *Epulopiscium*-like Symbionts. PhD Thesis, Department of Biology. Indiana University, Bloomington, IN, p 196

Angert ER (2005) Alternatives to binary fission in bacteria. Nat Rev Microbiol 3:214–224

Angert ER, Clements KD (2004) Initiation of intracellular offspring in *Epulopiscium*. Mol Microbiol 51:827–835

Angert ER, Losick RM (1998) Propagation by sporulation in the guinea pig symbiont *Metabacterium polyspora*. Proc Natl Acad Sci USA 95:10218–10223

Angert ER, Brooks AE, Pace NR (1996) Phylogenetic analysis of *Metabacterium polyspora*: clues to the evolutionary origin of daughter cell production in *Epulopiscium* species, the largest bacteria. J Bacteriol 178:1451–1456

Angert ER, Clements KD, Pace NR (1993) The largest bacterium. Nature 362:239–241

Blakemore RP, Canale-Parola E (1973) Morphological and ecological characteristics of *Spirochaeta plicatilis*. Arch Mikrobiol 89:273–289

Bohrmann B, Kellenberger E (2001) Cryosubstitution of frozen biological specimens in electron microscopy: use and application as an alternative to chemical fixation. Micron 32:11–19

Bohrmann B, Villiger W, Johansen R, Kellenberger E (1991) Coralline shape of the bacterial nucleoid after cryofixation. J Bacteriol 173:3149–3158

Bresler V, Montgomery WL, Fishelson L, Pollak PE (1998) Gigantism in a bacterium, *Epulopiscium*, correlates with complex patterns in arrangement, quantity, and segregation of DNA. J Bacteriol 180:5601–5611

Chatton E, Pérard C (1913) Schizophytes du caecum du cobaye. II *Metabacterium polyspora* n.g., n.s. C R Hebd Soc Biol (Paris) 74:1232–1234

Clements KD (1997) Fermentation and gastrointestinal microorganisms of fishes. In: Mackie RI, White BA (eds) Gastrointestinal Microbiology: Gastrointestinal Ecosystems and Fermentations. Chapman & Hall, New York, p 156–198

Clements KD, Bullivant S (1991) An unusual symbiont from the gut of surgeonfishes may be the largest known prokaryote. J Bacteriol 173:5359–5362

Clements KD, Sutton DC, Choat JH (1989) Occurrence and characteristics of unusual protistan symbionts from surgeonfishes Acanthuridae of the Great Barrier Reef Australia. Marine Biol 102:403–412

Dame RT (2005) The role of nucleoid-associated proteins in the organization and compaction of bacterial chromatin. Mol Microbiol 56:858–870
Errington J (2003) Regulation of endospore formation in *Bacillus subtilis*. Nat Rev Microbiol 1:117–126
Fenchel T (1994) Motility and chemosensory behaviour of the sulfur bacterium *Thiovulum majus*. Microbiol 140:3109–3116
Fishelson L, Montgomery WL, Myrberg AA (1985) A unique symbiosis in the gut of a tropical herbivorous surgeonfish (Acanthuridae: Teleostei) from the Red Sea. Science 229:49–51
Fitz-James PC (1962) Morphology of spore development in *Clostridium pectinovorum*. J Bacteriol 84:104–114
Fitz-James P, Young E (1969) Morphology of sporulation. In: Gould GW, Hurst A (eds) The Bacterial Spore. Academic Press, London, p 39–72
Flint JF, Drzymalski D, Montgomery WL, Southam G, Angert ER (2005) Nocturnal production of endospores in natural populations of *Epulopiscium*-like surgeonfish symbionts. J Bacteriol 187:7460–7470
Fuerst JA (2005) Intracellular compartmentalization in planctomycetes. Annu Rev Microbiol 59:299–328
Hashimoto M, Ichimura T, Mizoguchi H, Tanaka K, Fujimitsu K, Keyamura K, Ote T, Yamakawa T, Yamazaki Y, Mori H, Katayama T, Kato J (2005) Cell size and nucleoid organization of engineered *Escherichia coli* cells with a reduced genome. Mol Microbiol 55:137–149
Hilbert DW, Piggot PJ (2004) Compartmentalization of gene expression during *Bacillus subtilis* spore formation. Microbiol Mol Biol Rev 68:234–262
Jannasch HW, Nelson DC, Wirsen CO (1989) Massive natural occurence of unusually large bacteria (*Beggiatoa* spp.) at a hydrothermal deep-sea vent site. Nature 342:834–836
Komaki K, Ishikawa H (1999) Intracellular bacterial symbionts of aphids possess many genomic copies per bacterium. J Mol Evol 48:717–722
Komaki K, Ishikawa H (2000) Genomic copy number of intracellular bacterial symbionts of aphids varies in response to developmental stage and morph of their host. Insect Biochem Mol Biol 30:253–258
Kunstyr I, Schiel R, Kaup FJ, Uhr G, Kirchhoff H (1988) Giant gram-negative noncultivable endospore-forming bacteria in rodent intestines. Naturwissenschaften 75:525–527
Lewis PJ (2004) Bacterial subcellular architecture: recent advances and future prospects. Mol Microbiol 54:1135–1150
Losick R, Shapiro L (1999) Changing views on the nature of the bacterial cell: from biochemistry to cytology. J Bacteriol 181:4143–4145
Mackey BM, Morris JG (1971) Ultrastructural changes during sporulation in *Clostridium pasteurianum*. J Gen Microbiol 66:1–13
Macnab RM (2003) How bacteria assemble flagella. Annu Rev Microbiol 57:77–100
Makarova KS, Aravind L, Wolf YI, Tatusov RL, Minton KW, Koonin EV, Daly MJ (2001) Genome of the extremely radiation-resistant bacterium *Deinococcus radiodurans* viewed from the perspective of comparative genomics. Microbiol Mol Biol Rev 65:44–79
Montgomery WL, Pollak PE (1988) *Epulopiscium fishelsoni* n.g., n.s., a protist of uncertain taxonomic affinities from the gut of an herbivorous reef fish. J Protozool 35:565–569
Norris V, Turnock G, Sigee D (1996) The *Escherichia coli* enzoskeleton. Mol Microbiol 19:197–204

Powell JF, Strange RE (1953) Biochemical changes occurring during the germination of bacterial spores. Biochem J 54:205–209

Powell JF, Strange RE (1956) Biochemical changes occurring during sporulation in *Bacillus* species. Biochem J 63:661–668

Reams AB, Neidle EL (2004) Selection for gene clustering by tandem duplication. Annu Rev Microbiol 58:119–142

Robinow CF (1953) Spore structure as revealed by thin sections. J Bacteriol 66:300–311

Robinow CF (1957) Kurzer Hinweis auf *Metabacterium polyspora*. Z Tropenmed Parasitol 8:225–227

Robinow C, Angert ER (1998) Nucleoids and coated vesicles of "Epulopiscium" spp. Arch Microbiol 170:227–235

Robinow C, Kellenberger E (1994) The bacterial nucleoid revisited. Microbiol Rev 58:211–232

Schulz HN, Brinkhoff T, Ferdelman TG, Marine MH, Teske A, Jorgensen BB (1999) Dense populations of a giant sulfur bacterium in Namibian shelf sediments. Science 284:493–495

Shapiro L, Losick R (1997) Protein localization and cell fate in bacteria. Science 276:712–718

Sherratt DJ (2003) Bacterial chromosome dynamics. Science 301:780–785

Shively JM (1974) Inclusion bodies of prokaryotes. Annu Rev Microbiol 28:167–187

Stevens CE, Hume ID (1995) Comparative Physiology of the Vertebrate Digestive System. Cambridge University Press, Cambridge, UK

Stragier P, Losick R (1996) Molecular genetics of sporulation in *Bacillus subtilis*. Annu Rev Genet 30:297–341

Thar R, Fenchel T (2005) Survey of motile microaerophilic bacterial morphotypes in the oxygen gradient above marine sulfidic sediment. Appl Environ Microbiol 71:3682–3691

Tower J (2004) Developmental gene amplification and origin regulation. Annu Rev Genet 38:273–304

van Niftrik LA, Fuerst JA, Sinninghe Damste JS, Kuenen JG, Jetten MS, Strous M (2004) The anammoxosome: an intracytoplasmic compartment in anammox bacteria. FEMS Microbiol Lett 233:7–13

Wu LJ (2004) Structure and segregation of the bacterial nucleoid. Curr Opin Genet Dev 14:126–132

Zimmerman SB (2002) Toroidal nucleoids in *Escherichia coli* exposed to chloramphenicol. J Struct Biol 138:199–206

Zufall RA, Robinson T, Katz LA (2005) Evolution of developmentally regulated genome rearrangements in eukaryotes. J Exp Zoolog B Mol Dev Evol 304:448–455

Part II
Additional Complex Intracellular Structures

Microbiol Monogr (2)
J.M. Shively: Complex Intracellular Structures in Prokaryotes
DOI 10.1007/7171_029/Published online: 20 May 2006

Cytoskeletal Elements in Prokaryotes

Waldemar Vollmer

Mikrobielle Genetik, Universität Tübingen,
Auf der Morgenstelle 28, 72076 Tübingen, Germany
waldemar.vollmer@uni-tuebingen.de

Abstract In the recent decade, our view of the sub-cellular organization of bacterial cells has been revolutionized by the application of modern cell biology methods. Cytoskeletal proteins that are precisely targeted to specific cellular locations and that assemble in dynamic filamentous structures have been discovered. Moreover, most bacterial species contain homologous proteins of eukaryotic tubulin and actin, which, like in eukaryotes, have functions in cell shape maintenance, cell division, DNA movement or alignment of organelles. Additional cytoskeletal elements are present in bacterial species with more complex cell shapes, e.g., in the curved *Caulobacter crescentus*. Cell wall-free Mollicutes contain cytoskeletal proteins for maintaining their cell shape and for motility.

1
Introduction

Eukaryotic cells have a complex and dynamic cytoskeleton required for the determination and maintenance of cell shape, for the movement of DNA and organelles, and for cell division. The eukaryotic cytoskeleton is composed of three major components, tubulin, actin and intermediate filaments (IF), each of which can form long filamentous structures and/or sheets. Until recently, the absence of a cytoskeleton has been regarded as a hallmark of prokaryotes. A landmark publication in 1991 reported about the formation of an intracellular, ring-like structure by the cell division protein FtsZ (Bi and Lutkenhaus 1991). One year later, it was (correctly) predicted that several bacterial proteins (DnaK, FtsA, ParM and MreB) should adopt a similar fold as the ATPase domain of actin (Bork et al. 1992). Since then, the combination of sequence comparison, modern cell biology methods for intracellular protein localization, and structural biology have led to the discovery of bacterial tubulins, actins and an intermediate filament like protein, that not only share key structural features with their eukaryotic analogues, but also have similar physiological roles (Graumann 2004; Löwe et al. 2004). In addition, certain cytoskeleton elements have been identified only in bacteria and (so far) not in eukaryotes. This chapter provides a brief summary about the prokaryotic cytoskeleton, a rapidly expanding research field.

2
Prokaryotic Tubulins

FtsZ is the bacterial tubulin and is widely conserved in prokaryotes, chloroplasts and some mitochondria. FtsZ is an essential cell division protein. Self-assembly of FtsZ leads to the formation of a membrane-associated ring-structure (the Z-ring) at the site of cell or organelle division (Fig. 1). The Z-ring is stabilized by the actin-like FtsA (see below) and serves as a platform for the assembly of more than 10 other essential cell division proteins in a sequential manner. Contraction of the Z-ring complex (the divisome) at the leading edge of constriction, together with new cell envelope synthesis, leads to fission of the cell or organelle.

FtsZ is a structural homologue of tubulins, although they share < 20% amino acid sequence identity (Löwe and Amos 1998). Like tubulins, FtsZ is able to bind to and hydrolyze GTP and to assemble into protofilaments in

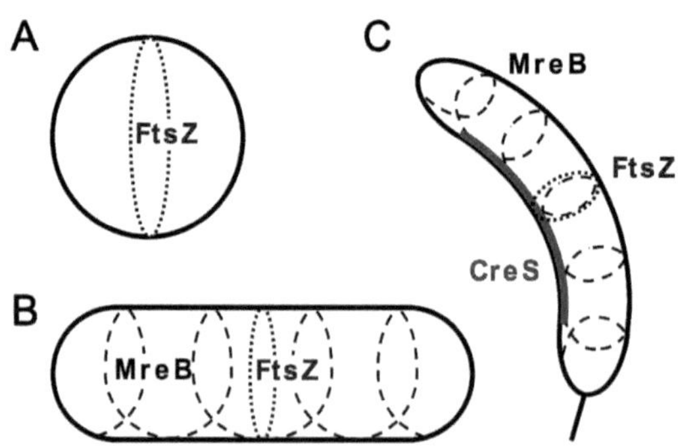

Fig. 1 Cytoskeletal elements and cell shape. During cell division, the tubulin-like FtsZ forms a ring at mid-cell in round (**A**, *Staphylococcus aureus*), rod-shaped (**B**, *Escherichia coli*), and crescent-shaped (**C**, *Caulobacter crescentus*) bacteria. The actin-like MreB forms intracellular helical structures in most rod-shaped bacteria. The intermediate filament-like CreS locates at one side of the Caulobacter cell and is responsible for its crescent-shape

vitro. The nucleotide is bound to the interface of two FtsZ molecules, by this forming the active site for hydrolysis. Unlike tubulin, FtsZ does not form microtubules, and the precise structure of the Z-ring inside the cell is unknown. It is also unknown whether the Z-ring generates an inwards directed force. During division, only about 30% of the cellular FtsZ is assembled into the Z-ring. The ring is surprisingly dynamic with a half-life of about 30 s or less, due to the constant release of FtsZ molecules from the ring and incorporation of new monomers. Surprisingly, FtsZ (outside the Z-ring) oscillates very rapidly in helical waves through the cell by a yet unknown mechanism (Thanedar and Margolin 2004).

Some members of the phylum Verrucomicrobia contain intracellular microtubule-like structures of unknown function. In *Prosthecobacter dejongeii*, two bacterial tubulins (BtubA and BtubB) were identified that have a higher sequence similarity to eukaryotic tubulins than to FtsZ. The structures of BtubA and BtubB are highly similar to tubulin, and BtubA/B form tubulin-like protofilaments (Schlieper et al. 2005). Presumably, *Prosthecobacter* has obtained the corresponding genes from a eukaryotic cell by horizontal gene transfer.

3
Prokaryotic Actins

FtsA is an essential cell division protein and interacts with the *C*-terminus of FtsZ. FtsA stabilizes the Z-ring and tethers it to the cell membrane via a conserved membrane targeting sequence. The structure of FtsA from *Thermotoga maritima* shows the typical architecture of actin, with the exception of one subdomain, that is moved to the other side of the molecule (Löwe et al. 2004). FtsA binds ATP, but a hydrolytic activity could not be demonstrated yet, possibly because of a missing binding partner. So far, only FtsA from *Streptococcus pneumoniae* was shown to polymerize in vitro to protofilaments in a nucleotide-dependent manner (Lara et al. 2005).

Besides FtsA, cells of *Bacillus subtilis* contain three other actin homologues: MreB, Mbl and MreBH. MreB-like proteins are found almost exclusively in rod-shaped bacteria and not in coccal species, and they appear to have a major role in establishment of a rod-shape (Jones et al. 2001). The structure of MreB is very similar to that of actin (van den Ent et al. 2001), and MreB polymerizes in an ATP-dependent manner. MreB and Mbl form independent dynamic, helical filaments on the inner surface of the cytoplasmic membrane, coiling along the long axis of the rod-shaped cell (Fig. 1). Mbl filaments are more extended, ranging from pole-to-pole, whereas MreB filaments have a shorter lead and are absent from the cell poles. The filaments of MreB-like proteins are associated with four different cellular functions: growth of the rod-shaped cell, movement of DNA (segregation of chromosomes, partitioning of low-copy number plasmids), polar positioning of proteins, and alignment of organelles.

The shape of a bacterial cell is determined and maintained by an exoskeleton, the murein (or peptidoglycan) sacculus, that is multi-layered in Gram-positive bacteria and mainly single-layered in the Gram-negative *Escherichia coli.* During the elongation of (most) rod-shaped bacteria, the sacculus is enlarged by incorporation of precursors at the side wall by murein synthases, the penicillin-binding proteins (PBPs). Growth of the murein sacculus requires the well-coordinated activities of both, murein synthases and hydrolases, which probably assemble into multi-enzyme complexes (Vollmer and Höltje 2001). In *B. subtilis*, the sites of murein incorporation appear to follow a helical path on the cell surface (Daniel and Errington 2003), and this pattern is dependent on the presence of the intracellular Mbl helical filaments. Therefore, it is possible that the insertion of new murein precursors into the sacculus is directed by the Mbl filaments, maybe via the MreC and MreD proteins. MreB, MreC and MreD form a membrane-bound complex in *E. coli*, and the *Caulobacter crescentus* MreC, that has a major exocytoplasmic domain, localizes in a helical pattern along the cell length and interacts with PBPs. The molecular mechanism(s) by which Mbl and MreB control wall growth during elongation of rod-shaped bacteria are still unknown.

Bacterial chromosomes are highly condensed into so-called nucleoids, located in the centre of the cell. Before cell division takes place, the DNA is replicated and the sister chromosomes are segregated into both cell halves. Depletion of MreB in *B. subtilis* results in a defect in chromosome segregation before a change in cell shape is visible. The duplication of the chromosome occurs at a stationary DNA replicase complex, localized at the cell centre, through which the DNA travels. The dynamic MreB filaments could drive segregation by pushing each of the sister chromosomes towards another cell pole (Graumann and Soufo 2004). This (yet to be proven) mechanism would be analogous to the known partitioning system of low copy number plasmids, employing the actin-like ParM protein, the ParC DNA-binding protein, and the *parC* locus on the plasmid (Moller-Jensen et al. 2002). ParC binds to the *parC* region and pairs the plasmid after replication and prior to segregation. Plasmid-bound ParC induces filament formation of ParM, which pushes the plasmids away from each other towards the opposite poles of the cell.

Another function of the helical MreB filament appears to be the targeting of certain proteins to the poles of the rod-shaped cells. Among them are the aspartate chemoreceptor Tar and the *Shigella* virulence factor IcsA that, interestingly, promotes bacterial motility in the cytosol of an infected eukaryotic cell by inducing polymerization of actin of the host cell. In the absence of MreB, polar targeting of IcsA and Tar is impaired (Nilsen et al. 2005). The molecular mechanism, by which proteins are targeted to the cell poles, with the direct or indirect requirement for MreB, is not known.

Yet another function of actin-like proteins was found recently in a magnetotactic bacterium, *Magnetospirillum gryphiswaldense*, that contains unique intracellular, nano-sized organelles (magnetosomes) required to navigate along the magnetic field of the earth (see Chap. 7). For full functioning, the magnetosomes need to be assembled into a straight chain, despite their tendency to agglomerate. For this, the magnetosomes are aligned along cytoskeleton filaments of the actin-like MamK protein (Komeili et al. 2005), via the acidic protein MamJ, that interacts with both, the surface of the magnetosomes and the MamK filaments (Scheffel et al. 2005).

4
The Intermediate Filament Protein CreS

C. crescentus is a crescent-shaped bacterium and a model for bacterial polarity and development. The cytoskeleton protein CreS is presently the only known bacterial equivalent to eukaryotic intermediate filament proteins. Like those, CreS has large coiled-coil segments and assembles spontaneously into filaments. CreS localizes at the inner cell curvature underneath the cytoplasmic membrane (Fig. 1C). *Caulobacter* mutant cells lacking CreS are not crescent-shaped but grow as straight rods. The mechanisms by which CreS localizes to one side of the cell and by which it induces curved cell growth are not known (Ausmees et al. 2003).

5
The Oscillating Min Proteins

The *E. coli* MinCDE system, which has (so far) no analogue in eukaryotic organisms, prevents cell division near the cell poles, which would lead to round, DNA-free minicells (Raskin and de Boer 1999). Surprisingly, a MinC–MinD complex oscillates from pole to pole every 30 to 50 s, driven by the following mechanism: MinC–MinD (ATP-form) binds to the cytoplasmic membrane at one pole in a cooperative manner, such that all cellular MinCD localizes in one cell halve. Close to mid-cell, MinE assembles to a membrane-bound ring structure that moves towards the pole, replacing MinCD from the membrane by increasing the ATP hydrolase activity of MinD (the ADP form of MinD does not bind to the membrane). After the replacement of ADP by ATP, the MinC–MinD (ATP) complex assembles at the opposite cell pole, and the cycle starts again. Recent findings indicate that the Min proteins are organized in and oscillate along extended coiled structures associated with the cytoplasmic membrane (Shih et al. 2003). MinC is an inhibitor of Z-ring formation; therefore, the pole-to-pole oscillation of MinC prevents cell division near the poles and allows division to take place only at mid-cell, where the MinC concentration is lowest. In *B. subtilis* MinE is replaced by DivIVA and the system does not oscillate. DivIVA localizes at the cell poles and permanently recruits the MinCD cell division inhibitor to prevent polar cell division.

6
Cytoskeleton Elements in Cell-wall-less Prokaryotes

Mollicutes (*Mycoplasma*, *Spiroplasma* and *Acholeplasma*) are phylogenetically related to Gram-positive bacteria, but they lack a cell wall exoskeleton and instead have a cholesterol-containing cell membrane. Yet, these organisms have defined cell shapes, ranging from coccal (*Acholeplasma*) to spiral shape (*Spiroplasma*). Ultra-structural studies revealed that Mollicutes have internal cytoskeletal structures for cell shape maintenance and motility (Trachtenberg 1998). *Mycoplasma pneumoniae* contains a highly complex asymmetric multi-component cytoskeleton, that is only poorly characterized. In *Spiroplasma* species, the cytoskeleton is made of a unique fibril protein that forms a flat, helical ribbon. The architecture of the helical cytoskeleton elements of *Spiroplasma melliferum*, probably including MreB filaments, is described in detail in Chap. 13.

References

Ausmees N, Kuhn JR, Jacobs-Wagner C (2003) The bacterial cytoskeleton: an intermediate filament-like function in cell shape. Cell 115:705–713

Bi E, Lutkenhaus J (1991) FtsZ ring structure associated with division in *Escherichia coli*. Nature 354:161–164

Bork P, Sander C, Valencia A (1992) An ATPase domain common to prokaryotic cell cycle proteins, sugar kinases, actin, and hsp70 heat shock proteins. Proc Natl Acad Sci USA 89:7290–7294

Daniel RA, Errington J (2003) Control of cell morphogenesis in bacteria: two distinct ways to make a rod-shaped cell. Cell 113:767–776

Graumann PL (2004) Cytoskeletal elements in bacteria. Curr Opin Microbiol 7:565–571

Graumann PL, Soufo HJD (2004) An intracellular actin motor in bacteria? BioEssays 26:1209–1216

Jones LJ, Carballido-Lopez R, Errington J (2001) Control of cell shape in bacteria: helical, actin-like filaments in *Bacillus subtilis*. Cell 104:913–922

Komeili A, Newman DK, Jensen GJ (2005) Magnetosomes are cell membrane invaginations organized by the actin-like protein MamK. Science 311:242–245

Lara B, Rico AI, Petruzzelli S, Santona A, Dumas J, Biton J, Vicente M, Mingorance J, Massidda O (2005) Cell division in cocci: localization and properties of the *Streptococcus pneumoniae* FtsA protein. Mol Microbiol 55:699–711

Löwe J, Amos LA (1998) Crystal structure of the bacterial cell-division protein FtsZ. Nature 391:203–206

Löwe J, van den Ent F, Amos LA (2004) Molecules of the bacterial cytoskeleton. Annu Rev Biophys Biomol Struct 33:177–198

Moller-Jensen J, Jensen RB, Löwe J, Gerdes K (2002) Prokaryotic DNA segregation by an actin-like filament. Embo J 21:3119–3127

Nilsen T, Yan AW, Gale G, Goldberg MB (2005) Presence of multiple sites containing polar matrial in spherical *Escherichia coli* cells that lack MreB. J Bacteriol 187:6187–6196

Raskin DM, de Boer PA (1999) Rapid pole-to-pole oscillation of a protein required for directing division to the middle of *Escherichia coli*. Proc Natl Acad Sci USA 96:4971–4976

Scheffel A, Gruska M, Faivre D, Linaroudis A, Plitzko JM, Schüler D (2005) An acidic protein aligns magnetosomes along a filamentous structure in magbetotactic bacteria. Nature 440:110–114

Schlieper D, Oliva MA, Andreu JM, Löwe J (2005) Structure of bacterial tubulin BtubA/B: evidence for horizontal gene transfer. Proc Natl Acad Sci USA 102:9170–9175

Shih YL, Le T, Rothfield L (2003) Division site selection in *Escherichia coli* involves dynamic redistribution of Min proteins within coiled structures that extend between the two cell poles. Proc Natl Acad Sci USA 100:7865–7870

Thanedar S, Margolin W (2004) FtsZ exhibits rapid movement and oscillation waves in helix-like patterns in *Escherichia coli*. Curr Biol 14:1167–1173

Trachtenberg S (1998) Mollicutes - wall-less bacteria with internal cytoskeleton. J Struct Biol 124:244–256

van den Ent F, Amos LA, Löwe J (2001) Prokaryotic origin of the actin cytoskeleton. Nature 413:39–44

Vollmer W, Höltje J-V (2001) Morphogenesis of *Escherichia coli*. Curr Opin Microbiol 4:625–633

Microbiol Monogr (2)
J.M. Shively: Complex Intracellular Structures in Prokaryotes
DOI 10.1007/7171_030/Published online: 20 May 2006

Cryo-electron Tomography Reveals the Architecture of a Bacterial Cytoskeleton

Julia Kürner · Wolfgang Baumeister (✉)

Department of Structural Biology, Max Planck Institute of Biochemistry,
Am Klopferspitz 18, 82152 Martinsried, Germany
baumeist@biochem.mpg.de

Abstract Cryo-electron tomography (cryo-ET) is an emerging imaging technique that combines the power of three-dimensional (3-D) imaging of large pleomorphic structures such as cells or organelles with a close-to-life preservation of the sample. At the present resolution of approximately 4 nm, supramolecular structures can be studied in unperturbed cellular environments. Application of this method to prokaryotic cells has provided new insights into the structural organization of the bacterial cytoskeleton. In *Spiroplasma melliferum*, it is built of three parallel ribbons of thicker and thinner filaments that span the cell from one end to the other just underneath the cell membrane, thereby giving the cell its helical shape and enabling it to swim by processive change in body helicity.

1 Introduction

Structural investigations of biological materials at levels from molecules to cells provide important contributions to the understanding of their function. X-ray crystallography gives structural insights into molecular complexes at atomic resolution. Some macromolecular complexes are robust enough to withstand biochemical isolation and purification procedures, others assemble only transiently or are held together by weak forces. Ideally, such structures should be studied in their native environment, i.e., the intact cell. Although conventional electron microscopic methods based on fixation, dehydration or plastic embedding of cells allow insights into the cellular context, they are prone to artifacts and the interpretation of the images at the molecular level is very problematic. Therefore, new approaches are needed to analyze the molecular interaction patterns and networks that underlie cellular behavior. Electron tomography (ET) in combination with cryo-sample preparation technique allows a non-invasive three-dimensional (3-D) observation of the molecular architecture of complexes and assemblies in a close-to-native state (Baumeister et al. 1999). Hence, cryo-ET can bridge the gap between structural studies at the molecular and cellular level (Baumeister 2005). Cryotomograms of whole cells with molecular resolution are essentially images of the cell's entire proteome and can be used to reveal and interpret supramolecular architectures or protein networks. For example, cryo-ET has been used to explore in 3-D the actin cytoskeleton of *Dictyostelium discoideum* cells (Medalia et al. 2002). Here, we describe the structural organization of the bacterial cytoskeleton of *Spiroplasma melliferum* as revealed by cryo-ET of whole cells.

2 Structural Elements in Prokaryotes

For decades, the cytoskeleton has been seen as a distinctive feature of eukaryotes, but evidence has accumulated recently, that bacteria can also have a primordial cytoskeleton that determines and maintains their shape (see e.g., Ausmees et al. 2003; Jones et al. 2001; Trachtenberg 1998). Since, e.g., Mollicutes (*Mycoplasma*, *Acholeplasma* and *Spiroplasma*) (Tully 1992) are enveloped by only a cell membrane and still show distinct morphologies, the cell wall can obviously not be the primary determinant of cell shape. Furthermore, the peculiar mode of movement of these bacteria in the absence of appendages normally implicated in motility (e.g., flagella or secretion organelles) makes the existence of an internal bacterial cytoskeleton likely.

There has been an ongoing controversy about the existence of actin or actin-like proteins in prokaryotes, including Mollicutes (for review, see Neimark 1983). Bacterial homologues of actin were first identified in *Bacillus subtilis* when the cell-shape determinants MreB and Mbl (MreB-like) were shown to assemble into helical filamentous structures that run in spirals around the periphery of the cell under the cytoplasmic membrane in *B. subtilis* (Jones et al. 2001). The major role of MreB lies in the control of cell width, as depletion of MreB induces the formation of rounded, inflated cells which is ultimately lethal (Formstone and Errington 2005). In *Caulobacter crescentus*, MreB forms again spirals that encircle the whole cell (Figge et al. 2004; Gitai et al. 2004). Additionally, these cells possess the intermediate filament-like protein crescentin, which in the helical stationary phase cells builds a helix along the cell membrane. Without these crescentin filaments, the cells become rod-shaped (Ausmees et al. 2003).

3 The Cytoskeleton of *Spiroplasma melliferum* in 3-D

The isolation of a cytoskeleton of different *Spiroplasma* species by various methods was reported by Williamson (1974) and Townsend et al. (1980). Using electron microscopy and SDS gel electrophoresis, it was shown that *S. melliferum* possesses a cytoskeletal ribbon comprising ~ 4–5 nm wide fibrils with an axial repeat of ~ 9 nm; these fibrils form pairs within the ribbon and are composed of the fibril protein (*fib*) (Townsend et al. 1980, Williamson et al. 1991). Townsend and Plaskitt (1985) further used immunogold staining of thin sections with anti-p55-antibody to localize the ribbon built of the fibril protein within *S. melliferum* cells. A detailed description of the structure and localization of the cytoskeleton of *S. melliferum* was presented by Trachtenberg and Gilad (Trachtenberg 1998; Trachtenberg and Gilad 2001). They described this cytoskeleton as a flat, monolayered, membrane-bound ribbon comprising six or seven pairs of fibrils following the shortest helical path through the whole cell. Since so far, however, only 2-D images of whole cells, sections or isolated filaments have been obtained, these results provided only part of the truth.

Tomographic studies of intact vitrified *S. melliferum* cells revealed a more complex cytoskeletal structure (Kürner et al. 2005). Two types of filaments (thin and thick) form three parallel ribbons that span the entire cell just underneath the cell membrane (Fig. 1a,b). Profiles of the filament grey values showed that the two outer ribbons are built of five thick filaments with a spacing of about 11 nm, while the region in between is composed of nine thin filaments with a spacing of about 4 nm (Fig. 1c). Although the ribbons cannot be visualized as continuous bands along the cell membrane of the whole cell due to the "missing wedge" problem (Koster et al. 1997), it becomes clear that the ribbons follow, in parallel, a helical path from the blunt to the tapered end of the cell (Fig. 2). The two different filamentous structures presumably comprise the fibril protein (Townsend et al. 1980) and the actin-like protein MreB, which were shown to exist in this bacterium by Western Blot and sequence analysis. Furthermore, the motility modes of these swimming cells could be explained and modeled in the light of these new structural data. One possibility is to change the handedness of the cell in order to move it forward. Shaevitz et al. recently verified this mechanism by measuring cell kinematics during free swimming of *Spiroplasma* cells (Shaevitz et al. 2005). They showed that propulsion is generated by the propagation of kink pairs down the length of the cell body. At these kinks, the handedness of the cell changes (Fig. 3). This handedness change can be achieved by differentially changing the length of one of the two outer cytoskeletal ribbons as a result of conformational changes of its subunits. A similar mechanism has already been proposed by Trachtenberg and Gilad (2001). Our assumptions for this movement are that the ribbons are con-

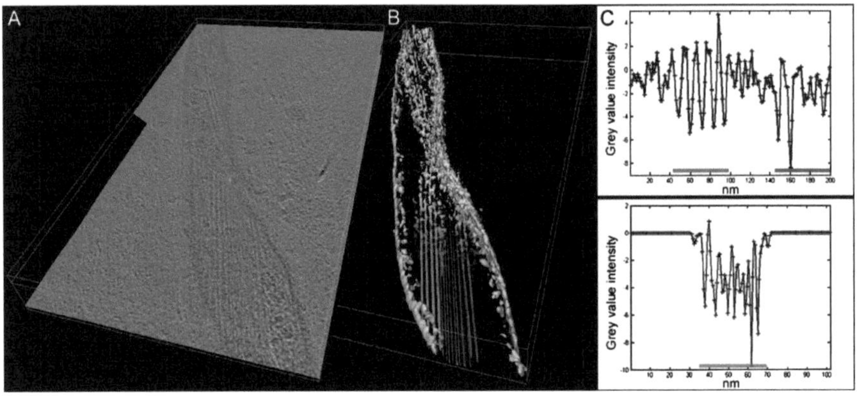

Fig. 1 The fine structure of the bacterial cytoskeleton of *S. melliferum*: (**A**) Superimposed slices from a tomogram, (**B**) corresponding 3-D visualization and (**C**) profiles of the filament grey values. Cryo-electron tomography not only reveals the helical course of the cytoskeleton but also its composition of thicker (*red and green*) and thinner (*purple*) filaments that form three parallel ribbons underneath the cell membrane (*blue*). The two outer ribbons are built of five thick filaments (**C**, *top*) with a spacing of 11 nm, while the region in between comprises nine thin filaments (**C**, *bottom*) with a spacing of 4 nm

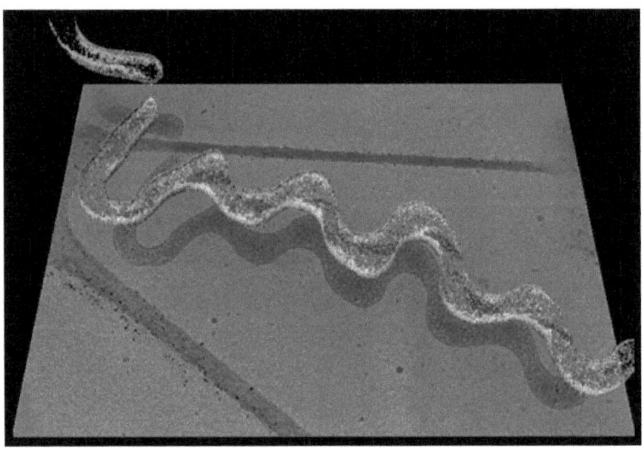

Fig. 2 Tomogram slice with superimposed 3-D visualization of a whole *S. melliferum* cell, which illustrates the localization and course of the two outer cytoskeletal ribbons (*red and green*) underneath the cell membrane (*blue*). It shows that the ribbons follow, in parallel, a helical path through the cell

nected to each other and to the cell membrane, and that the inner ribbon functions elastically during cell movement. So far, the resolution of our tomograms does not allow predictions about any anchoring point. However, Kruse and his colleagues (Kruse et al. 2005) recently suggested a model, how MreB filaments are attached to the cell membrane via a membrane-bound complex consisting of the three proteins MreB, MreC and MreD. Further-

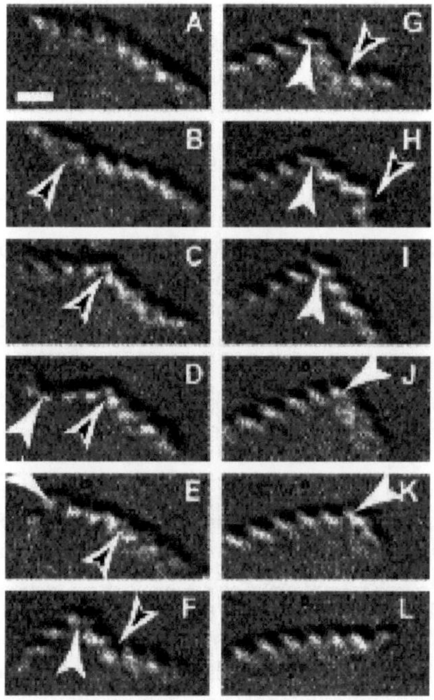

Fig. 3 Video frames of a swimming *S. melliferum* cell showing the propagation of a pair of kinks (*black and white arrows*) along the helical cell body, which allows directional movement in liquid medium. The kinks are generated by a progressive change of handedness. Time duration from A to L: 858 ms. *Scale bar*: 1 μm. Reprinted from Shaevitz et al. 2005, Copyright, with permission from Elsevier

more, this complex was shown to be essential to maintain cell morphology in *Escherichia coli*.

4
Conclusion

Regarding the limits and problems of traditional specimen preparation and imaging techniques, it is obvious that new approaches were needed to analyze the complex architecture and intracellular networks of cells. Cryo-ET enables us to explore supramolecular architecture like the cytoskeleton in an unperturbed cellular context. However, it remains difficult to capture specific states of this highly dynamic system, which could also further elucidate the motility mechanism of *Spiroplasma*. This could be accomplished by combining correlative light and electron microscopy techniques. Furthermore, the structural analysis of mutants lacking certain proteins might help to identify the individual contributions of the different filaments with respect to cell shape and motility. Additionally, it would allow one to clearly assign the two types of filaments to specific proteins. For this, however, genetic tools for Mollicutes have to be developed first.

References

Ausmees N, Kuhn JR, Jacobs-Wagner C (2003) The bacterial cytoskeleton: An intermediate filament-like function in cell shape. Cell 115:705–713

Baumeister W (2005) From proteomic inventory to architecture. FEBS Lett 579:933–937

Baumeister W, Grimm R, Walz J (1999) Electron tomography of molecules and cells. Trends Cell Biol 9:81–85

Figge RM, Divakaruni AV, Gober JW (2004) MreB, the cell shape-determining bacterial actin homologue, co-ordinates cell wall morphogenesis in *Caulobacter crescentus*. Mol Microbiol 51:1321–1332

Formstone A, Errington J (2005) A magnesium-dependent mreB null mutant: implications for the role of mreB in *Bacillus subtilis*. Mol Microbiol 55:1646–1657

Gitai Z, Dye N, Shapiro L (2004) An actin-like gene can determine cell polarity in bacteria. Proc Natl Acad Sci USA 101:8643–8648

Jones LJF, Carballido-López R, Errington J (2001) Control of cell shape in bacteria: Helical, actin-like filaments in *Bacillus subtilis*. Cell 104:913–922

Koster AJ, Grimm R, Typke D, Hegerl R, Stoschek A, Walz J, Baumeister W (1997) Perspectives of molecular and cellular electron tomography. J Struct Biol 120:276–308

Kruse T, Bork-Jensen J, Gerdes K (2005) The morphogenetic MreBCD proteins of *Escherichia coli* form an essential membrane-bound complex. Mol Microbiol 55:78–89

Kürner J, Frangakis AS, Baumeister W (2005) Cryo-electron tomography reveals the cytoskeletal structure of *Spiroplasma melliferum*. Science 307:436–438

Medalia O, Weber I, Frangakis AS, Nicastro D, Gerisch G, Baumeister W (2002) Macromolecular architecture in eukaryotic cells visualized by cryoelectron tomography. Science 298:1209–1213

Neimark H (1983) Mycoplasma and bacterial proteins resembling contractile proteins: A review. Yale J Biol Med 56:419–423

Shaevitz JW, Lee JY, Fletcher DA (2005) Spiroplasma swim by a processive change in body helicity. Cell 122:941–945

Townsend R, Plaskitt KA (1985) Immunogold localization of p55-fibril protein and p25-spiralin in Spiroplasma cells. J Gen Microbiol 131:983–992

Townsend R, Archer DB, Plaskitt KA (1980) Purification and preliminary characterization of spiroplasma fibrils. J Bacteriol 142:694–700

Trachtenberg S (1998) Mollicutes – Wall-less bacteria with internal cytoskeletons. J Struct Biol 124:244–256

Trachtenberg S, Gilad R (2001) A bacterial linear motor: cellular and molecular organization of the contractile cytoskeleton of the helical bacterium *Spiroplasma melliferum* BC3. Mol Microbiol 41:827–848

Tully JG (1992) Mollicutes (Mycoplasmas). In: Lederberg J (ed) Encyclopedia of microbiology, vol 3. Academic Press, New York, pp 181–191

Williamson DL (1974) Unusual fibrils from the spirochete-like sex ratio organism. J Bacteriol 117:904–906

Williamson DL, Renaudin J, Bové J-M (1991) Nucleotide sequence of the *Spiroplasma citri* fibril protein gene. J Bacteriol 173:4353–4362

Microbiol Monogr (2)
J.M. Shively: Complex Intracellular Structures in Prokaryotes
DOI 10.1007/7171_031/Published online: 25 May 2006

Organization and Assembly of the *Mycoplasma pneumoniae* Attachment Organelle

Mitchell F. Balish

Department of Microbiology, Miami University, Oxford, OH 45056, USA
BalishMF@MUOhio.edu

Abstract *Mycoplasma pneumoniae* synthesizes a complex polar structure, the attachment organelle (AO), which is required for productive adherence to host cells. Because in nature *M. pneumoniae* cannot survive outside the host, this structure is essential to the *M. pneumoniae* cell. While it is understood that the AO is the site at which adhesin proteins are concentrated, how that localization is mediated is unknown. However, the presence at the AO of a set of novel, detergent-insoluble proteins and structures which are indirectly required for normal function of this structure has provided insight into its assembly and architecture. Biochemical, genetic, and immunocytochemical studies of these cytadherence-accessory proteins and their interrelationships have revealed remarkable complexity in this supposedly minimal bacterial cell.

1 Introduction to *Mycoplasma pneumoniae* and its Ultrastructure

Mycoplasma pneumoniae is a common etiological agent of community-acquired tracheobronchitis and primary atypical pneumonia, and is also implicated in chronic asthma and causes disease at a variety of sites other than the respiratory tract (Waites and Talkington 2004). Like related mucosal pathogens, commensals, and opportunists of the bacterial class *Mollicutes*, *M. pneumoniae* is characterized by small size, genome reduction, absence of a cell wall, and fastidiousness resulting in obligate host association in nature. Nonetheless, *M. pneumoniae* is culturable in isolation in complex growth media.

Despite considerable pleomorphy, discrete regions are readily visible in a typical *M. pneumoniae* cell (Fig. 1). The trailing filament, so called because of its position behind the cell body during gliding motility (Bredt 1968), is the most variable portion of the cell; its significance is uncertain. The attachment organelle (AO), which leads during gliding and is implicated in the mechanism of that activity, is regular in dimensions, approximately 300 nm long and 40 nm wide. Some cells have multiple AOs; these cells are likely in the process of division, in which the AO appears to duplicate prior to cytokinesis (Seto et al. 2001). *M. pneumoniae* cells typically attach to host cells via the AO tip (Collier and Clyde 1971).

Extraction of *M. pneumoniae* cells attached to substrates with the detergent Triton X-100 reveals a rod (Fig. 1) from which fibrous structures are sometimes seen to emanate. This rod, also called the electron-dense core (Biberfeld and Biberfeld 1970), is located within the cytoplasmic space of the AO (Meng and Pfister 1980; Gobel et al. 1981) and is readily seen in cross-section (Hegermann et al. 2002) (Fig. 2). It is perpendicularly striated at non-uniform but characteristic intervals and consists of two parallel bar-shaped structures (Hegermann et al. 2002; Willby and Krause 2002). The distal end of the core is enlarged, forming the terminal button (Biberfeld and Biberfeld 1970). At the proximal end is a double-ring called the wheel-like structure (Hegermann et al. 2002). In addition, fine filaments are suggested to link the core to the inner surface of the plasma membrane (Hegermann et al. 2002). Although the proteins constituting the Triton X-100-insoluble fraction have been identified (Regula et al. 2001), little information is available concerning the composition of individual elements of the visible structures.

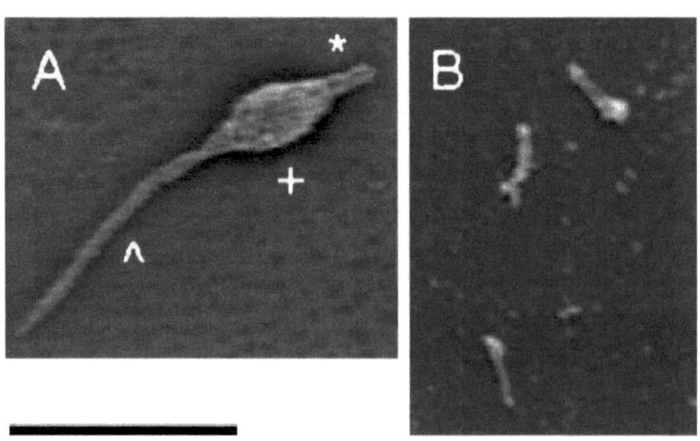

Fig. 1 Scanning electron micrographs of *M. pneumoniae* grown attached to glass. *Panel A*, a single, typical, intact cell with three parts: $^\wedge$, trailing filament; $^+$, cell body; *, AO. *Panel B*, cells extracted with 2% TX-100. Only the electron-dense cores of the AOs remain. Three are visible. *Scale bar*, 1 μm

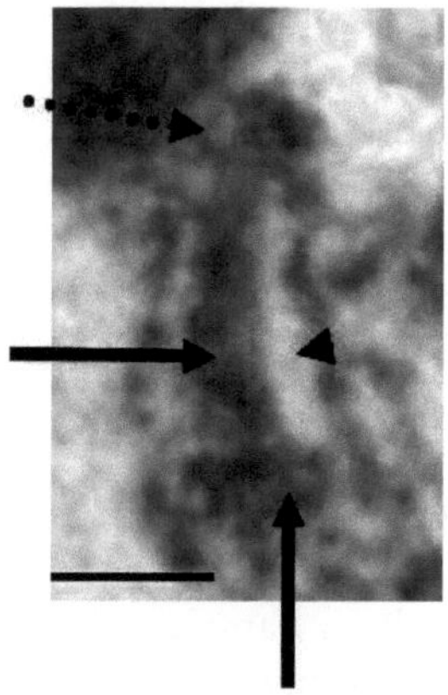

Fig. 2 Transmission electron micrograph of a thin section through an AO in a pellet of *M. pneumoniae* cells. The electron-dense core (*horizontal arrow*) is located within an electron-lucent area of cytoplasm (*arrowhead*) bounded by the membrane of the attachment organelle. At the distal end is an enlarged terminal button (*dotted arrow*), and at the proximal end is a wheel-like base (*vertical arrow*)

2 Identification of AO Components

The *M. pneumoniae* AO contains proteins involved directly in adherence as well as proteins associated with architecture of the structure.

2.1 P1 Adhesin

Protein P1, which is required for association of *M. pneumoniae* with host cells (Hu et al. 1977), is located preferentially at the AO (Baseman et al. 1982; Feldner et al. 1982; Collier et al. 1983), though it is also distributed less densely over the entire mycoplasma cell surface. It traverses the membrane (Su et al. 1987) and can be chemically cross-linked to other AO components (Layh-Schmitt and Herrmann 1994; Layh-Schmitt et al. 2000). Although determinants within the structure of P1 for clustering at the AO have not been identified, analyses of avirulent *M. pneumoniae* mutants have resulted in identification of other proteins required for this process.

2.2 Selection of Avirulent *M. pneumoniae* Mutants

M. pneumoniae colonies adsorb erythrocytes (Sobeslavsky et al. 1968). This hemadsorption (HA) capability correlates with cytadherence and virulence, since *M. pneumoniae* cells appear to bind the same determinants on both erythrocytes and host cells. HA-impaired *M. pneumoniae* mutants, which are likely to harbor mutations in adherence genes, have been screened and selected in several ways: high-passage mutants (Lipman et al. 1969); mutants created by chemical mutagenesis (Hansen et al. 1979); spontaneous mutants among a wild-type population (Krause et al. 1982; Layh-Schmitt and Harkenthal 1999); transposon insertion mutants (Krause et al. 1997); and mutants with reduced plastic-binding capacity (Willby and Krause 2002).

2.3 Identification of Mutant Genes

Sodium dodecyl sulfate polyacrylamide gel electrophoresis (Hansen et al. 1979; Krause et al. 1982) and immunoblotting (Baseman et al. 1987; Krause et al. 1997; Layh-Schmitt and Harkenthal 1999; Willby and Krause 2002; Willby et al. 2004) have been used to identify proteins absent or at reduced levels in HA mutants. In addition to P1, a small set of proteins has been identified in this manner. These are: A, B, and C (Hansen et al. 1979), of which B and C have been identified as proteins P90 and P40 from other studies (Sperker et al. 1991; Waldo et al. 2005); HMW1, HMW2, and HMW3 (Krause et al. 1982); and P30 (Baseman et al. 1987). Additionally, proteins P65, P41, and P24, whose coding genes are cotranscribed with HMW2 have been implicated in AO function (Krause et al. 1997; Jordan et al. 2001; Kenri et al. 2004), but unlike the others, strains with mutations in their coding genes have not been described. B and C are derived from a precursor that is structurally similar to P1 and encoded by a gene that is cotranscribed with the gene for P1 (Sperker et al. 1991; Layh-Schmitt and Herrmann 1992); they are likely to function in concert with P1. A is unidentified but appears upon reacquisition of B and C (Krause et al. 1983). P30 is an adhesin as well, as antibodies against it prevent *M. pneumoniae* cytadherence (Morrison-Plummer et al. 1986). None of these proteins shares sequence homology with proteins from unrelated bacteria.

3 Interactions Among AO Components

The components of the *M. pneumoniae* AO are likely to interact with each other, but it has been difficult to identify these interactions directly. However, there is plentiful circumstantial evidence suggesting interactions among these proteins. Immunocytochemical data indicate that P1, B, and C are clustered on the AO surface but are also present at other sites on the cell at lower concentrations (Baseman et al. 1982; Feldner et al. 1982; Collier et al. 1983; Seto et al. 2001). In contrast, P30, which is also an adhesin, is principally restricted to the AO tip AO (Baseman et al. 1987; Seto and Miyata 2003).

3.1 Stabilization

Studies of mutants reveal an intricate web of probable interactions among AO proteins that are predominantly or entirely Triton X-100-insoluble. HMW1 and HMW2 are mutually stabilizing, and both are required to prevent post-translational loss of HMW3 and P65 (Popham et al. 1997; Jordan et al. 2001; Willby et al. 2004). In the case of HMW1, failure of an HMW2-dependent stabilization event that is possibly linked to HMW1 export results in its accumulation in the Triton X-100-soluble fraction, where it is degraded (Balish et al. 2001). HMW1 stabilization requires both the central part of HMW2 and the C-terminal part of HMW1 (Balish et al. 2003a; Willby et al. 2004). P65 stabilization is directly dependent upon both P30 and HMW3 (Jordan et al. 2001; Willby and Krause 2002). P30 levels are also slightly reduced in the absence of HMW2 (Jordan et al. 2001). HMW1 and HMW2 are both required for core formation (Seto and Miyata 2003); in the absence of HMW3, cores appear split, suggesting a role for HMW3 in core stabilization (Willby and Krause 2002). Stability of P1, B, and C is independent of the other proteins (Balish and Krause 2002).

3.2 Localization

Immunocytochemistry and green fluorescent protein fusion technology have yielded considerable information about the locations of AO components despite the small size of the structure (Fig. 3). P65 (Proft et al. 1995), HMW1 (Stevens and Krause 1991; Balish et al. 2001), and HMW3 (Stevens and Krause

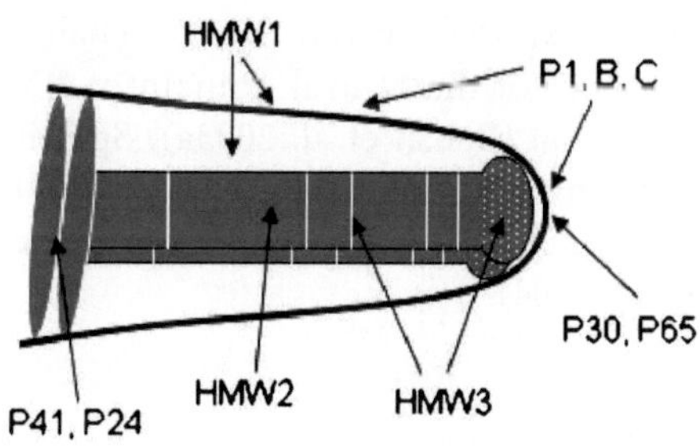

Fig. 3 Schematic model of structures and localization of proteins in the *M. pneumoniae* AO. Two striated rods constitute the electron-dense core; HMW2 is a likely component of this structure, distinct from P41 and P24, whose proximal location within the AO might represent localization to the wheel-like structure. HMW3 is localized to the terminal button and perhaps the striations. HMW1 appears to be present on both the outer surface and the interior, but not at the most distal part of the AO. P1, B, C, and P30 are on the surface, but P30, perhaps in association with P65, is only at the distal portion

1992) have all been reported to be located partly on the cell surface despite the absence of recognizable transmembrane domains, though for at least HMW3 this is thought to be artifactual (Balish and Krause 2002). HMW3 is associated with both the terminal button and fine filaments associated with the core (Stevens and Krause 1992). Like P30, P65 is located at the distal end of the AO (Seto and Miyata 2003; Kenri et al. 2004). The same is true for HMW2 (Balish et al. 2003b; Kenri et al. 2004), although HMW2 is believed to be located exclusively in the cell interior, where it might be a component of the electron-dense core (Balish et al. 2002, 2003b). HMW1 and HMW3 are more cell-proximal, and P41 and P24 are frankly at the base of the AO (Seto and Miyata 2003; Kenri et al. 2004), suggesting that the last two are associated with the wheel-like structure.

P1 clustering at the AO occurs only in the presence of HMW1, HMW2, HMW3, and B and C (Seto and Miyata 2003; unpublished data). Moreover, the form of HMW1 that results from stabilization by HMW2 is specifically required for P1 clustering (Balish et al. 2003a). P30, whose localization to the AO requires HMW1, HMW2, and HMW3, is not involved in P1 clustering (Romero-Arroyo et al. 1999; Seto and Miyata 2003; unpublished data). HMW1 and HMW2 are mutually required for AO localization (Balish et al. 2003a).

3.3
AO Assembly

Consideration of stabilization and localization data from mutants has led to a model for how the components of the AO are organized and perhaps assembled (Krause and Balish 2004). In order for both P30 and the presumptive P1/B/C complex to localize to the AO, a stable core must form. HMW3 is essential for the stability of the core (Willby and Krause 2002), which itself requires both HMW1 and HMW2 for its formation (Seto and Miyata 2003). A stabilized and surface-exposed form of HMW1 (Balish et al. 2001), which exists only in the presence of the central domain of HMW2, is additionally required for P1 localization (Balish et al. 2003a). Splitting of the two parallel core components during AO duplication (Hegermann et al. 2002) results in two half-cores, each of which serves as a template for the construction of a full core (Krause and Balish 2004).

4
Conclusion

The achievement of a full understanding of the organization and assembly of the *M. pneumoniae* AO is progressing, but is slowed by some of the properties of the organism, including fastidiousness, small size, and limited repertoire of molecular techniques developed for use in it. Nevertheless, the importance of *M. pneumoniae* as a pathogen, as a model for related mycoplasma species, and as an example of complex intracellular structure formation in prokaryotes ensures that research in this area will continue to progress.

Acknowledgements I thank J. Hatchel, M. Duley, S. Ross, and D. Krause for providing electron micrographs. This work was supported by Miami University startup funds.

References

Balish MF, Hahn TW, Popham PL, Krause DC (2001) Stability of *Mycoplasma pneumoniae* cytadherence-accessory protein HMW1 correlates with its association with the triton shell. J Bacteriol 183:3680–3688

Balish MF, Krause DC (2002) Cytadherence and the cytoskeleton. In: Razin S, Herrmann R (eds) Molecular biology and pathogenesis of mycoplasmas. Kluwer Academic/Plenum Publishers, New York, pp 491–518

Balish MF, Ross SM, Fisseha M, Krause DC (2003a) Deletion analysis identifies key functional domains of the cytadherence-associated protein HMW2 of *Mycoplasma pneumoniae*. Mol Microbiol 50:1507–1516

Balish MF, Santurri RT, Ricci AM, Lee KK, Krause DC (2003b) Localization of *Mycoplasma pneumoniae* cytadherence-associated protein HMW2 by fusion with green fluorescent protein: implications for attachment organelle structure. Mol Microbiol 47:49–60

Baseman JB, Cole RM, Krause DC, Leith DK (1982) Molecular basis for cytadsorption of *Mycoplasma pneumoniae*. J Bacteriol 151:1514–1522

Baseman JB, Morrison-Plummer J, Drouillard D, Puleo-Scheppke B, Tryon V, Holt SC (1987) Identification of a 32-kilodalton protein of *Mycoplasma pneumoniae* associated with hemadsorption. Isr J Med Sci 23:474–479

Biberfeld G, Biberfeld P (1970) Ultrastructural features of *Mycoplasma pneumoniae*. J Bacteriol 102:855–861

Bredt W (1968) Motility and multiplication of *Mycoplasma pneumoniae*. A phase contrast study. Pathol Microbiol 32:321–326

Collier AM, Clyde WA Jr (1971) Relationships between *Mycoplasma pneumoniae* and human respiratory epithelium. Infect Immun 3:694–701

Collier AM, Hu PC, Clyde WA Jr (1983) Location of attachment moiety on *Mycoplasma pneumoniae*. Yale J Biol Med 56:671–677

Feldner J, Gobel U, Bredt W (1982) *Mycoplasma pneumoniae* adhesin localized to tip structure by monoclonal antibody. Nature 298:765–767

Gobel U, Speth V, Bredt W (1981) Filamentous structures in adherent *Mycoplasma pneumoniae* cells treated with nonionic detergents. J Cell Biol 91:537–543

Hansen EJ, Wilson RM, Baseman JB (1979) Isolation of mutants of *Mycoplasma pneumoniae* defective in hemadsorption. Infect Immun 23:903–906

Hegermann J, Herrmann R, Mayer F (2002) Cytoskeletal elements in the bacterium *Mycoplasma pneumoniae*. Naturwissenschaften 89:453–458

Hu PC, Collier AM, Baseman JB (1977) Surface parasitism by *Mycoplasma pneumoniae* of respiratory epithelium. J Exp Med 145:1328–1343

Jordan JL, Berry KM, Balish MF, Krause DC (2001) Stability and subcellular localization of cytadherence-associated protein P65 in *Mycoplasma pneumoniae*. J Bacteriol 183:7387–7391

Kenri T, Seto S, Horino A, Sasaki Y, Sasaki T, Miyata M (2004) Use of fluorescent-protein tagging to determine the subcellular localization of *Mycoplasma pneumoniae* proteins encoded by the cytadherence regulatory locus. J Bacteriol 186:6944–6955

Krause DC, Balish MF (2004) Cellular engineering in a minimal microbe: structure and assembly of the terminal organelle of *Mycoplasma pneumoniae*. Mol Microbiol 51:917–924

Krause DC, Leith DK, Baseman JB (1983) Reacquisition of specific proteins confers virulence in *Mycoplasma pneumoniae*. Infect Immun 39:830–836

Krause DC, Leith DK, Wilson RM, Baseman JB (1982) Identification of *Mycoplasma pneumoniae* proteins associated with hemadsorption and virulence. Infect Immun 35:809–817

Krause DC, Proft T, Hedreyda CT, Hilbert H, Plagens H, Herrmann R (1997) Transposon mutagenesis reinforces the correlation between *Mycoplasma pneumoniae* cytoskeletal protein HMW2 and cytadherence. J Bacteriol 179:2668–2677

Layh-Schmitt G, Harkenthal M (1999) The 40 and 90kDa membrane proteins (ORF6 gene product) of *Mycoplasma pneumoniae* are responsible for the tip structure formation and P1 (adhesin) association with the Triton shell. FEMS Microbiol Lett 174:143–149

Layh-Schmitt G, Herrmann R (1992) Localization and biochemical characterization of the ORF6 gene product of the *Mycoplasma pneumoniae* P1 operon. Infect Immun 60:2906–2913

Layh-Schmitt G, Herrmann R (1994) Spatial arrangement of gene products of the P1 operon in the membrane of *Mycoplasma pneumoniae*. Infect Immun 62:974–979

Layh-Schmitt G, Podtelejnikov A, Mann M (2000) Proteins complexed to the P1 adhesin of *Mycoplasma pneumoniae*. Microbiology 146:741–747

Lipman RP, Clyde WA Jr, Denny FW (1969) Characteristics of virulent, attenuated, and avirulent *Mycoplasma pneumoniae* strains. J Bacteriol 100:1037–1043

Meng KE, Pfister RM (1980) Intracellular structure of *Mycoplasma pneumoniae* revealed after membrane removal. J Bacteriol 144:390–399

Morrison-Plummer J, Leith DK, Baseman JB (1986) Biological effects of anti-lipid and anti-protein monoclonal antibodies on *Mycoplasma pneumoniae*. Infect Immun 53:398–403

Popham PL, Hahn TW, Krebes KA, Krause DC (1997) Loss of HMW1 and HMW3 in noncytadhering mutants of *Mycoplasma pneumoniae* occurs post-translationally. Proc Natl Acad Sci USA 94:13979–13984

Proft T, Hilbert H, Layh-Schmitt G, Herrmann R (1995) The proline-rich P65 protein of *Mycoplasma pneumoniae* is a component of the Triton X-100-insoluble fraction and exhibits size polymorphism in the strains M129 and FH. J Bacteriol 177:3370–3378

Regula JT, Boguth G, Gorg A, Hegermann J, Mayer F, Frank R, Herrmann R (2001) Defining the mycoplasma 'cytoskeleton': the protein composition of the Triton X-100 insoluble fraction of the bacterium *Mycoplasma pneumoniae* determined by 2D gel electrophoresis and mass spectrometry. Microbiology 147:1045–1057

Romero-Arroyo CE, Jordan J, Peacock SJ, Willby MJ, Farmer MA, Krause DC (1999) *Mycoplasma pneumoniae* protein P30 is required for cytadherence and associated with proper cell development. J Bacteriol 181:1079–1087

Seto S, Layh-Schmitt G, Kenri T, Miyata M (2001) Visualization of the attachment organelle and cytadherence proteins of *Mycoplasma pneumoniae* by immunofluorescence microscopy. J Bacteriol 183:1621–1630

Seto S, Miyata M (2003) Attachment organelle formation represented by localization of cytadherence proteins and formation of the electron-dense core in wild-type and mutant strains of *Mycoplasma pneumoniae*. J Bacteriol 185:1082–1091

Sobeslavsky O, Prescott B, Chanock RM (1968) Adsorption of *Mycoplasma pneumoniae* to neuraminic acid receptors of various cells and possible role in virulence. J Bacteriol 96:695–705

Sperker B, Hu P, Herrmann R (1991) Identification of gene products of the P1 operon of *Mycoplasma pneumoniae*. Mol Microbiol 5:299–306

Stevens MK, Krause DC (1991) Localization of the *Mycoplasma pneumoniae* cytadherence-accessory proteins HMW1 and HMW4 in the cytoskeletonlike Triton shell. J Bacteriol 173:1041–1050

Stevens MK, Krause DC (1992) *Mycoplasma pneumoniae* cytadherence phase-variable protein HMW3 is a component of the attachment organelle. J Bacteriol 174:4265–4274

Su CJ, Tryon VV, Baseman JB (1987) Cloning and sequence analysis of cytadhesin P1 from *Mycoplasma pneumoniae*. Infect Immun 55:3023–3029

Waites KB, Talkington DF (2004) *Mycoplasma pneumoniae* and its role as a human pathogen. Clin Microbiol Rev 17:697–728

Waldo RH III, Jordan JL, Krause DC (2005) Identification and complementation of a mutation associated with loss of *Mycoplasma pneumoniae* virulence-specific proteins B and C. J Bacteriol 187:747–751

Willby MJ, Balish MF, Ross SM, Lee KK, Jordan JL, Krause DC (2004) HMW1 is required for stability and localization of HMW2 to the attachment organelle of *Mycoplasma pneumoniae*. J Bacteriol 186:8221–8228

Willby MJ, Krause DC (2002) Characterization of a *Mycoplasma pneumoniae* hmw3 mutant: implications for attachment organelle assembly. J Bacteriol 184:3061–3068

Microbiol Monogr (2)
J.M. Shively: Complex Intracellular Structures in Procaryotes
DOI 10.1007/7171_032/Published online: 25 May 2006

The Junctional Pore Complex: Molecular Motor of Microbial Motility

Egbert Hoiczyk

W. Harry Feinstone Department of Molecular Microbiology and Immunology, Johns Hopkins Bloomberg School of Public Health, Baltimore, MD 21205, USA
ehoiczyk@jhsph.edu

Abstract Many diverse bacteria move in the absence of any visible locomotory organelle by a mechanism termed gliding. Gliding motility requires contact with a solid substrate and occurs in the direction parallel to the long axis of the bacterial cell. Recent research indicates that bacteria use at least two different mechanisms to glide. One mechanism, termed social motility, is based on the extension and retraction of pili. The other mechanism, termed adventurous motility, involves the secretion of slime from a specialized organelle called the junctional pore complex. This work discusses the possible role of the junctional pore complex in gliding motility of cyanobacteria, myxobacteria, and the alphaproteobacterium *Gluconacetobacter*.

1 Introduction

Bacterial motility is perplexingly complex. To swim through liquids, most bacteria use "propeller"-like flagella that are driven by small rotary motors (Berg 2003). Other swimming bacteria, however, lack flagella and use for example internal contractile cytoskeletons like *Spiroplasma* (Gilad et al. 2003) or completely unknown mechanisms like certain marine *Synechococcus* strains (Waterbury et al. 1985). In contrast, bacteria that are living on surfaces often employ other forms of locomotion. One such form is twitching or social motility, which is powered by the extension and retraction of type IV pili (Merz et al. 2000; Skerker and Berg 2001). Other bacteria are able to crawl over surfaces, even though they lack any detectable locomotory organelle.

One of the most interesting of these diverse, and sometimes mysterious, types of surface-associated locomotion is gliding. Gliding is widespread among bacteria and found in many cyanobacteria, flexibacteria, myxobacteria, mycoplasma species, the *Chloroflexaceae*, and the *Beggiatoaceae* (McBride 2001). Compared to swimming (10–100 μm/s; Macnab and Aizawa 1984), gliding is slow with speeds ranging from 0.04 to 11 μm/s (Halfen 1979).

Although the molecular mechanism of gliding is currently unknown for any bacterium, numerous studies have shown that virtually all gliders share a number of distinct features: (*i*) Usually their cells are greater in length then width and they move in only one direction, parallel to the cell axis. (*ii*) During motility, the cells frequently reverse direction resulting in characteristic back-and-forth movements. (*iii*) In order to glide, the cells require contact with a solid surface. Once they lose contact, motility stops. (*iv*) The source of energy for gliding is the proton motive force rather than ATP. (*v*) Finally, as bacteria move, they leave behind trails of polymeric carbohydrate slime. Although the function of the slime is not known, it has been suggested that it either acts as a passive adhesive or as an active force-generating hydrogel that propels the cells. More recently, electron microscopy has revealed that some gliding bacteria share yet another feature, a novel organelle termed the junctional pore complex (JPC), that might be the molecular motor for this type of microbial motility. Ultrastructural and cell behavioral studies of several gliding bacteria indicate that the JPC secretes slime during gliding and it has been suggested that this process generates the necessary propulsive force (Hoiczyk and Baumeister 1998; Wolgemuth et al. 2002). In this work we will discuss the structure and occurrence of this novel organelle and how hydration-driven swelling of the secreted slime could power gliding motility in these bacteria.

2
The Junctional Pore Complex of Cyanobacteria

Cyanobacteria are a large group of photosynthetic microorganisms of diverse morphology ranging from simple unicellular species to complex filamentous forms capable of cellular differentiation. When motile, most multicellular filamentous cyanobacteria such as *Oscillatoria*, *Phormidium*, *Anabaena*, and *Nostoc* move by gliding (Castenholz 1982); while some unicellular species are capable of twitching motility using type IV pili (*Synechocystis*; Bhaya et al. 1999) or swim by an unidentified mechanism (*Synechococcus*; Waterbury et al. 1985).

Gliding in filamentous cyanobacteria is a slow, well-coordinated movement at speeds of up to 10 μm/s. Some species, such as *Oscillatoria* and *Phormidium* rotate during gliding, while others such as *Nostoc* and *Anabaena* do not rotate but are capable of lateral movements. About every 5–8 minutes the multicellular filaments stop and reverse their direction. Since signals such as light modulate the frequency of filament reversals, gliding motility allows the cells to actively search for an optimal niche in the environment.

Light microscopy, using India ink particles or fluorescently-labeled lectins, indicates that during gliding, all cyanobacteria secrete slime that either completely ensheathes the filaments or is secreted in multiple thin slime bands (Fig. 1A). If slime bands are sheared off the surface of the filaments by the flow of fluid, they remain attached at the site of their origin in close proximity of the cross walls between the cells in the filament. Observations of such detached slime bands in *Phormidium* and *Anabaena* filaments reveal two

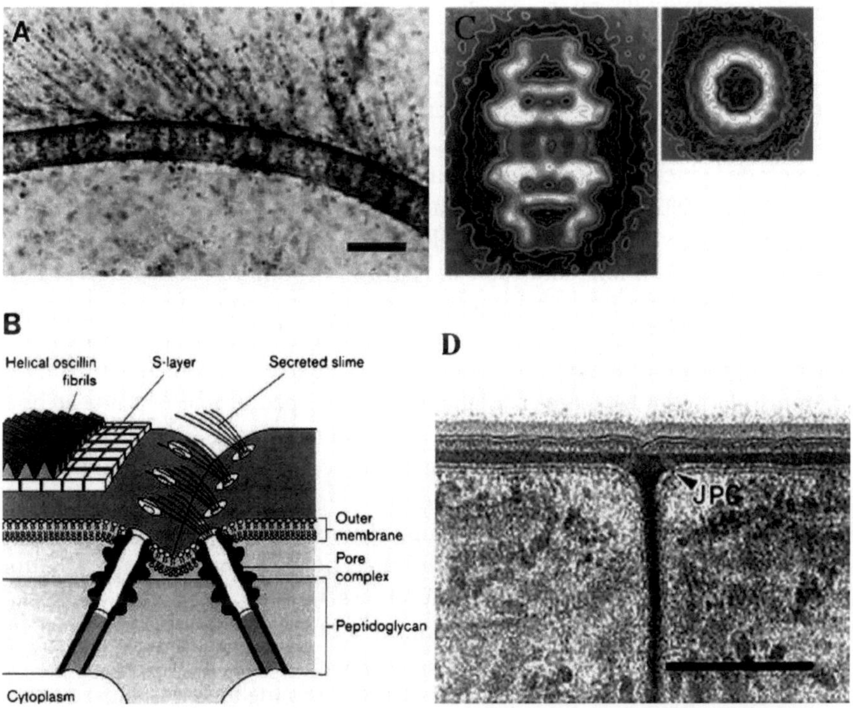

Fig. 1 JPC organelles and slime secretion in the gliding filamentous cyanobacterium *Phormidium uncinatum*. **A** Secretion of multiple thin slime bands that are sheared off the cell surface by the flow of fluid. The slime bands emanate in close proximity to the cross walls where the JPCs are located and they are elongated at a speed comparable to the speed of gliding. *Bar* 5 µm. **B** Illustration of the cell envelope of *P. uncinatum* showing the arrangement of the JPCs and the helical oscillin fibrils. The organism can control the direction of locomotion by using alternate sets of nozzles for secretion. **C** Averages of side- and top-views of the barrel-shaped nozzle part of the JPC of *P. uncinatum*. **D** Electron microscopic cross section of the same portion of the multicellular filament shown in the schematic illustration. *JPC* junctional pore complex. *Bar* 200 nm. Pictures taken from Hoiczyk and Baumeister (1998) with kind permission of Elsevier

important aspects of the secretion process (Hoiczyk and Baumeister 1998): First, in both species the slime bands are elongated at rates equal to the rates of gliding (up to 3 μm/s); and second, the direction of slime extrusion is always opposite to the direction of the movement. Both observations strongly suggest that the secretion of slime plays an active rather than passive function in cyanobacterial gliding motility. Finally, electron microscopic analysis of the cell wall architecture of both *Phormidium* and *Anabaena* showed that these cyanobacteria possess pore-like organelles that are located at the cross walls where the slime emanates (Hoiczyk and Baumeister 1998; Fig. 1B, D). These organelles, termed JPC, are arranged in rows that encircle both sides of the cross walls. While *Phormidium* has one row of pores, *Anabaena* possesses multiple rows and this difference might explain *Anabaena*'s ability to move laterally. To detect the JPCs, isolated cell walls can be treated with acid exposing empty trans-peptidoglycan channels. Using this method more than twenty years ago, Guglielmi and Cohen-Bazire (1982) had shown for more than two dozen cyanobacterial species that motility invariably correlates with the presence of the JPC. However, their findings were initially disregarded since the visualized channels did not penetrate the entire cell wall (Castenholz 1982). Careful reexamination of the cell walls showed that the trans-peptidoglycan channels in the living cell harbor the JPC organelle. The JPCs are 70–80 nm long and span the entire cell wall. At least two large substructures form the JPC. The first is a tube-like structure 13 nm in diameter, which traverses the peptidoglycan, and the second is a nozzle-like structure located in the outer membrane. The nozzle is a barrel-shaped structure about 32 nm long and 14–16 nm in diameter. It has openings of 6–8 nm at both ends and a central bulge measuring about 14 nm (Fig. 1C). Each JPC is inclined at an angle of 30–40° relative to the cell axis (Fig. 1B, D) pointing in an opposite direction on either side of the cross wall. This arrangement is thought to have two important implications for the mechanism of gliding. The inclination of the JPC provides directionality for the slime secretion and the control of the JPC's activity allows the cell to reverse the direction of its movement by using alternate sets of JPCs. The importance of the JPC for gliding is further substantiated by the observation that cells that become non-motile during cultivation lose the organelles and, subsequently, no longer secrete slime.

In addition, the ultrastructural analysis of the cell walls also offered a structural explanation for the fact that some cyanobacteria such as *Phormidium* and *Oscillatoria* rotate during locomotion while others such as *Anabaena* do not (Hoiczyk and Baumeister 1997). In all rotating cyanobacteria the filament surface is covered with a helically arranged Ca^{2+}-binding glycoprotein termed oscillin (Fig. 1B). If the continuously secreted slime flows along these protein helices the multicellular filament will rotate. In comparison, non-rotating species have radially arranged surface proteins and, therefore, do not rotate during locomotion.

3
The Junctional Pore Complex of Myxobacteria

Myxobacteria are ubiquitous rod-shaped unicellular soil bacteria that use gliding motility for two important aspects of their unique life style: predation and fruiting body formation (Kaiser 2003). Interestingly, myxobacteria possess two different propulsion mechanisms for gliding that they use simultaneously and synergistically. Extension and retraction of polar type IV pili is used for social- or (S-) motility and is limited to cells within pilus-contact distance of each other (Wall and Kaiser 1999). Adventurous- or (A-) motility allows movement of individual cells and is suggested to be powered by the secretion of slime from JPC-like organelles (Wolgemuth et al. 2002).

To test whether myxobacteria possess JPCs, negatively stained isolated cell envelopes were examined in the electron microscope. Using this technique, nozzle-like structures were found in *M. xanthus* that are similar to, although somewhat smaller than, the JPC organelles of cyanobacteria (Fig. 2A, B). In *M. xanthus*, the individual nozzles have an outer diameter of 12–14 nm and a central pore of about 6.5 nm. Up to 250 of these organelles are clustered at each of the two opposite cell poles, while only a few scattered pores are located in-between. Since so far no side-views of the isolated JPC organelles exist, it is currently impossible to determine how similar the morphology of these pore complexes in myxobacteria and cyanobacteria really is. Despite this incomplete morphological information, the examination of slime secretion in *M. xanthus* further substantiates the similarity of gliding motility in these two different groups of bacteria. Using acridine orange to stain the secreted slime, it can be shown that each *Myxococcus* cell leaves a slime trail behind that originates from its rear cell pole (Fig. 2C). To improve the resolution of the visualization, cells were allowed to glide over electron microscopic grids and were negatively stained. Examination of such cells in the electron microscope show that the slime trails are made up of a number of individual slime bands that originate at the cell poles where the JPCs are clustered (Fig. 2D). The observed correlation between the arrangement of the JPCs and the pattern of slime secretion strongly suggest that this process is the molecular motor of A-motility in myxobacteria. Various S-motility mutants still possess the JPCs, confirming that these structures are not part of the S-motility machinery. Furthermore, the observation that *M. xanthus* cells follow lines of stress in an agar surface provides additional evidence for a link between A-motility and slime secretion (Fontes and Kaiser 1999). This behavior is termed "elasticotaxis" and is thought to arise from the tendency of the secreted slime polymer to align with the polymer chains of the substrate. Thus, A-mutant strains that are deficient in slime secretion, fail to show elasticotaxis, while S-mutants are not affected. Finally, theoretical calculations show that the force produced by the expansion of the slime within the JPCs is large enough to produce a propulsive force

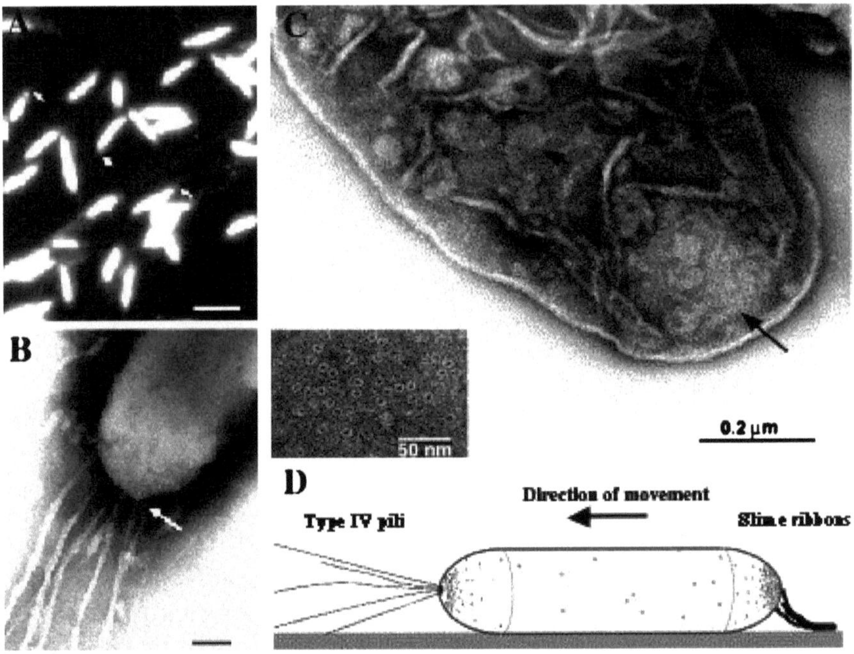

Fig. 2 Slime secretion and JPC organelle structure in *M. xanthus*. **A** Light micrograph of acridine orange labeled slime trails of *M. xanthus*. **B** Electron micrograph of a cell pole of a gliding *M. xanthus* cell showing the secretion of multiple slime bands that emerge at the cell pole where the JPC organelles are located. **C** Electron micrograph of a negatively stained isolated cell envelope of *M. xanthus* showing the polar cluster of the ring-shaped JPCs. The *inset* shows a detail of the JPC cluster at higher magnification. **D** Illustration of the two motors of *Myxococcus*. Type IV pili pull the cell at the front pole, while slime secretion from the JPCs at the rear pole push the cell. Both motors work in coordination and during a reversal of movement the pili are disassembled and reassembled at the opposite pole, while the cluster of active JPCs switches and starts secreting slime from the opposite pole. Pictures taken from Wolgemuth et al. (2002) with kind permission of Elsevier

sufficient to propel both myxobacteria and cyanobacteria at the observed speeds (Wolgemuth et al. 2002).

4 The Junctional Pore Complex of *Gluconacetobacter*

Gluconacetobacter xylinum (formerly known as *Acetobacter*) is a rod-shaped gram-negative bacterium that synthesizes cellulose in static culture (Brown et al. 1976). Even though this process has been extensively studied to understand cellulose biogenesis, relatively little attention has been paid to the fact that these bacteria move as a consequence of cellulose secretion. Closer analysis of these movements show that they possess all the hallmarks of gliding motility: They occur at an interface (air–liquid); the cells move without changing their shape in the direction of the cell's long axis; and they secrete a polymer during motility (cellulose). Intriguingly, *Gluconacetobacter* rotates during locomotion like some cyanobacteria. However, the rotation is not the result of slime flow along helically arranged surface fibrils but is caused by the twisting of the secreted cellulose microfibrils (Fig. 3A–D). The speed of locomotion of *G. xylinum* is about 0.04 μm/s and is comparable to the speed of other unicellular gliders such as *M. xanthus*. Electron microscopic studies show that each cell uses a linear array of about 50 JPC organelles for the secretion of individual cellulose microfibrils (Zarr 1979). Like in cyanobacteria, in vivo measurements confirm that the speed with which the cellulose microfibrils are secreted is comparable to the speed of gliding (Fig. 3A,B). Moreover, freeze fractures of the cell envelope demonstrate that the individual JPC organelles are formed by ring-shaped complexes similar to the ones found in cyanobacteria and myxobacteria (Fig. 3E,F). Even though no further morphologic characterization has been carried out, the biochemical mechanism of cellulose synthesis by this JPC is remarkably well understood (Ross et al. 1991). Among other details, the actual secretion capacity of a single JPC has been calculated, based on the speed and number of extruded cellulose microfibrils. Each organelle secretes up to 4000 glucose molecules per second totaling 200 000 glucose molecules per cell per second that are polymerized into a cellulose ribbon. Like in other JPC-based motility systems, it has been suggested that the propulsive force for locomotion is generated by both the incorporation of new polymer subunits as well as polymer aggregation and swelling. Even though it is not known whether motility is important for the survival of *G. xylinum* or only a by-product of cellulose secretion, the study of this system documents the enormous secretion capacities of these organelles and their potential to drive secretion-based motility even in heterotrophic microorganisms.

5
Outlook and Conclusion

Observations in several groups of gliding bacteria strongly suggest that the JPC organelle is the molecular motor powering gliding motility in these bacteria and that polymer secretion provides the propulsive force for this type of motility. While this concept provides a framework for future research, many aspects of this mechanism remain to be investigated. In particular, are the JPC organelles of these different gliding microorganisms morphologically similar? How do the cells control and coordinate the activity of these nano-machines? Given the great phylogenetic diversity and the different metabolism, morphology, and lifestyle among gliding bacteria, it is reason-

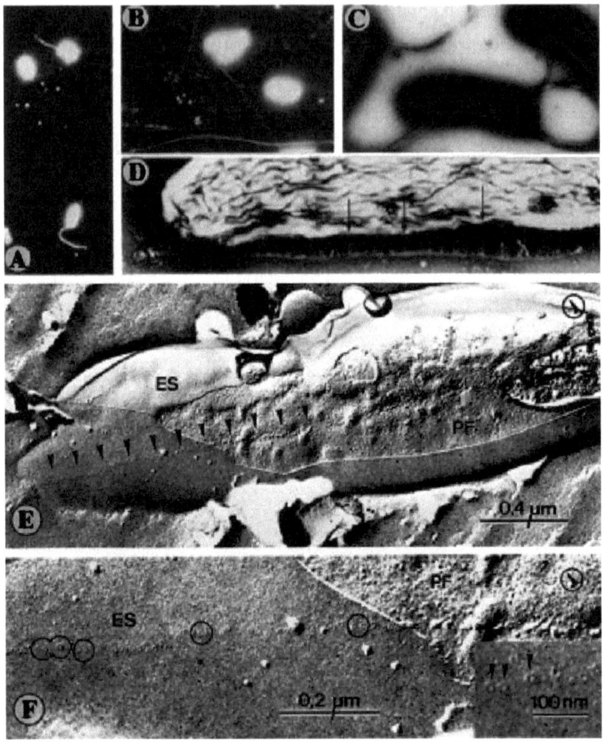

Fig. 3 JPC organelles and cellulose secretion in the gliding bacterium *Gluconacetobacter xylinum*. **A,B** Darkfield light micrograph of bacterial cells showing growing cellulose ribbons of various length attached to the bacteria. The twisting of the cellulose ribbons causes the bacteria to rotate during locomotion. **C** Electron micrograph of negatively stained bacteria showing the cellulose ribbons at higher magnification. **D** Separation of the cellulose ribbon reveals individual microfibrils that are secreted from the individual junctional pore complexes in the bacterial cell envelope. **E** Freeze fracture of a cellulose-secreting *G. xylinum* cell. The cellulose ribbon has been peeled away and the row of junctional pore complexes is visible both on the cell surface (ES) and their extension in the membrane (PF). **F** Enlarged portion of Fig. **E** demonstrating the alignment between the pores on the cell surface and their corresponding structures in the membrane. The inset in the lower right corner shows the ring-like appearance of the pores at higher magnification. *ES* etched surface, *PF* P-fracture face of the membrane. Pictures **A–D** taken from Brown et al. (1976) with kind permission of the National Academy of Sciences and pictures **E,F** taken from Zarr (1979) with kind permission of Rockefeller University Press

able to assume that there exist variations in the design and working of these propulsive organelles. Now that the complete genome of several gliding organisms is in hand and the first parts of the gliding machinery of *Myxococcus* have been isolated, it should soon be possible to start answering some of these questions. Although this work has only begun, it is likely to help us answer the simple sounding question of how gliding motility works.

References

Berg HC (2003) The rotary motor of bacterial flagella. Ann Rev Biochem 72:19–54

Bhaya D, Watanabe N, Ogawa T, Grossman AR (1999) The role of an alternative sigma factor in motility and pilus formation in the cyanobacterium *Synechocystis* sp. strain PCC6803. Proc Natl Acad Sci USA 96:3188–3193

Brown RM Jr, Willison JHM, Richardson CM (1976) Cellulose biosynthesis in *Acetobacter xylinum*: visualization of the site of synthesis and direct measurement of the in vivo process. Proc Natl Acad Sci USA 73:4565–4569

Castenholz RW (1982) The movements of cyanobacteria. In: Carr NG, Whitton BA (eds) The Biology of Cyanobacteria. Blackwell, London, pp 413–439

Fontes M, Kaiser D (1999) *Myxococcus* cells respond to elastic forces in their substrate. Proc Natl Acad Sci USA 96:8052–8057

Gilad R, Porat A, Trachtenberg S (2003) Motility modes of *Spiroplasma melliferum* BC3: a helical, wall-less bacterium driven by a linear motor. Mol Microbiol 47:657–669

Guglielmi G, Cohen-Bazire G (1982) Structure and distribution des pores et des perforations de l'enveloppe de peptidoglycane chez quelques cyanobacteries. Protistologica 18:151–165

Halfen LN (1979) Gliding movements. In: Haupt W, Feinleib ME (eds) Encyclopedia of Plant Physiology. Vol 7. Physiology of Movements. Springer, Berlin Heidelberg New York, pp 250–267

Hoiczyk E, Baumeister W (1997) Oscillin, an extracellular, Ca^{2+}-binding glycoprotein essential for the gliding motility in cyanobacteria. Mol Microbiol 26:699–708

Hoiczyk E, Baumeister W (1998) The junctional pore complex, a prokaryotic secretion organelle, is the molecular motor underlying gliding motility in cyanobacteria. Curr Biol 8:1161–1168

Kaiser D (2003) Coupling cell movement to multicellular development in myxobacteria. Nat Rev Microbiol 1:45–54

Macnab RM, Aizawa S-I (1984) Bacterial motility and the bacterial flagellar motor. Ann Rev Biophys Bioeng 13:51–83

McBride MJ (2001) Bacterial gliding motility: multiple mechanisms for cell movement over surfaces. Ann Rev Biochem 55:49–75

Merz A, Sheetz M, So M (2000) Pilus retraction powers bacterial twitching motility. Nature 407:98–102

Ross P, Mayer R, Benziman M (1991) Cellulose biosynthesis and function in bacteria. Microbiol Rev 55:35–58

Skerker JM, Berg HC (2001) Direct observation of extension and retraction of type IV pili. Proc Natl Acad Sci USA 98:6901–6904

Waterbury JB, Willey JM, Franks DG, Valois, FW, Watson SW (1985) A cyanobacterium capable of swimming motility. Science 230:74–76

Wall D, Kaiser D (1999) Type IV pili and cell motility. Mol Microbiol 32:1–10

Wolgemuth C, Hoiczyk E, Kaiser D, Oster G (2002) How myxobacteria glide. Curr Biol 12:369–377

Zarr K (1979) Visualization of pores (export sites) correlated with cellulose production in the envelope of the gram-negative bacterium *Acetobacter xylinum*. J Cell Biol 80:773–777

Microbiol Monogr (2)
J.M. Shively: Complex Intracellular Structures in Prokaryotes
DOI 10.1007/7171_033/Published online: 10 June 2006

Type III Secretion Systems: Bacterial Injection Devices for Microbe–Host Interactions

Kirill Sergueev · Colleen A. McHugh · Egbert Hoiczyk (✉)

W. Harry Feinstone Department of Molecular Microbiology and Immunology,
Johns Hopkins Bloomberg School of Public Health, Baltimore, MD 21205, USA
ehoiczyk@jhsph.edu

Abstract Many pathogenic and symbiotic bacteria interact with their eukaryotic host using a complex protein secretion apparatus termed the type III secretion system. Type III secretion systems are multiprotein machineries that form a complex syringe-like organelle spanning the entire Gram-negative bacterial envelope. The bacteria use these machineries to translocate effector proteins into the host cells that either enable them to evade the host's immune system or help establish a close relationship that is necessary to engage in a mutualistic symbiosis. This work discusses the occurrence, structure, and regulation of these novel and important protein translocation systems.

1 Introduction

Gram-negative bacteria possess a wide range of cell envelope-associated organelles that are involved in microbe–host interaction (Hueck 1998), biogenesis of cell surface structures such as pili or flagella (Aizawa 2004), and the uptake or release of various macromolecules such as DNA, proteins, carbohydrates, nutrients, etc. (Kostakioti et al. 2005). In each of these cases the bacteria use highly sophisticated transport machines, which are formed by large macromolecular protein assemblies. The following sections will discuss the occurrence, structure, function, and regulation of one such transport machine, a protein secretion apparatus termed the type III secretion system

(TTSS) that is used by many pathogenic and symbiotic bacteria to interact with their eukaryotic host.

2
The Occurrence of Type III Secretion Systems in Bacteria

With an increasing number of bacterial genomes sequenced, ever more TTSSs have been recognized indicating the widespread occurrence of these important pathogenicity determinants (Pallen et al. 2005). While the TTSS of human pathogenic *Yersinia* is often considered the prototype of these injection devices, similar systems have been found in many other pathogens such as enteropathogenic *Escherichia coli, Salmonella, Shigella, Pseudomonas, Bordetella*, and *Chlamydia* species. Intriguingly, TTSSs are also found in many plant pathogenic bacteria such as *Xanthomonas* and *Erwinia* species that cause spot disease or soft rot in important crop plants. Most pathogenic bacteria possess only one TTSS, but some species have more than one. For example, *Salmonella typhimurium* utilizes one TTSS to invade intestinal epithelial cells, while the second system is activated upon cell entry to enable the bacterium to establish a persistent infection. TTSSs have also been found in symbiotic bacteria such as *Rhizobium* species or the insect endosymbiont *Sodalis glossinidius*. Although these bacteria are not pathogenic, they use the TTSS to establish close relationships that are necessary for a mutualistic symbiosis with their host. Finally, TTSSs have recently been found in non-pathogenic bacteria such as *Myxococcus, Chromobacterium, Desulfovibrio*, and *Verrumicrobium* suggesting that these machines have yet undiscovered functions beyond pathogenicity.

3
The Structure of Bacterial Type III Secretion Systems

TTSSs consist of about 25 different proteins that form a complex syringe-like organelle (Fig. 1), which spans the whole multi-layered Gram-negative bacterial envelope (Kubori et al. 1998; Blocker et al. 1999). Approximately ten of these proteins share sequence similarities with proteins present in the basal body of the bacterial flagellum, pointing to a shared ancestry of these two organelles. Each TTSS organelle is formed by a large cylindrical basal body (Marlovits et al. 2004) and ends in a 60–80 nm long needle-like structure protruding outwards from the bacterium (Hoiczyk and Blobel 2001). This architecture is shared by all TTSSs, with the exception of phyto-pathogenic bacteria, which possess, instead of a short needle, a 2 μm long pilus to penetrate the thick walls of plant cells (He and Jin 2003). The basal body portion of the TTSS consists of two unequally-sized concentric rings that are con-

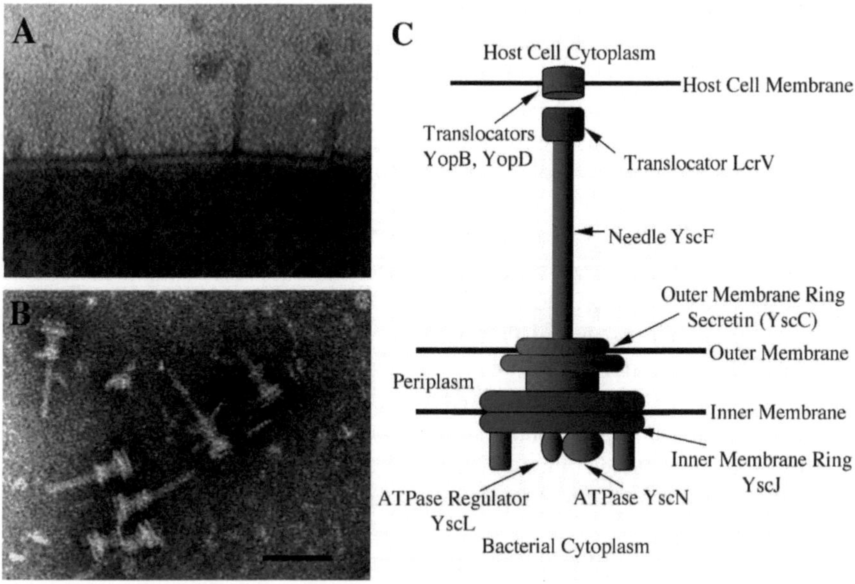

Fig. 1 Structure of the type III secretion system of animal-pathogenic bacteria. **A** Appearance of the needle-like portion of the TTSS on the surface of a *Yersinia enterocolitica* cell. *Scale bar* 50 nm. **B** Isolated needle complexes of the TTSS of *Salmonella typhimurium* showing the needle and the inner and outer ring of the secretion apparatus. *Scale bar* 100 nm. **C** Schematic illustration of the TTSS encoded by the pYV plasmid of human-pathogenic *Yersinia* species. The model is based on published data and images of the TTSSs from *Yersinia* species and other related enteric bacteria. (Picture A taken from Hoiczyk and Blobel (2001) with kind permission of the National Academy of Sciences; picture B taken from Kimbrough and Miller (2002) with kind permission of Elsevier)

nected through a central rod-like structure. The larger of these rings, termed the "inner" ring, is anchored in the bacterial inner membrane and the smaller "outer" ring is embedded in the outer membrane. Isolated outer rings have an outer diameter of about 15–20 nm and an inner diameter of 5–7 nm (Koster et al. 1997). They are formed by a single protein that belongs to a large family of bacterial outer membrane proteins called secretins. Secretins form homo-oligomeric ring-like structures that function as apertures to allow large structures such as pili, filamentous phages, or TTSS needles to traverse the outer membrane without perturbing its integrity. Studies of isolated rings indicate that they are formed by 12–13 subunits and that a small TTSS-encoded lipoprotein is necessary for their proper oligomerization and localization (Burghout et al. 2004). At least two proteins have been identified to be part of the inner ring in *Salmonella* (PrgK, PrgH) and *Shigella* (MxiJ, MxiG). One of these two, PrgK/MxiJ, is homologous to the flagellar basal body protein FliF, which forms a ring-shaped passive core of the flagellar motor. However, neither PrgH nor MxiG share any homology with the proteins of the flagellar basal body. Electron microscopic studies of the *Salmonella* inner ring indicate that this structure has a 21 fold rotational symmetry (Marlovits et al. 2004), while the 1.8 Å crystal structure of the inner ring formed by the homolog protein EscJ of enteropathogenic *E. coli* appears to possess a 24 fold symmetry (Yip et al. 2005). Current models of TTSS biogenesis suggest that these large ring-shaped supramolecular structures form a platform on which the assembly of the complete needle complex occurs. More than ten additional proteins have been suggested to be part of the complete basal body-like structure of the TTSS. Although these proteins have not yet been recovered in TTSS preparation, yeast two- and three-hybrid analyses indicate their association with the TTSS (Jackson and Plano 2000). One of these proteins, a conserved ATPase, shows homology to the α and β subunits of the F_0F_1 ATPases and is thought to energize the translocation of the effector proteins similar to FliI, an ATPase that energizes flagellar transport (Woestyn et al. 1994). Studies with the overexpressed purified ATPases of *Yersinia* and *Salmonella* suggest that they form hexameric complexes that might be located at the base of the TTSS apparatus.

All TTSSs have needle-like structures that protrude from the bacterial surface. Analyses of these needles in different pathogenic bacteria show that they are composed of several hundred subunits of a single small (< 10 kDa) hydrophilic protein (Hoiczyk and Blobel 2001). Individual needles are straight hollow tubes ca. 60–80 nm long, 6–7 nm wide, with an inner diameter of about 2 nm. Once assembled, it is generally assumed that the structure forms a channel through which the effectors are directly translocated into the cytosol of the host cells. Recently, it has been shown in *Yersinia* that another TTSS-related protein, LcrV, forms a distinct structure on the tip of the needle (Mueller et al. 2005). Intriguingly, LcrV is one of the earliest discovered antigens in *Y. pestis* and has been shown to be crucial for effector translocation (Bacon and Burrows 1956). The localization of LcrV at the tip of the needle strongly suggests a role in host cell membrane interaction. Two more hydrophobic secreted proteins, YopB and D, have been implicated to play a role in the same process by forming a translocation pore in the host membrane (Rosqvist et al. 1994). Even though genetic and electrophysiologic data support the possibility that YopB and D could insert into host membranes, attempts to visualize this pore have so far been unsuccessful.

4
Toxic Ammunition: Effectors and Chaperones

TTSSs are protein translocation machineries and the secreted effectors are the ammunition used by the bacterium to target basic cellular functions in order to survive, multiply, and establish infection. The reaction of the host cells to this attack will vary depending on the bacterial species and the nature of the injected effectors. For example, the effectors injected by *Yersinia* species prevent the phagocytic killing of the bacteria and allow the pathogen to proliferate extracellularly, whereas the effectors of *Salmonella* species promote the uptake of bacteria and allow them to disseminate intracellularly. Despite these differences, the effectors of most pathogenic bacteria target similar key cellular processes such as cytoskeleton dynamics, signal transduction, or cell survival in order to subvert the host cell. Here we will discuss the Yop effector proteins of *Yersinia* species to highlight common mechanisms among effector proteins. For a more detailed discussion of this extensive topic, the reader is referred to the literature (Hueck 1998).

Human pathogenic *Yersinia* species inject six different Yop effectors; YopE, H, T, O/YpkA, YopP/J, YopM (Cornelis 2002). Four of these Yop-effectors, namely YopE, H, T, and O/YpkA, target cytoskeleton dynamics highlighting the central role of this structural element in mounting an effective host cell defense. The mechanisms by which these proteins interfere with the cytoskeleton are different, but appear to be synergistic. YopH is a phosphatase that is recruited to focal adhesions and dephosphorylates proteins such as p130Cas, SKAP-HOM, and focal adhesion kinase. This dephosphorylation disrupts the formation of actin filaments at the focal adhesions. YopE, T and O/YpkA counteract phagocytosis by interacting with small GTPases that regulate cytoskeleton dynamics. In particular, YopE is a GTPase-activating protein that accelerates GTP-hydrolysis and thereby inactivates Rac1, RhoA, and Cdc42. YopT is a cysteine protease that inactivates the same GTPases by proteolytic cleavage. Finally, YopO/YpkA is similar to RhoA-binding kinases and seems to interfere with the cytoskeleton by phosphorylating actin directly. Other major targets for Yop effectors are signaling pathways. YopP/J inhibits activation of the transcription factor NF-κB and as a result reduces the release of tumor-necrosis factor alpha by macrophages and interleukin-8 by epithelial and endothelial cells. This process down-regulates the inflammatory response and prevents the recruitment of macrophages to the site of infection. In addition, YopP/J induces apoptosis in macrophages. Finally, the target of the sixth Yop effector, YopM, is less clear and might be related to gene transcription.

Analyses of the Yop effectors show that they often resemble eukaryotic proteins rather than bacterial house-keeping proteins, suggesting that at least some of these proteins have been obtained from the host. However, the mechanism by which pathogenic bacteria acquire new effector proteins is currently unknown. Another unsolved problem is how the TTSS recognizes the effector proteins. Two hypotheses have been suggested. The secretion signal might be constituted by the first 15 N-terminal amino acids or, alternatively, by the corresponding 5′ end of the messenger RNA (Cornelis 2002).

Translocation of the effector proteins depends on the activity of specific chaperones that bind the effectors inside the bacterial cytoplasm. In the absence of its chaperone, the secretion of the cognate effector is severely reduced. TTSS chaperones are small (< 15 kDa) highly acidic (pI < 5) proteins that share a similar three-dimensional structure even though they do not possess significant homology in their amino acid sequences (Page and Parsot 2002; Stebbins and Galan 2003). Three functions have been proposed for TTSS chaperones: stabilization and prevention of premature associations of the effector; piloting of the effector to the TTSS apparatus; and regulation of the transcription of some effectors.

5
The Genetic Organization of Type III Secretion Systems

The genetic organization of the TTSSs is principally similar in different bacterial species. Genes coding for the TTSS are clustered in operons, which are located in some bacteria, such as human pathogenic *Yersinia* and *Shigella* species, on large single copy plasmids. In other bacteria, such as enteropathogenic *E. coli* and *Salmonella* species, the genes of the TTSS are clustered on the chromosome in DNA regions termed pathogenicity islands (Kaper and Hacker 1999). The sizes of the plasmids and pathogenicity islands vary considerably, ranging from 10 kb to over 200 kb. In addition to the structural genes of the TTSS, these genetic elements code for the secreted effector proteins, their chaperones, regulatory proteins, and proteins which fulfill other important pathogenic functions such as host cell adherence, toxin production, cell invasion, and iron uptake. Interestingly, these genetic elements are present in pathogenic bacteria but absent from non-pathogenic members of the same or closely related species. Additionally, these pathogenicity elements are often characterized by the presence of different G+C% content, alternative codon usage, and the presence of cryptic or functional genes encoding mobility factors like integrases or transposases. This indicates that these pathogenicity-related genetic elements represent foreign DNA that has been acquired horizontally by the bacterium. The mobility of these elements is also indicated by the observation that pathogenicity islands are often associated with tRNA genes that act as target sites for recombination events. For example, in *Y. pseudotuberculosis* it has been shown that pathogenicity islands can relocate within days from one tRNA site to the next (Buchrieser et al. 1998). Therefore, TTSSs and their surrounding genetic regions can spread horizontally between different bacterial strains or even species, potentially increasing the plasticity of the genomes of these bacteria and their overall fitness.

6
Regulation of Type III Secretion Systems

The complex nature of microbe–host interactions is reflected in an equally complex network of control mechanisms that ensure the proper temporal and spatial regulation of TTSS gene expression. In all studied systems, the regulation of the TTSS occurs in at least two separate steps: the expression of genes required for the assembly of the machine occurs first, and is then followed by the expression of genes that code for the secreted effector proteins (Francis et al. 2002). To regulate the genes of the first group, animal pathogenic bacteria utilize AraC-like thermal inducible transcriptional activators that sense the increase in temperature to 37 °C upon host contact. Plant pathogenic bacteria, on the other hand, seem to respond to either plant specific compounds such as flavonoids or the unique osmolarity of their host environment. The expression of the second group of genes in both systems is thought to be activated by physical contact with a host cell. In this context it has recently been suggested that the tip of the needle could act as a contact-dependant sensor. However, the mechanism by which this physical contact could trigger a change of gene expression is currently not understood. One possibility would be that the system is controlled by a feedback mechanism. The expression of effector genes might depend upon the contact-mediated release of a previously synthesized protein that represses effector gene expression. Although this model explains many aspects of the observed regulation, there exists so far only limited experimental evidence. To complicate matters, many small molecules can selectively activate or repress TTSS gene expression. For example, millimolar concentrations of calcium block secretion in *Yersinia* and *Pseudomonas*, while the small amphiphilic dye Congo red activates the TTSS in *Shigella*. Even though these molecules have been extensively used to study the TTSS in vitro, their molecular mode of action is unknown. Finally, numerous extracellular factors such as pH, cell density, media composition, oxygen availability, and intracellular factors such as nutritional status, RNA stability, and growth phase have been implicated in the regulation of TTSS gene expression in different bacteria.

7
Outlook and Conclusion

Recent work has identified bacterial TTSSs as an important feature of many Gram-negative bacteria that orchestrate pathogenic and symbiotic interactions between the bacteria and animals, plants or even insects. The observation that these secretion systems are present in many disease-causing bacteria but are absent from non-pathogenic species has made them the focus of intense research. Although these efforts have already provided many insights into the overall function, architecture, and regulation of these fascinating microinjection devices, many basic questions remain. What is the phylogenetic origin of these secretion organelles? Since many TTSSs are encoded on plasmids or mobile genetic elements like pathogenicity islands, their bacterial carrier must have acquired them from another source. What mechanisms control these acquisitions or the horizontal transfer between different bacterial species? Other unsolved questions are related to the general mechanism of action of the TTSSs. How does the machinery recognize the substrate proteins that are translocated? What drives the translocation of the effector proteins? Which signals trigger the deployment of the TTSS? The answer to these questions will undoubtedly not only improve our understanding of bacterial infections, but might ultimately help to identify appropriate targets for novel antimicrobial agents and vaccine development.

References

Aizawa S-I (2004) Flagella. In: Schaechter M (ed) The Desk Encyclopedia of Microbiology. Elsevier, Amsterdam, pp 470–479

Bacon GA, Burrows TW (1956) The basis of virulence in *Pasteurella pestis*: an antigen determining virulence. Br J Exp Pathol 37:481–493

Blocker A et al. (1999) The tripartite type III secreton of *Shigella flexneri* inserts IpaB and IpaC into host membranes. J Cell Biol 147:683–693

Buchrieser C, Brosch R, Bach S, Guiyoule A, Carniel E (1998) The high pathogenicity island of *Yersinia tuberculosis* can be inserted into any of the three chromosomal *asn* tRNA genes. Mol Microbiol 30:965–978

Burghout P et al. (2004) Role of the pilot protein YscW in the biogenesis of the YscC secretin in *Yersinia enterocolitica*. J Bacteriol 186:5366–5375

Cornelis GR (2002) *Yersinia* type III secretion: send in the effectors. J Cell Biol 158:401–408

Francis MS, Wolf-Watz H, Forsberg Å (2002) Regulation of type III secretion systems. Curr Opin Microbiol 5:166–172

He SY, Jin Q (2003) The Hrp pilus: learning from flagella. Curr Opin Microbiol 6:15–19

Hoiczyk E, Blobel G (2001) Polymerization of a single protein of the pathogen *Yersinia enterocolitica* into needles punctures eukaryotic cells. Proc Natl Acad Sci USA 98:4669–4674

Hueck CJ (1998) Type III protein secretion systems in bacterial pathogens of animals and plants. Microbiol Mol Biol Rev 62:379–433

Jackson MW, Plano GV (2000) Interactions between type III secretion apparatus components from *Yersinia pestis* detected using the yeast two-hybrid system. FEMS Microbiol Lett 186:85–90

Kaper JB, Hacker J (1999) Pathogenicity islands and other mobile virulence elements. ASM, Washington, DC

Kimbrough TG, Miller SI (2002) Assembly of the type III secretion needle complex of *Salmonella typhimurium*. Microbes Infect 4:75–82

Kostakioti M, Newman CL, Thanassi DG, Stathopoulos C (2005) Mechanisms of protein export across the bacterial outer membrane. J Bacteriol 187:4306–4314

Koster M, Bitter W, de Cock H, Allaoui A, Cornelis GR, Tommassen J (1997) The outer membrane component, YscC, of the Yop secretion machinery of *Yersinia enterocolitica* forms a ring-shaped multimeric complex. Mol Microbiol 26:789–797

Kubori T et al. (1998) Supramolecular structure of the *Salmonella typhimurium* type III protein secretion system. Science 280:602–605

Marlovits TC, Kubori T, Sukhan A, Thomas DR, Galan JE, Unger VM (2004) Structural insights into the assembly of the type III secretion needle complex. Science 306:1040–1042

Mueller CA et al. (2005) The V-antigen of *Yersinia* forms a distinct structure at the tip of injectisome needles. Science 310:674–676

Page A-L, Parsot C (2002) Chaperones of the type III secretion pathway: jacks of all trades. Mol Microbiol 46:1–11

Pallen MJ, Beatson SA, Bailey CM (2005) Bioinformatics, genomics and evolution of non-flagellar type-III secretion systems: a Darwinian perspective. FEMS Microbiol Rev 29:201–229

Rosqvist R, Magnusson K-E, Wolf-Watz H (1994) Target cell contact triggers expression and polarized transfer of *Yersinia* YopE cytotoxin into mammalian cells. EMBO J 13:964–972

Stebbins CE, Galan JE (2003) Priming virulence factors for delivery into the host. Nat Rev Mol Cell Biol 4:738–743

Woestyn S, Allaoui A, Wattiau P, Cornelis GR (1994) YscN, the putative energizer of the *Yersinia* Yop secretion machinery. J Bacteriol 176:1561–1569

Yip CK et al. (2005) Structural characterization of the molecular platform for type III secretion system assembly. Nature 435:702–707

Microbiol Monogr (2)
J.M. Shively: Complex Intracellular Structures in Procaryotes
DOI 10.1007/7171_034/Published online: 25 May 2006

Gas Vesicles in Actinomycetes: Not Simply a Case of Flotation in Water-Logged Soil

Geertje van Keulen

Department of Molecular Microbiology, John Innes Centre, Norwich Research Park, Norwich NR47UH, UK
geertje.van-keulen@bbsrc.ac.uk

Abstract Gas vesicles are gas-filled prokaryotic organelles that provide buoyancy in many planktonic (cyano)bacteria and halophilic archaea. Remarkably, more and more genomes of soil bacteria, especially those of actinomycetes, show gas vesicle gene (*gvp*) clusters often encoding homologues of at least the eight *gvp* genes essential for gas vesicle formation in archaea. Here, I discuss characteristics specific to actinomycete Gvp proteins, their expression under stress conditions, and the apparent absence of a buoyancy phenotype in *gvp* mutant strains. Alternative functions for gas vesicles in actinomycetes are discussed in relation to their unusually complex developmental life cycle.

1 Introduction

Gas vesicles are gas-filled prokaryotic organelles that provide buoyancy in many planktonic (cyano)bacteria and halophilic archaea [see Pfeifer 2006, in this volume and Walsby (1994) for reviews and references]. Eight genes, *gvpFGJKLM-gvpAO*, are essential for wild-type gas vesicle formation in halophilic archaea (Pfeifer 2006, in this volume) (Offner et al. 2000). Intriguingly, orthologues of the eight essential *gvp* genes in haloarchaea occur in the genomes of soil bacteria, such as *Bacillus megaterium* (Li and Cannon 1998) and members of the actinomycete genera *Streptomyces*, *Rhodococcus*, *Frankia*, *Saccharopolyspora*, and *Amycolatopsis* (Bentley et al. 2002; Ikeda et al. 2003; Offner et al. 2000; Van Keulen et al. 2005). Actinomycetes are often developmentally complex, mycelial organisms with the remarkable capacity to produce a vast array of secondary metabolites, including antibiotics. The unexpected discovery of *gvp* genes in many actinomycetes suggests that gas vesicle formation is a relatively general trait in them, raising interesting questions concerning their functions.

This work focuses on the unusual characteristics of actinomycete (-specific) Gvp proteins, discussing the results of expression studies of *S. coelicolor gvp* genes under stress conditions, and the lack of a buoyancy phenotype in mutant strains. Alternative functions for gas vesicles in actinomycetes will be discussed in relation to their complex developmental life cycle.

2
Gas Vesicle Genes in Actinomycetes

All streptomycetes have at least two *gvp* clusters of similar gene order *gvpO, A, F, G, Y, Z, J, L, S, K* (Van Keulen et al. 2005). This is in contrast to cyanobacteria and haloarchaea that usually contain one *gvp* cluster, except for *Halobacterium* sp. NRC-1, which carries two clusters on two plasmids that are expressed differentially and result in gas vesicles of different shapes (Pfeifer et al. 1997). The other *gvp*-containing actinomycetes usually contain one *gvp* cluster, although *Rhodococcus* sp. RHA1 has a second, shorter cluster. It is noteworthy that the genome sequences of mycobacteria and corynebacteria, with smaller genomes than most actinomycetes and a very different habitat and life cycle, lack *gvp* genes (Van Keulen et al. 2005). Two novel genes, *gvpY* and *gvpZ*, lie either between *gvpG* and *gvpJ* or at the beginning of the *gvp* clusters in all except *gvp1* of *S. avermitilis* and *gvp1* of *S. peucetius*. Remarkably, *S. coelicolor* contains a third *gvpYZ* pair (*sco3687-3688*), which is not associated with other *gvp* genes. It is probably also significant that none of the actinomycete clusters encode GvpC, which is required for gas vesicles to withstand higher pressures, particularly in deep lakes (Beard et al. 2000; Kinsman et al. 1995). The absence of GvpC may correspond with the natural soil habitat of actinomycetes in which high external pressures do not occur.

Eight of the ten Gvp proteins encoded by the actinomycete *gvp* clusters show significant similarities to their haloarchaeal and cyanobacterial counterparts. Yet, some actinomycete Gvp proteins or domains are unusual. The (unusual) characteristics of *S. coelicolor* Gvp proteins are summarized in Table 1 and discussed below. It must be noted that the putative functions are derived from their functions in haloarchaea and cyanobacteria. Actinomycete GvpJ, -S and -K proteins do not show significant differences and will therefore not be discussed.

Table 1 Characteristics of Gvp proteins of *S. coelicolor*

S. coelicolor Gvp proteins	pI	Archaeal homologues	Characteristics and putative functions
GvpO/O2	5.6/7.3	GvpO	Regulatory protein; GvpO1 contains N-terminal 9*x* degenerate QXS repeat followed by HTH motif (aa 53–74);
GvpA/A2	5.4/6.0	GvpA	Major gas vesicle protein with long C-terminal extension; long C-terminus rich in R, E, (P);
GvpF/F2	5.0/10.3	GvpF	Minor gas vesicle protein with coiled-coil domain; GvpF2 has unusually high pI; Archaea have core deletion of 39–41 aa;
GvpG/G2	4.2/7.5	GvpG	Minor gas vesicle protein; GvpG2 contains a probable coiled-coil (aa40 to 67); GvpG2 has higher pI possibly due to Arg-rich C-terminus;
GvpY/Y2	9.5/10.3		GvpY contains a highly acidic repetitive stretch and a highly basic repetitive stretch; High pI;
GvpZ/Z2	4.6/4.5		Contains a highly acidic repetitive region; GvpZ1 contains 6x degenerate EPEQ repeat;
GvpJ/J2	5.1/4.6	GvpJ	Minor gas vesicle protein orthologous to GvpA, GvpM and GvpS;
GvpL/L2	5.2/9.4	GvpL	Minor gas vesicle protein with coiled-coil domain; SCOGvpL2/SAVGvpL3 have high pI (core); GvpL1 has short polyE tail;
GvpS/S2	4.7/4.5	GvpM	Minor gas vesicle protein orthologous to GvpA, GvpJ and GvpM;
GvpK/K2	4.7/4.5	GvpK	Low similarity to GvpA protein family

2.1 GvpO

The protein encoded by the first gene in the *S. coelicolor gvp* clusters is GvpO. A helix-turn-helix motif is apparent, e.g. in ScoGvpO1 at aa 53–74, suggesting that it has a regulatory function. Interestingly, the only intergenic regions of significant length in the *gvp* clusters of *S. coelicolor* lie between *gvpO1-A1* and *gvpO2-A2* (72 and 37 bp, respectively) and between *sco6498-gvpO1* and *sco0648-gvpO2* (66 and 43 bp, respectively).

An amino acid alignment of GvpOs showed that the *C*-termini are well conserved (data not shown) with a notable arginine pattern: Y-E/R-R-X-R-R-Y-X-R-G-X-X-D-X_{0-6}. The GvpO consensus sequence of the *C*-termini of archaea is different from the actinomycete GvpO *C*-termini but, in archaea, is also well conserved. In contrast, the *N*-termini of GvpOs are hypervariable. The *N*-terminus of ScoGvpO1 is unusual in having a 9x degenerate QXS repeat, which is not present in other GvpOs. Furthermore, the predicted isoelectric points (pI) of ScoGvpO2 and SavGvpO3 are unusually high (Table 1). The higher pI most likely results from a higher percentage of arginine residues. If GvpO is indeed a regulatory protein, differences in pI and amino acid sequence of the *N*-terminus may result in different regulatory patterns.

2.2 GvpA

The second *gvp* gene encodes the major structural protein of gas vesicles, GvpA. An amino acid alignment revealed a long *C*-terminal extension on actinomycete GvpAs but not on other GvpAs (Fig. 1A) (Van Keulen et al. 2005). This extension almost doubles the length of actinomycete GvpA proteins and is different from other known sequences in the databases. Furthermore, the percentage of glutamate and arginine residues in this domain is extraordinarily high. These residues are often present in long, alternating acidic and basic tracts. It is implied that the *C*-terminus of GvpA faces the cytosolic side of the gas vesicle (Belenky et al. 2004; Stuart et al. 2001) and is most likely not incorporated into the rigid gas vesicle membrane structure.

The *N*-terminal half of actinomycete GvpAs aligns well with the full-length GvpAs of non-actinomycetes, enabling phylogenetic analysis (Van Keulen et al. 2005). GvpAs from actinomycetes, cyanobacteria and archaea each form a distinct clade. Separate actinomycete subclades are formed by GvpAs of *Frankia* and *Streptomyces*. Within the latter subclade, ScoGvpA2 and SavGvpA3 form a deep-branching group. In the third actinomycete subclade, GvpAs of rhodococci group together with GvpA of *Sac. erythraea*.

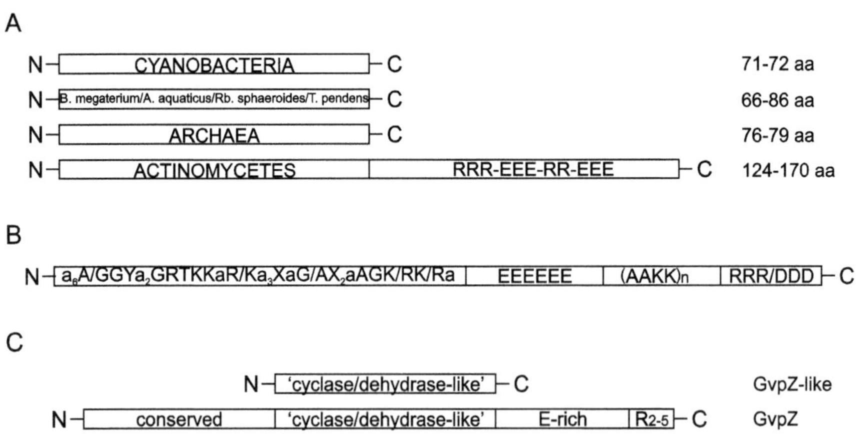

Fig. 1 Domain organization of GvpA (**A**) (Reprinted from Van Keulen et al., 2005 with permission from Elsevier); GvpY (**B**); and GvpZ (**C**). *Capital letters* indicate amino acid residues, "a" indicates an aliphatic amino acid residue

2.3
GvpF/GvpL and GvpG

The paralogues GvpF and GvpL are encoded, respectively, by the third and eighth genes of the streptomycete *gvp* clusters. An amino acid alignment of all GvpFs showed that archaeal GvpFs have a deletion of 39–41 residues in the core of the protein (data not shown). Remarkably, the GvpFs and GvpLs encoded by *gvp2* of *S. coelicolor* and *gvp3* of *S. avermitilis* showed a higher predicted pI, this is also seen with the pI for GvpG2 of *S. coelicolor* (Table 1). Again, the pI most likely reflects a higher proportion of arginine residues.

Phylogenetic analyses of Gvp proteins found in all *gvp*-containing organisms show a consistent branching and grouping pattern (data not shown) similar to GvpA (Van Keulen et al. 2005). The separate actinomycete branches of Gvp proteins indicate that the actinomycetes did not obtain *gvp* genes through lateral gene transfer. Furthermore, a global comparison of the genomes of *F. alni* and *S. coelicolor* showed significant overall synteny that includes *gvp* of *F. alni* and *gvp1* of *S. coelicolor* (Van Keulen et al. 2005), suggesting that a common actinomycete ancestor must already have contained *gvp* gene clusters.

2.4
GvpY

GvpY represents a novel, actinomycete-specific set of Gvp proteins encoded by genes found only directly upstream of *gvpZ* genes. The amino acid sequence and domain structure of GvpYs is quite extraordinary (Fig. 1B). The very well-conserved *N*-terminus (a_6-A/G-G-Y-a_2-G-R-T-K-K-a-R/K- a_3-X-a-G/A-X_2-a-A-G-K/R-K/R-a; a is an aliphatic residue) is followed by an extremely acidic and repetitive core region mainly consisting of glutamate residues, which is then followed by a highly basic repetitive tract commonly consisting of AAKK repeats. The extreme *N*-terminus often ends with either three arginine or three aspartate residues. GvpY2 of *S. coelicolor* is the exception to the rule in lacking the acidic core region.

All GvpY proteins have unusually high pI predictions of over 9.5 (Table 1 and data not shown). The only other known Gvp protein family that also has high pI values is the GvpI family of haloarchaea (Pfeifer 2006, in this volume). It is thought that GvpI is involved in the termination of GvpA incorporation into gas vesicles (Pfeifer 2006, in this volume) (Offner et al. 2000). Whether GvpY is a functional homologue of GvpI remains to be established.

2.5
GvpZ

gvpZ is the second novel gene absent from other *gvp* clusters. However, when ScoGvpZ1 was used to search the databases, significant hits arose with protein sequences from actinomycetes and other bacteria (data not shown). An amino acid alignment showed that the core region of GvpZs aligned well with the full-length sequence of the GvpZ-like proteins (Fig. 1C). This domain belongs to the InterPro domain IPR005031 "Streptomyces cyclase/dehydrase" involved in polyketide synthesis. GvpZ-like proteins lacked the well-conserved *N*-terminus of "true" GvpZ proteins as well as the unusual *C*-terminus that is repetitive in glutamate residues and ends with two to five arginine residues. In ScoGvpZ1, the glutamate residues are found in a 6x degenerate, EPEQ repeat. The presence of poly-E tracts in GvpZ resembles the poly-E tracts of GvpA and GvpG.

Phylogenetic analysis revealed that the GvpZ-like proteins form a separate branch from the "true" GvpZ proteins (data not shown). It is noteworthy that the third GvpZ-like protein, SCO3688, groups with "true" GvpZs and also contains the conserved *N*-terminus and glutamate-rich *C*-terminus. Interestingly, the corresponding gene is next to the third *gvpY* gene.

3
Expression of *gvp1* of *S. coelicolor* is Induced by (Osmotic) Stress Conditions

Transcript profiling and low-resolution S1-mapping studies have shown that *gvp1* and neighbouring genes are transiently induced in expression in shaken liquid cultures of *S. coelicolor* after osmotic, phosphate or temperature up-shifts (Karoonuthaisiri et al. 2005; Lee et al. 2005). Induction of *gvp1* expression was indirectly dependent on the alternative sigma factor, σ^{B}, which is a master regulator of the osmotic stress response (Lee et al. 2005). However, low-level induction of *gvp1* was observed in a *sigB* null mutant, indicating that *gvp1* expression is not obligately dependent on σ^{B}. Lee et al. (2005) suggested that gas vesicles might serve a function related to the response against hyperosmotic stress, although they did not propose a molecular mechanism. Furthermore, recent evidence indicates that osmo-adaptation is also important in the erection of aerial hyphae and secondary metabolite production (Bishop et al. 2004; Cho et al. 2001; Viollier et al. 2003).

The presence of Gvp proteins or gas vesicles unfortunately was not shown under osmotic up-shift, which leaves the interesting question open as to whether streptomycete Gvp proteins form typical gas vesicles or some other kind of novel protein assemblies. Interestingly, induction of expression of *gvp2* was not observed, which indicates that the two *gvp* clusters are differentially regulated, and may serve different functions.

4 Gvp Proteins are not Essential for Flotation of *S. coelicolor*

The habitats of saprophytic actinomycetes range from soils to sediments, although actinomycetes have now also been discovered in marine environments (Stach et al. 2003). Recently, it was shown that the non-motile *S. coelicolor* forms floating colonies in standing liquid cultures (Van Keulen et al. 2003), a condition that resembles flooded, water-logged soils. These environments readily become anoxic, so flotation and subsequent sporulation at the air–water interface provides an excellent escape from the unfavourable, oxygen-poor environment. Surprisingly, however, *S. coelicolor* mutants lacking one or both *gvp* clusters still floated and reached the air interface in standing liquid cultures (Van Keulen et al. 2005). This indicates that gas vesicle proteins, including the intriguing *C*-terminal extension on GvpA as well as the other unusual Gvp proteins, are not essential for flotation. This suggests the existence of other, unknown mechanism(s) of buoyancy as well as other roles for the Gvp proteins.

5 Alternative Functions for Gas Vesicle Proteins in Actinomycetes

Whether the properties conferred by the glutamate and/or arginine-rich regions in many of the Gvp proteins have a role in the actinomycete (osmo)stress response and/or differentiation remains to be established. One possibility is that the E- and R-rich regions of GvpA, -Y and -Z might have a role in binding these proteins to nucleic acids. Similarly charged amino acid tracts are observed in virus coat proteins and RNA-binding proteins (Tan and Frankel 1995). Additionally, arginine-rich peptides have been reported to assist translocation of other peptides through membranes (Futaki et al. 2001), which presents a further putative function for actinomycete Gvp proteins rich in arginine. The AAKK repeats in the *C*-terminus of GvpY are equally intriguing and resemble AAKK repeat regions in bacterial nucleoid proteins such as HU and IHF, e.g. HupB of *Mycobacterium tuberculosis* (Cohavy et al. 1999; Swinger and Rice 2004).

Does the different composition of *gvp* clusters in the various actinomycetes perhaps suggest different functions for these multiprotein assemblies in these strains? Why do streptomycetes have more than one *gvp* cluster? It will also be fascinating to identify the roles of the separate (acidic and/or basic) domains of GvpA, GvpY and GvpZ in Gvp protein assembly and function(s). And what is the function of the Gvp proteins with high pI that are encoded by *gvp2* of *S. coelicolor*? What is the role of gas vesicles in actinomycetes if it is not for buoyancy? Although never observed in streptomycete spores, do gas vesicles perhaps confer a wider (aerial) distribution of spores as the gas vesicle-containing caps on endospores of a few sporulating, strictly anaerobic clostridia and sulphate-reducing bacteria, such as *Clostridium corinoforum*, result in wider distribution (Duda and Makar'eva 1977; Widdel and Pfennig 1981)? These exciting questions need to be addressed to fully understand the role of gas vesicle proteins in the complex biology of these microbes.

Acknowledgements I thank Gary Sawers and David Hopwood for helpful comments on the manuscript. This work is supported by a Marie Curie Postdoctoral Fellowship (MEIF-CT-2004-506056).

References

Beard SJ, Davis PA, Iglesias-Rodríguez D, Skulberg OM, Walsby AE (2000) Gas vesicle genes in *Planktothrix* spp. from Nordic lakes: strains with weak gas vesicles possess a longer variant of GvpC. Microbiology 146:2009–2018

Belenky M, Meyers R, Herzfeld J (2004) Subunit structure of gas vesicles: a MALDI-TOF mass spectrometry study. Biophys J 86:499–505

Bentley SD, Chater KF, Cerdeño-Tárraga A-M, Challis GL, Thomson NR, James K.D. ea (2002) Complete genome sequence of the model actinomycete *Streptomyces coelicolor* A3(2). Nature 417:141–147

Bishop A, Fielding S, Dyson P, Herron P (2004) Systematic insertional mutagenesis of a Streptomycete genome: a link between osmoadaptation and antibiotic production. Genome Res 14:893–900

Cho YH, Lee EJ, Ahn BE, Roe J-H (2001) SigB, an RNA polymerase σ factor required for osmoprotection and proper differentiation of *Streptomyces coelicolor*. Mol Microbiol 42:205–214

Cohavy O, Harth G, Horwitz M, Eggena M, Landers C, Sutton C, Targan SR, Braun J (1999) Identification of a Novel Mycobacterial Histone H1 Homologue (HupB) as an Antigenic Target of pANCA Monoclonal Antibody and Serum Immunoglobulin A from Patients with Crohn's Disease. Infect Immun 67:6510–6517

Duda VI, Makar'eva ED (1977) Morphogenesis and function of gas caps on spores of anaerobic bacteria of the genus *Clostridium*. Mikrobiologiya 46:689–694

Futaki S, Suzuki T, Ohashi W, Yagami T, Tanaka S, Ueda K, Segiura Y (2001) Arginine-rich peptides: An abundant source of membrane-permeable peptides having potential as carriers for intracellular protein delivery. J Biol Chem 276:5836–5840

Ikeda H, Ishikawa J, Hanamoto A, Shinose M, Kikuchi H, Shiba T, Sakaki Y, Hattori M, Omura S (2003) Complete genome sequence and comparative analysis of the industrial microorganism *Streptomyces avermitilis*. Nat Biotechnol 21:526–531

Karoonuthaisiri N, Weaver D, Huang J, Cohen SN, Kao CM (2005) Regional organization of gene expression in *Streptomyces coelicolor*. Gene 353:53–66

Kinsman R, Walsby AE, Hayes PK (1995) GvpCs with reduced numbers of repeating sequence elements bind to and strenghten cyanobacterial gas vesicles. Mol Microbiol 17:147–154

Lee EJ, Karoonuthaisiri N, Kim HS, Park JH, Cha CJ, Kao CM, Roe J-H (2005) A master regulator σ^B governs osmotic and oxidative response as well as differentiation via a network of sigma factors in *Streptomyces coelicolor*. Mol Microbiol 57:1252–1264

Li N, Cannon MC (1998) Gas vesicle genes identified in *Bacillus megaterium* and functional expression in *Escherichia coli*. J Bacteriol 180:2450–2458

Offner S, Hofacker A, Wanner G, Pfeifer F (2000) Eight of fourteen *gvp* genes are sufficient for formation of gas vesicles in halophilic archaea. J Bacteriol 182:4328–4336
Pfeifer F, Krüger K, Röder R, Mayr A, Ziesche S, Offner S (1997) Gas vesicle formation in halophilic Archaea. Arch Microbiol 167:259–268
Stach JE, Maldonado LA, Ward AC, Goodfellow M, Bull AT (2003) New primers for the class Actinobacteria: application to marine and terrestrial environments. Environ Microbiol 5:828–841
Stuart ES, Morshed F, Sremac M, DasSarma S (2001) Antigen presentation using novel particulate organelles from halophilic archaea. J Biotechnol 88:119–128
Swinger KK, Rice PA (2004) IHF and HU: flexible architects of bent DNA. Curr Opin Struct Biol 14:28–35
Tan R, Frankel AD (1995) Structural variety of arginine-rich RNA-binding peptides. Proc Natl Acad Sci USA 92:5282–5286
Van Keulen G, Hopwood DA, Dijkhuizen L, Sawers RG (2005) Gas vesicles in actinomycetes: old buoys in novel habitats? Trends Microbiol 13:350–354
Van Keulen G, Jonkers HM, Claessen D, Dijkhuizen L, Wösten HAB (2003) Differentiation and anaerobiosis in standing liquid cultures of *Streptomyces coelicolor*. J Bacteriol 185:1455–1458
Viollier PH, Kelemen GH, Dale GE, Nguyen KT, Buttner MJ, Thompson CJ (2003) Specialized osmotic stress response systems involve multiple SigB-like sigma factors in *Streptomyces coelicolor*. Mol Microbiol 47:699–714
Walsby AE (1994) Gas vesicles. Microbiol Rev 58:94–144
Widdel F, Pfennig N (1981) Sporulation and further nutritional characteristics of *Desulfotomaculum acetoxidans*. Arch Microbiol 129:401–402

Microbiol Monogr (2)
J.M. Shively: Complex Intracellular Structures in Prokaryotes
DOI 10.1007/7171_035/Published online: 13 April 2006

Bacterial Endosymbionts in Prokaryotes

Daniele Corsaro[1] (✉) · Danielle Venditti[2]

[1]CHLAREAS, Chlamydia Research Association, 12 rue du Maconnais, 54500 Vandoeuvre-lès-Nancy, France
corsaro@voila.fr

[2]Tredi Research Department, Faculty of Medecine, 9 avenue Foret de Haye, BP189, 54505 Vandoeuvre-lès-Nancy, France

Abstract We have summarized herein the available data concerning either parasitic or mutualistic bacteria having as host cell another prokaryote, with particular emphasis on those cases in which endosymbiosis emerges. These types of associations are central to the endosymbiotic theory for the origin of the eukaryotic cell. Endosymbionts of prokaryotes (today restricted to Gram-negative bacteria) may be clustered into two categories: the intraperiplasmic, long time known as bdellovibrios, and the more invading intracytoplasmic, described more recently.

1 Introduction

Symbiosis is the condition in which different organisms live together, and endosymbiosis is the case in which one partner, the endosymbiont, penetrates the cells or tissues of the other, the host. In the large sense, (endo)symbiotic associations include both mutualistic and parasitic forms, and also organelle-based ones, as plastids and mitochondria are today recognized as the remnants of ancient bacterial endosymbionts. Endosymbioses are widespread in the living world, involving either prokaryotic or eukaryotic organisms, and play a key role in ecology and evolution (Corsaro et al. 1999; Corsaro and Venditti 2002; Margulis 1993; Margulis et al. 2000).

By considering the size ratio between the partners, intracellular endosymbioses are expected to have frequently, if not exclusively, eukaryotic cells as hosts (the larger partner), while the endosymbiont (the smaller partner) may be prokaryotic or eukaryotic. Very large bacteria exist (Schulz and Jørgensen 2001), but prokaryote as host has been very rarely reported, nor this condition seems to be linked effectively to cell dimensions.

In this review we summarized the few but very interesting data available on the endosymbioses in which both partners are prokaryotic organisms. Major examples of such associations may be classified in two categories: i) predators/parasites and ii) non-predators/mutualists.

2
Bacterial Endosymbiosis and the Origin of the Eukaryotic Cell

It is currently widely accepted that the eukaryotic cell originated from the merger of bacterial symbioses.

2.1
Serial Endosymbiotic Theory

The serial endosymbiotic theory supposes successive endosymbiotic events between different prokaryotic organisms to lead finally to the emergence of modern eukaryotic cells (Margulis 1993). The successive endosymbiosis of an intracellular parasite, phylogenetically linked to the rickettsiae (Fitzpatrick et al. 2006), gave origin to the mitochondrion. Endosymbiosis of cyanobacteria by one eukaryote lineage led to Plantae (i.e. eukaryotes containing plastids from primary symbiosis), while other photosynthetic eukaryotes originated by events of secondary symbioses of eukaryotic algae (Bhattacharya et al. 2003).

2.2
Archeal-eubacterial Merger

A key event was the merger of archeal and eubacterial symbionts, giving rise to the chimeric eukaryotic genome structure, that is archeal-type for information genes, and eubacterial-type for metabolic genes. Various models have been proposed, differing essentially on the first step. For example, upon phylogenomic analyses, Horiike et al. (2004) suggested that the eukaryotic cell originated from the endosymbiosis of a *Pyrococcus*-like (*Euryarchaeota*, i.e.

a major branch of *Archae*) into a γ-proteobacterium. Margulis et al. (2000) proposed the merger between archeal (*Thermoplasma*-like) and eubacterial (spirochaeta-like) cells into a motility symbiosis, forming a "*Thiodendron*"-like stage, similar to the extant sulfur syntrophic consortium of spirochaetes and sulfidogenic bacteria known as "*Thiodendron latens*" (Dubinina et al. 1993). From this *Thiodendron*-like stage emerged first a nucleated cell with flagella (karyomastigont), and then true eukaryotes (Margulis et al. 2000). Other models also postulated the epi/endobiosis (either mutualistic or parasitic) between archeal and eubacterial cells, but from different lineages. For example, the "hydrogen hypothesis" of Martin and Müller (1998) hypothesized the engulfment by an anaerobic methanogen (the future nucleocytoplasm) of a hydrogen-producing (α-)Proteobacterium. The latter developed into a hydrogenosome under anaerobic conditions, but evolved into the mitochondrion within the aerobic eukaryotes. Most of these models postulated a "hypertrophic" state of the membrane system of a host (preferably the archeal cell) to envelop the partner, and in some way to form a nuclear body. Some archaebacteria showed to have complex cell structures, with outer membrane and periplasmic vesicle budding (Rachel et al. 2002). However, an autogenous origin of the nucleus cannot be discarded, as some peptidoglycan-less, Gram-negative eubacteria (Planctomycetes) possess nuclear-like structures (see also Fuerst 2006, in this volume).

3 Predatory Prokaryotes

Bacteriophages and protistan grazers are important factors influencing the dynamic of prokaryotic populations. However, predatory prokaryotes also may play a role in modulating microbial population structure: they can be categorized following one of the four basic predation strategies observed so far (Martin 2002): i) group predation or wolfpack; ii) epibiotic or epicellular; iii) periplasmic; and iv) direct invasion or intracytoplasmic.

3.1 Extracellular Predatory Strategies

Wolfpack and epibiotic strategies are commonly encountered in the prokaryotic world. Classical examples of wolfpack are myxobacteria (δ-Proteobacteria), which release degrading enzymes to lyse the prey cell.

The epibiotic strategy implies a cell-to-cell contact between the predator and the prey. Thus, epibiosis may be an evolutive step towards an invasive stage. Care should be made to distinguish between surface colonization, a widespread feature among prokaryotes, and true epibiosis, the latter implying a change of nutrients. Additionally, evidence that the prey cell is degrad-

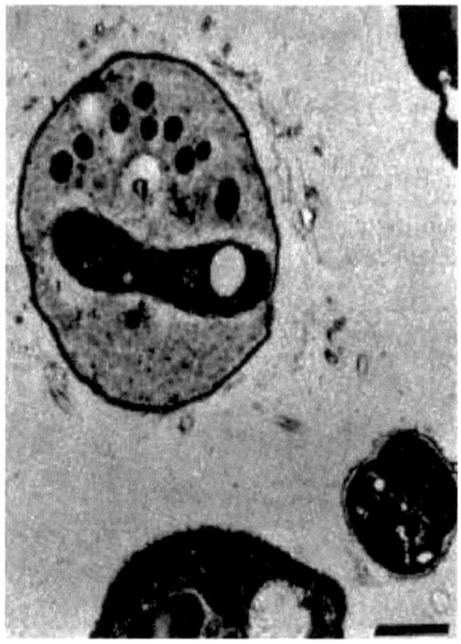

Fig. 1 A *Chromatium* cell with partially degraded cytoplasm harboring a *Daptobacter* cell in division. In the *lower right*, another *Chromatium* cell is attacked by an epicellular *Vampirococcus*. *Scale bar* = 0.5 μm. Reprinted from Guerrero et al. (1986) with permissions of Prof. Margulis and Dr. Esteve

ing should differentiate predatory from non-predatory/mutualistic epibiosis. *Vampirococcus* is a small (0.6 μm), ovoidal, anaerobic, aflagellate bacterium, requiring obligatorily *Chromatium* species as preys to reproduce by binary fission (Fig. 1). Prey cytoplasm is finally degraded (Esteve et al. 1983; Guerrero et al. 1986). Another epibiotic predator is the vibroid (0.2–0.4 × 0.6–1 μm), flagellated *Micavibrio* which adheres to and destroys various Gram-negative bacteria (Lambina et al. 1983). *Vampirovibrio* (Gromov and Mamkaeva 1980) is often cited as an additional example of epibiotic predator, but it "vampirizes" eukaryotic *Chlorella* algae. *Nanoarchaeum* is a very small (0.4 μm) spherical archaebacterium, epibiontic of another archeal, *Ignicoccus* (Huber et al. 2002), but its mutualistic or parasitic nature has not been clearly established yet.

3.2
Periplasmic Strategy: *Bdellovibrio*-And-Like-Organisms (BALO)

BALO, once regrouped into a single genus *Bdellovibrio*, are at present recognized as two distinct species-rich families within the δ-Proteobacteria: the *Bdellovibrionaceae*, comprising *B. bacteriovorus*, and the *Bacteriovoracaceae*, comprising *Bacteriovorax* (e.g., *B. stolpii*) and *Peredibacter* (e.g., *P. starrii*) species (Davidov and Jurkevitch 2004). Both lineages comprise uniflagellated, small to medium (0.2–0.5 × 0.5–2.5 μm) vibroids preying on Gram-negative bacteria, and living in a variety of habitats including fresh and salt waters,

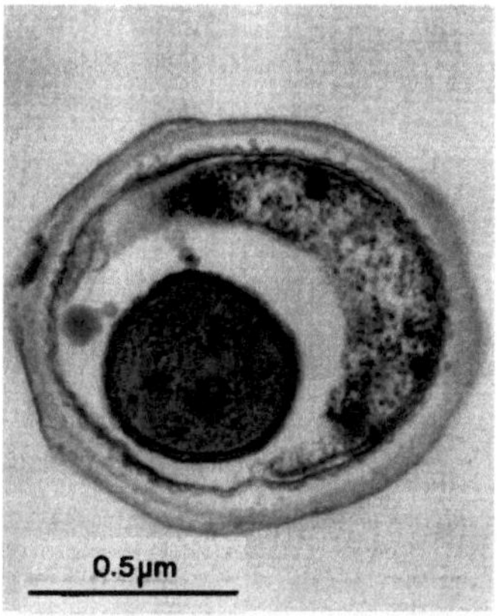

Fig. 2 Transverse section of a *Thiothrix* cell, showing an intraperiplasmic symbiont. Reprinted from Larkin et al. (1990) with permission of ASM Journals Department

soils, sewages, and the intestinal tracts of animals. The latter localisation, coupled with the inability of BALO to invade mammalian cells, promises for a possible development of (some) BALO as a "living antibiotics". The predatory strategy includes the invasion of the periplasm, but not of the cytoplasm, of the prey by active penetration, followed by the modification of the prey cell into a so-called bdelloplast, which becomes resistant to additional invasion. Within the bdelloplast, BALO exploit prey's solutes, then elongate in a filamentous cell that consecutively produces multiple BALO cells by septation. The new generation of BALO is released by the prey cell lysis. The whole genome sequencing of *B. bacteriovorus* (Rendulic et al. 2004) will allow the study at a molecular level of such a unique life cycle.

Another periplasmic predator has been described as growing within *Thiothrix* filaments (trichomes) covering the exoskeleton of mayfly larvae (*Drunella*, Insecta: Ephemeroptera) in a sulfur spring effluent (Larkin et al. 1990). This bacterium, described as morphotype 3 (to distinguish from two intracytoplasmic morphotypes, see Sect. 3.3), appears to reside within the periplasm of the prey, pushing aside the prey cytoplasm (Fig. 2). Finally, it probably escapes by a hole that is drilled from the inside of the cell sheath of the prey. This bacterium is aflagellated and non-comma-shaped, and apparently has no septation stage. Therefore, it is not a BALO, but molecular data are lacking.

3.3 Cytoplasmic Invasion

Guerrero et al. (1986) described *Daptobacter*, the first predatory prokaryote able to penetrate the cytoplasm of its bacterial preys, Gram-negative bacteria

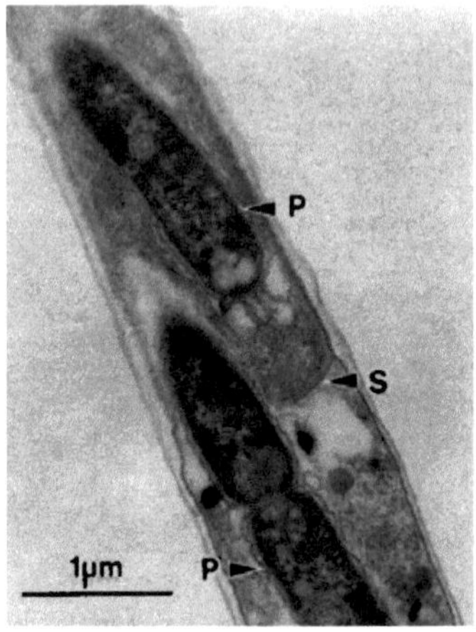

Fig. 3 *Thiothrix* trichome harboring intracytoplasmically *Daptobacter*-like parasites (P). The lower organism is in division and is penetrating the septum (S) between two *Thiothrix* cells. Reprinted from Larkin et al. (1990) with permission of ASM Journals Department

of the *Chromatium* group. *Daptobacter* is a rod (0.5 × 1.5 μm) Gram-negative, facultative anaerobic, uniflagellated, isolated from Spanish sulfurous lakes. Division by binary fission occurs within the prey cytoplasm (Fig. 1) (Guerrero et al. 1986).

Additional intracytoplasmic parasites have been described by Larkin et al. (1990) as two morphotypes infecting *Thiotrix* spp. covering mayfly larvae. These *Daptobacter*-like are rods (0.4 × 1–4.5 μm), they divide by binary fission, penetrate the septum between the individual cells (about 1 μm width) of the *Thiothrix* trichome (Fig. 3), and are supposed to leave the prey by passing through the open end of the sheathed trichome.

Despite their very interesting nature, being the unique intracytoplasmic parasites of bacteria, no other study has been performed on these organisms, and their phylogenetic affiliations (i.e., 16S rDNA sequence) remain unknown.

3.4 Bacterial Parasites of Organelles

Mitochondria being the remnants of ancient bacterial endosymbionts their bacterial parasites deserve to be briefly described.

Bacterial parasites of mitochondria have been described in a mycetozoan (Duval 1966), two ciliates (de Puytorac and Grain 1972; Yamataka and Hayashi 1970), and the tick *Ixodes ricinus* (Lewis 1979; Zhu et al. 1992). Recently, the parasite of the tick mitochondria has been further characterised and named IricES1 (Sacchi et al. 2004; Beninati et al. 2004). IricES1 is present in the cytoplasm of the ovary cells of adult tick females, generally enclosed within a vacuole, and seems to be transovarially transmitted to the next tick generation. IricES1 cells are in close contact with mitochondria, and some bacteria are clearly within the organelles, presumably residing between the outer and the inner mitochondrial membranes (Sacchi et al. 2004). Mitochondria are progressively consumed as bacteria grow in number (Fig. 4), however the parasitism neither involves the entire mitochondrial population, nor seems to affect negatively the tick's fitness (Sacchi et al. 2004). Phylogenetic analyses based on the 16S rDNA (Beninati et al. 2004) showed that IricES1 belongs to the *Rickettsiales* (α-Proteobacteria), an important group of intracellular parasites of humans and animals (e.g., *Rickettsia*, *Ehrlichia*, *Wolbachia*), from which originated the ancestor of the

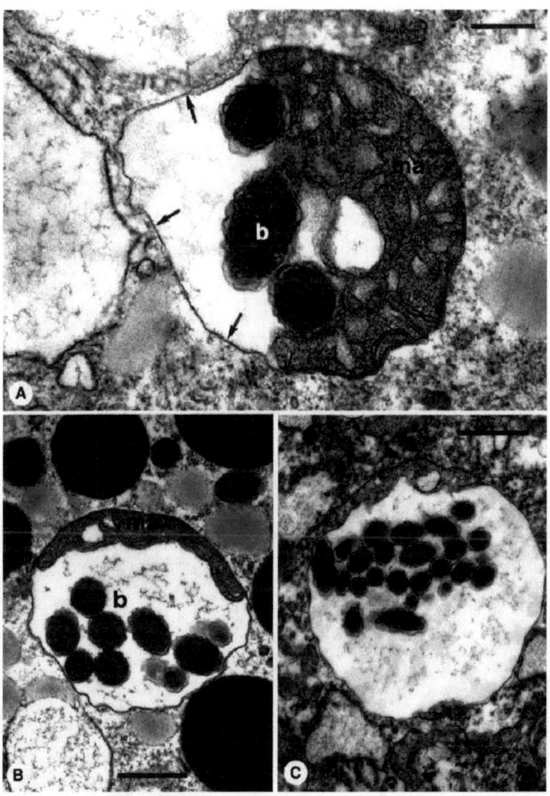

Fig. 4 Tick mitochondria parasitized by the pararickettsia IricES1 (**b**). The progressive consumption of the mitochondrial matrix (ma) is observable as the parasitic activity progresses (*panels* **a** *to* **c**), the mitochondrion ending as a swollen empty sac containing an increased population of bacteria (*panel* **c**). *Scale bar*: panel A = 0.25 μm; panel B = 0.8 μm; panel C = 1.2 μm. Reprinted from Tissue Cell (36)1:43–53, Sacchi et al. (2004), with permission from Elsevier

mitochondria (Fitzpatrick et al. 2006). In particular, IricES1 has as closest relatives the "pararickettsiae", intracytoplasmic, but not intramitochondrial, endosymbionts of *Acanthamoeba* (Fritsche et al. 1999), that, like mycetozoans, belong to the Amoebozoa.

4 Non-predator Endosymbioses

The only cases presently known of non-predator intrabacterial endosymbionts involve as hosts the sulfide-oxidizing *Beggiatoa* (Larkin and Henk 1996), and the primary symbiont of mealybug insects (von Dohlen et al. 2001).

4.1 *Beggiatoa* Endosymbionts

Beggiatoa spp. (γ-Proteobacteria) are sulfide-oxidizing bacteria, present in freshwater and marine habitats, as mats at the sediment–water interface. *Beggiatoa* cells form motile filaments, 1 to ~ 200 μm in diameter, but often up to 1 or more centimeter in length. Large *Beggiatoa* filaments (40–200 μm in diameter) have been reported from systems with high sulfide concentrations like hydrothermal vents and hydrocarbon seeps (Jannasch et al. 1989; Larkin and Henk 1996). As for many other big bacteria (Schulz and Jørgensen 2001), the single cells possess a central vacuole, with the cytoplasm pushed at the periphery. By studying microbial mats associated to hydrocarbon seeps in the Gulf of Mexico, Larkin and Henk (1996) described the

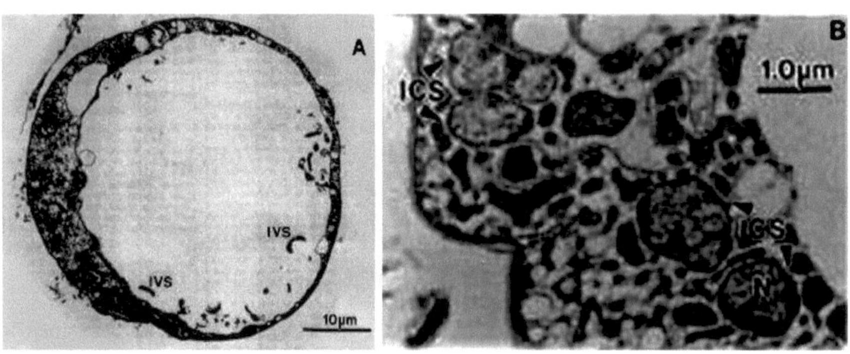

Fig. 5 Thin cross-section of a *Beggiatoa* filament, showing intravacuolar (IVS) (*panel A*) and intracytoplasmic (ICS) (*panel B*) prokaryotes. Reprinted from Larkin and Henk, Microsc Res Tech 33:23–31, Copyright (1996), with permission of Wiley-Liss, Inc., a subsidiary of Wiley

largest *Beggiatoa* filaments (up to 200 μm in diameter) having the particularity to harbor various prokaryotic endosymbionts. Those were observed both into the peripheric cytoplasm and in the central vacuole (Fig. 5), some of which also in binary fission. The authors described two morphotypes, based on ultrastructure: i) type II methanotrophic bacteria, remembering to bacterial symbionts present in the gills of mussels inhabiting methanogenic areas; and ii) carboxysome-like harboring prokaryotes. According to the authors, the known metabolic variety of *Beggiatoa* spp., ranging from various types of organotrophy to autotrophy, could depend on the variability of the prokaryotic endosymbionts.

4.2 Mealybug Endosymbionts

Mealybugs (superfamily Coccoidea) belong to the suborder Sternorrhyncha (Insecta: Hemiptera), that comprises mainly plant sap-sucking species, some being major agricultural plant pests. As many other insects, hemipterans also harbor specific bacterial P-symbionts (P for Primary) within particular cells, the bacteriocytes. They are transmitted vertically to the progeny, and show an ancient coevolutionary history with their insect hosts. P-symbionts are reputed to supply nutrient factors lacking in the diet of the insect host. In contrast, S-symbionts (S for Secondary) have multiple independent origins, they are neither limited to particular cells, nor are essential to the host.

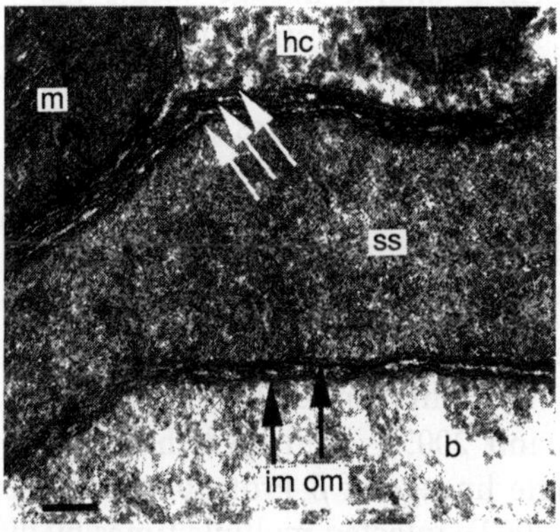

Fig. 6 High magnification of an electron micrograph showing a mealybug bacteriocyte, with the three-membraned (*white arrows*) symbiotic sphere including a two-membraned (*black arrows*; im, om, inner and outer membranes) bacterial S-symbiont (b). The matrix of the symbiotic sphere (ss) corresponds to the cytoplasm of the P-symbiont *Tremblaya*, while the three membranes (*white arrows*) correspond to the inner and outer membranes of *Tremblaya*, and to the membrane of the host cell vacuole. hc = insect cell cytoplasm, m = mitochondrion. *Scale bar* = 0.0706 μm. Reprinted by permission from Macmillan Publishers Ltd: Nature (von Dohlen et al. 412:433–436), Copyright (2001)

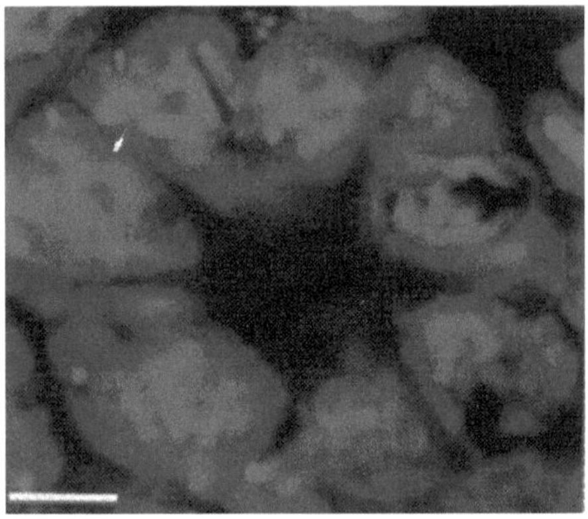

Fig. 7 Laser scanning confocal microscope image of in situ hybridization on a mealybug bacteriocyte section, combining 16S rRNA probe mix specific for beta- and gamma-proteobacteria. Fluorescence with beta (*blue*) and gamma (*red*) probes are from the matrix and the center of the symbiotic sphere, respectively. *Scale bar* = 4.5 μm. Reprinted by permission from Macmillan Publishers Ltd: Nature (von Dohlen et al. 412:433–436), Copyright (2001)

While in most hemipterans P-symbionts belong to the γ-Proteobacteria, e.g. *Buchnera* (aphids), in mealybugs early reports identified either γ- or β-symbionts. Fukatsu and Nikoh (2000) showed by in situ hybridization that bacteriocytes harbored both types of proteobacteria, and suggested that the β-symbionts are P-symbionts (16S rDNA sequences from different insect species form a unique lineage) while the γ-symbionts are unrelated S-symbionts. However, the structural relationship between P- and S-symbionts has been clarified by von Dohlen et al., who demonstrated that the three-membraned symbiotic spheres (about 20 μm in diameter) are in reality insect cell vacuoles containing the β-symbiont (the sphere "matrix") which harbors intracytoplasmic 3–6 μm-sized γ-symbionts (the center of the sphere) (Fig. 6). Such a localization was confirmed by in situ hybridization with β- and γ-specific 16S rRNA probes, labeling the matrix and the center of the symbiotic sphere, respectively (Fig. 7) (von Dohlen et al. 2001). Successive phylogenetic studies from a higher number of insect species (Thao et al. 2002; Baumann and Baumann 2005), confirmed the hypothesis of a P-symbiont represented by a single lineage of β-Proteobacteria, then named *Tremblaya princeps* (Thao et al. 2002), infected multiple times by different γ-symbionts.

5 Conclusions

Mutualistic epibiosis and consortia are widespread, and may be based on syntrophy and/or on motility symbiosis. Syntrophic associations have been described also in some invertebrates, in which bacterial partners live tightly together, either in the same host cell or extracellularly in the subcuticle (Corsaro and Venditti 2002). All these conditions may be the initial steps for a more "invasive" phase in the association, leading to i) predatory/parasitic species invading either the periplasm (BALO) or the cytoplasm of the prey (*Daptobacter* and like organisms) (Guerrero et al. 1986; Larkin et al. 1990), or to ii) likely mutualistic symbionts, today recognized within *Beggiatoa* (Larkin and Henk 1996) and within an insect (von Dohlen et al. 2001).

Whereas the eubacterial origin of mitochondria and plastids is today widely accepted, the absence of phagocytosis in prokaryotes remains the main criticism of the endosymbiogenesis of the eukaryote cell. However, even rare cases resumed herein demonstrate that some prokaryotes are able to invade the body of another prokaryote. The known cases of intracellular endosymbiosis all concern Gram-negative prokaryotes, involving various metabolic types and sizes. However, the rarity of cases may not exclude future discoveries concerning other types of prokaryotes: intrabacterial endosymbiosis by bacteria could be more common than presently known.

References

Baumann L, Baumann P (2005) Cospeciation between the primary endosymbionts of mealybugs and their hosts. Curr Microbiol 50:84–87

Beninati T, Lo N, Sacchi L, Genchi C, Noda H, Bandi C (2004) A novel alpha-proteobacterium resides in the mitochondria of ovarian cells of the tick *Ixodes ricinus*. Appl Environ Microbiol 70:2596–2602

Bhattacharya D, Yoon HS, Hackett JD (2003) Photosynthetic eukaryotes unite: endosymbiosis connects the dots. BioEssays 26:50–60

Corsaro D, Venditti D (2002) Endosymbiosis in ecology and evolution. In: Bitton G (ed) The Encyclopedia of Environmental Microbiology, vol 4. Wiley, New York, pp 1112–1130

Corsaro D, Venditti D, Padula M, Valassina M (1999) Intracellular life. Crit Rev Microbiol 25:39–79

Davidov Y, Jurkevitch E (2004) Diversity and evolution of *Bdellovibrio*-and-like organisms (BALOs), reclassification of *Bacteriovorax starrii* as *Peredibacter starrii* gen. nov., comb. nov., and description of the *Bacteriovorax-Peredibacter* clade as *Bacteriovoracaceae* fam. nov. Int J Syst Evol Microbiol 54:1439–1452

de Puytorac P, Grain J (1972) Bactéries intramitochondriales et particularités de l'ultrastructure cytostomo-pharyngienne chez le Cilié *Urotricha ovata* Kahl. CR Sc Soc Biol 4/5:604–607

Dubinina GA, Leshcheva NV, Grabovich MY (1993) The colorless sulfur bacterium *Thiodendron* is actually a symbiotic association of spirochetes and sulfidogenes. Microbiology (English translation of Mikrobiologiya) 62:432–444

Duval JC (1966) Mise en évidence de bactéries intramitochondriales chez *Didymium squamulosum* Fr. (Dydimiacée). CR Acad Sc Sér D 262:1441–1443

Esteve I, Guerrero R, Montesinos E, Abellà C (1983) Electron microscopy study of the interaction of epibiontic bacteria with *Chromatium minus* in natural habitats. Microb Ecol 9:57–64

Fitzpatrick DA, Creevey CJ, McInerney JO (2006) Genome phylogenies indicate a meaningful α-proteobacterial phylogeny and support a grouping of the mitochondria with the *Rickettsiales*. Mol Biol Evol 23(1):74–85

Fritsche TR, Horn M, Seyedirashti S, Gautom RK, Schleifer KH, Wagner M (1999) In situ detection of novel bacterial endosymbionts of *Acanthamoeba* spp. phylogenetically related to members of the order *Rickettsiales*. Appl Environ Microbiol 65:206–212

Fukatsu T, Nikoh N (2000) Endosymbiotic microbiota of the bamboo pseudococcid *Antonina crawii* (Insecta, Homoptera). Appl Environ Microbiol 66:643–650

Gromov BV, Mamkaeva KA (1980) [New genus of bacteria, Vampirovibrio, parasitizing Chlorella and previously assigned to the genus Bdellovibrio]. Mikrobiologyia 49:165–167

Guerrero R, Pedrós-Alió C, Esteve I, Mas J, Chase D, Margulis L (1986) Predatory prokaryotes: predation and primary consumption evolved in bacteria. Proc Natl Acad Sci USA 83:2138–2142

Horiike T, Hamada K, Miyata D, Shinozawa T (2004) The origin of eukaryotes is suggested as the symbiosis of *Pyrococcus* into gamma-proteobacteria by phylogenetic tree based on gene content. J Mol Evol 59:606–619

Huber H, Hohn MJ, Rachel R, Fuchs T, Wimmer VC, Stetter KO (2002) A new phylum of Archaea represented by a nanosized hyperthermophilic symbiont. Nature 417:63–67

Jannasch HW, Nelson DC, Wirsen CO (1989) Massive natural occurrence of unusually large bacteria (*Beggiatoa* sp.) at a hydrothermal deep-sea vent site. Nature 342:834–836

Lambina VA, Afinogenova AV, Romay Z, Konovalova SM, Andreev LV (1983) A new species of exoparasitic bacteria from the genus *Micavibrio* destroying Gram-negative bacteria. Mikrobiologiya 52:777–780

Larkin JM, Henk MC (1996) Filamentous sulfide-oxidizing bacteria at hydrocarbon seeps of the Gulf of Mexico. Microsc Res Tech 33:23–31

Larkin JM, Henk MC, Burton SD (1990) Occurrence of a *Thiothrix* sp. attached to mayfly larvae and presence of parasitic bacteria in the *Thiothrix* sp. Appl Environ Microbiol 56:357–361

Lewis D (1979) The detection of *Rickettsia*-like microorganism within the ovaries of female *Ixodes ricinus* ticks. Z Parasitenk 59:295–298

Margulis L (1993) Symbiosis in Cell Evolution. Freeman, New York

Margulis L, Dolan MF, Guerrero R (2000) The chimeric eukaryote: origin of the nucleus from the karyomastigont in amitochondriate protists. Proc Natl Acad Sci USA 97:6954–6959

Martin MO (2002) Predatory prokaryotes: an emerging research opportunity. J Mol Microbiol Biotechnol 4:467–477

Martin W, Müller M (1998) The hydrogen hypothesis for the first eukaryote. Nature 392:37–41

Rachel R, Wyschkony I, Riehl S, Huber H (2002) The ultrastructure of *Ignicoccus*: evidence for a novel outer membrane and for intracellular vesiscle budding in an archaeon. Archaea 1:9–18

Rendulic S, Jagtap P, Rosinus A, Eppinger M, Baar C, Lanz C, Keller H, Lambert C, Evans KJ, Goesmann A, Meyer F, Sockett RE, Schuster SC (2004) A predator unmasked: life cycle of *Bdellovibrio bacteriovorus* from a genomic perspective. Science 303:689–692

Sacchi L, Bigliardi E, Corona S, Beninati T, Lo N, Franceschi A (2004) A symbiont of the tick *Ixodes ricinus* invades and consumes mitochondria in a mode similar to that of the parasitic bacterium *Bdellovibrio bacteriovorus*. Tissue Cell 36:43–53
Schulz HN, Jørgensen BB (2001) Big bacteria. Annu Rev Microbiol 55:105–137
Thao ML, Gullan PJ, Baumann P (2002) Secondary (γ-*Proteobacteria*) endosymbionts infect the primary (β-*Proteobacteria*) endosymbionts of mealybugs multiple times and coevolve with their hosts. Appl Environ Microbiol 68:3190–3197
von Dohlen CD, Kohler S, Alsop SY, McManus WR (2001) Mealybug β-proteobacterial endosymbionts contain γ-proteobacterial symbionts. Nature 412:433–436
Yamataka S, Hayashi R (1970) Electron microscopic studies on the mitochondria and intramitochondrial microorganisms of *Halteria geleiana*. J Electron Microsc 19:50–62
Zhu Z, Aeschlimann A, Gern L (1992) *Rickettsia*-like microorganisms in the ovarian primordia of molting *Ixodes ricinus* (Acari: Ixodidae) larvae and nymphs. Ann Parasitol Hum Comp 67:99–110

Subject Index

MIX
Papier aus verantwortungsvollen Quellen
Paper from responsible sources
FSC® C105338
FSC
www.fsc.org

If you have any concerns about our products,
you can contact us on
ProductSafety@springernature.com

In case Publisher is established outside the EU,
the EU authorized representative is:
Springer Nature Customer Service Center GmbH
Europaplatz 3, 69115 Heidelberg, Germany

Printed by Libri Plureos GmbH
in Hamburg, Germany